Applied Latent Class Analysis

Applied Latent Class Analysis introduces several recent innovations in latent class analysis to a wider audience of researchers. Many of the world's leading innovators in the field of latent class analysis have contributed essays to this volume, each presenting a key innovation to the basic latent class model and illustrating how it can prove useful in situations typically encountered in actual research.

Professor Jacques A. Hagenaars is a full professor in Methodology of the Social Sciences; Chair of the Research Institute WORC; and Vice-Dean for Research, Faculty of Social Sciences, Tilburg University. His publications include *Categorical Longitudinal Data* (1990, Sage) and *Loglinear Models with Latent Variables* (1993, Sage).

Professor Allan L. McCutcheon is the Donald O. Clifton Chair of Survey Science, Chair of the Survey Research and Methodology graduate program, and Director of the Gallup Research Center at the University of Nebraska, Lincoln, and Senior Scientist with The Gallup Organization. He is the author of *Latent Class Analysis* (1987, Sage).

Applied Latent Class Analysis

Edited by

JACQUES A. HAGENAARS

Tilburg University

ALLAN L. McCUTCHEON

Gallup Research Center
University of Nebraska, Lincoln

CAMBRIDGE
UNIVERSITY PRESS

CAMBRIDGE UNIVERSITY PRESS
Cambridge, New York, Melbourne, Madrid, Cape Town, Singapore, São Paulo

Cambridge University Press
The Edinburgh Building, Cambridge CB2 2RU, UK

Published in the United States of America by Cambridge University Press, New York

www.cambridge.org
Information on this title: www.cambridge.org/9780521594516

First published 2002

A catalogue record for this publication is available from the British Library

Library of Congress Cataloguing in Publication data

Applied latent class analysis / edited by Jacques A. Hagenaars, Allan L. McCutcheon.
p. cm.
Includes bibliographical references and index.
ISBN 0-521-59451-0
1. Latent structure analysis. 2. Latent variables. I. Hagenaars, Jacques A., 1945–
II. McCutcheon, Allan L., 1950–
QA278.6 .A67 2002
519.5′35 – dc21 2001037649

ISBN-13 978-0-521-59451-6 hardback
ISBN-10 0-521-59451-0 hardback

Transferred to digital printing 2006

This volume is dedicated to the late Clifford C. Clogg, to honor his life and work

Contents

Contributors

Andries van der Ark, Tilburg University

Ulf Boekenholt, University of Illinois, Champaign-Urbana

Linda M. Collins, Pennsylvania State University

Marcel Croon, Tilburg University

C. Mitchell Dayton, University of Maryland, College Park

Wayne S. DeSarbo, Pennsylvania State University

Brian P. Flaherty, Pennsylvania State University

Anton K. Formann, University of Vienna

Leo A. Goodman, University of California, Berkeley

Jacques A. Hagenaars, Tilburg University

Peter G. M. van der Heijden, Utrecht University

Thomas Kohlman, University of Luebeck

Rolf Langeheine, University of Kiel

Allan L. McCutcheon, University of Nebraska, Lincoln

George B. Macready, University of Maryland, College Park

Jay Magidson, Statistical Innovations, United States

Robert D. Mare, University of California, Los Angeles

Ab Mooijaart, University of Leiden

Frank van der Pol, Central Bureau of Statistics, the Netherlands

Tamas Rudas, Eotvos Lorand University

Jeroen K. Vermunt, Tilburg University

John Robert Warren, University of Washington, Seattle

Michel Wedel, University of Groningen

Christopher Winship, Harvard University

Preface

Jacques A. Hagenaars and Allan L. McCutcheon

In two very important overviews of latent class modeling, Clifford Clogg discussed the advances made in the area of latent class analysis during the past two decades (Clogg, 1981, 1995). From a formal, statistical point of view, great progress has been made regarding the estimation and testing of latent class models. It also has become clear that particular developments in econometrics, biometrics, and mathematical statistics concerning (finite) mixture models, unobserved heterogeneity, frailty models, and random coefficient models are identical or at least have very close ties to latent class modeling, thus enhancing our insight into the potentialities of latent class analysis. Furthermore, in the social and behavioral sciences, close relationships between latent class and loglinear models and between latent class and latent trait (item response) models have been discovered, leading latent class analysis to be viewed as a very general latent variable model for categorical data. Finally, and perhaps most importantly, it has been shown that latent class analysis provides a very useful tool for answering many substantive questions in the social and behavioral sciences.

Nevertheless, and despite the present availability of user-friendly software with which latent class models can be easily and routinely applied, practicing social and behavioral researchers do not always consider latent class analysis a serious alternative for better-known techniques, such as factor analysis or linear structural equation modeling, even where it would be a more appropriate means to address their questions. In this volume, it will be shown that the latent class model is indeed a very useful and versatile tool and, in several important cases, the best or only tool. The chapters, all written by leading figures in the field, show how the basic latent class model and its variants can be applied to gain insight into important research problems.

In Chapter 1, an overview of the field is presented by Leo Goodman. The many faces of the latent class model are highlighted by means of simple but illuminating expositions and examples. Goodman first describes the latent class model as a model in which the observed association between two variables is regarded as spurious because the observed association is explained away by an indirectly observed latent variable. He then goes on to show how this basic property of conditional independence naturally leads to a measurement model for categorical data. He also demonstrates that the latent class model can be used to study a heterogeneous population, that is, a population consisting of several unidentified groups that behave differently regarding the problem at hand. Latent class analysis can make the existence of these groups apparent. In a separate section, Goodman explicitly describes the history of latent class models – a history that he himself wrote to a large extent. All these topics and many more (e.g., construction of latent class scores, estimation of the parameters, ways to solve the identifiability problem) are discussed in an integrative way, foreshadowing many issues discussed in the remainder of this book. Readers already familiar with the basic notions and applications of latent class analysis will find many new insights in this chapter. Some novices in the field may get the impression, at first sight, that the exposition here includes too much information. They are advised to read it just to get the flavor of the many possibilities of latent class analysis and, after having read the other chapters and having become readers already familiar with latent class analysis, return to it and digest it more thoroughly.

Allan McCutcheon provides an introduction into the standard latent class model in Chapter 2. As most people know it, the standard latent class model is a model for measuring one or more latent (unobserved) categorical variables by means of a set of observed categorical variables; these observed variables are considered to be indicators of the underlying concepts. The model is discussed in terms of two closely related parameterizations: (1) a parameterization in terms of probabilities of belonging to a particular latent class and of obtaining a particular scoring pattern on the observed variables, given the latent class one belongs to; and (2) a parameterization in terms of a loglinear model with a categorical latent variable. Because research problems often require particularly restricted models (e.g., the probability of giving a "wrong" response is the same for all indicators, or the relationship between the latent variable and the indicators is linear), attention is paid to imposing such restrictions by use of either parameterization. Another section in Chapter 2 is devoted to providing an outline of the basic procedures for estimating the parameters of restricted and unrestricted latent class models and for model selection. Finally, this chapter provides an introduction into multigroup latent class analysis, a technique that is useful to researchers who want to compare the outcomes of a latent class analysis among several populations (e.g., as in comparative or cross-cultural research) or for samples independently drawn from the same population but at different points in time (e.g., to study cross-temporal variations). Therefore, the discussion of the standard model is extended to include the principles of comparative latent class analysis. Altogether, in Chapter 2, the necessary basic concepts for being able to read the remaining chapters are provided, either by way of introduction for the newcomer to the field or as a refresher course for the more experienced reader.

The remaining chapters are divided into three parts according to the three main broad fields of application of latent class analysis: Classification and Measurement (Chapters 3–7), Causal Analysis and Dynamic Models (Chapters 8–11), and Unobserved Heterogeneity and Nonresponse (Chapters 12–15).

In the first type of application, Classification and Measurement, investigators essentially treat the latent variables as theoretical constructs and the observed variables as indicators of these constructs. They focus on the question of whether the (many) observed categories of one indicator or the observed scores on a number of indicators may be reduced to a small number of fundamental, theoretically meaningful latent categories or classes. The underlying classes may be treated as a classification, that is, as categories of nominal level latent variables (Chapters 3 and 4), or they can be regarded as ordered (Chapters 5 and 6), or as categories of interval level latent variables (Chapter 7).

When the categorical latent variables are regarded as nominal level variables, whose categories are not ordered, there is a close connection between the concepts of "cluster" and "latent class." In Chapter 3, Jeroen Vermunt and Jay Magidson introduce latent class cluster analysis and discuss the basic commonalities and differences between traditional clustering techniques and latent class analysis. Latent class cluster analysis is essentially a variant of what Gibson and Lazarsfeld in the 1950s called *latent profile analysis*, in which the underlying variable is treated as a nominal level latent variable, but the observed variables (the indicators) are treated as continuous (unlike the standard latent class model with categorical indicators). The latent profile model is extended here in several important ways, especially by introduction of the more complicated latent class cluster models for "mixed variables," that is, models in which some indicators are continuous but others are categorical, possibly with ordered categories. Finally, Vermunt and Magidson discuss how to introduce covariates into the latent class cluster model to find the "causes" of belonging to a particular cluster rather than to another. Examples are provided to illustrate each of their points.

In most applications of latent class analysis as a measurement model, the latent variable is seen as a "cause" of the indicators. In other words, the latent variable is seen as an antecedent variable that induces spurious relationships among the observed variables. Another way of looking at the latent class model is to view the latent variable as an intervening variable that completely explains (or interprets) the associations among the observed variables. This seemingly simple reconceptualization of the latent class model leads to many interesting applications. Clifford Clogg, for instance, used it to analyze social mobility tables: The observed variable Parents' Occupation "causes" the respondent (the child) to belong to one of the few unobserved latent (social) classes, which in turn influences the respondent's probability of obtaining a particular occupation. Peter van der Heijden, Andries van der Ark, and Ab Mooijaart refer to this model as the *latent budget model* (Chapter 4). A latent budget is essentially a latent class, a category of a nominal level latent variable. The term (latent) *budget* is reminiscent of financial or time budget studies, where (not directly observed) groups of people spend their money or time in different ways. Latent budget analysis seeks to characterize the kinds of people who belong to a particular latent group and to clarify what the budgets of these underlying groups look like (e.g., those who spend relatively more time on job-related activities than on leisure and those who spend relatively more time on leisure than on work). Thus the latent budget model approximates the observed budgets by a mixture of a small number of latent budgets. The versatility of the latent budget model (and its extensions) is reflected in a variety of distinct applications. Explicit attention is paid to the identification problem that is inherent in this approach; several interesting solutions are offered.

Following a good sociological tradition, the editors of this volume are inclined to denote *classification* as *measurement*; that is, the assignment of each and every research element to one and only one of a set of mutually exclusive and exhaustive nominal categories constitutes measurements of the persons' pertinent characteristic. Several other scientists, however, especially those with a more psychometric background, will reserve the term *measurement* for assigning respondents to categories that are at least ordered. There are several ways to deal with "ordered" latent variables in latent class analysis. Marcel Croon modifies the standard nominal level latent class model by assuming that all variables, both latent and observed, are "truly" ordinal level variables; that is, they are variables whose categories are ordered but that are not measured at intervals or a higher level of measurement (Chapter 5). Very often, models for ordered data involve analyses in which the relationships among the variables are restricted to being linear (or logistic). Such restrictions implicitly presuppose variables measured at the interval level. Croon, however, treats the data as truly ordinal, and he imposes inequality restrictions yielding (weakly) monotonic relationships between the latent and observed variables, rather than strictly linear relationships. Given the large amount of "truly ordered" data in social and behavioral research, this is a welcome extension of the latent class model. In Chapter 5, the main estimation and testing procedures (and problems) are extensively discussed for several variants of the basic ordinal latent class model (among other things, Guttman and Mokken scales).

Whereas Croon's contribution mainly focuses on indicators as "rating scales" with three or more ordered categories, other frequently occurring types of ordinal data are preference or ranking data. The latent class analysis of these latter kinds of data is the topic of Chapter 6. Ulf Böckenholt discusses several different kinds of ranking data, such as "pick j from J items," "pick any from J," incomplete and partial ranking data, and paired comparisons. The basic ideas underlying Coombs' unfolding model and Luce's choice model occupy a central place in these expositions. Böckenholt's explanations and applications clearly illustrate the usefulness and flexibility of latent class analysis for analyzing such ordered data. Unrestricted (or hardly restricted) latent class models can be meaningfully applied, but especially relevant are latent class models in which a particular response function describing the relationship between the latent and the observed variables is specified (e.g., within the context of the unfolding model, a symmetric, unimodal distribution exists). Moreover, models can be defined in which the latent classes are ordered along a particular continuum, but one can also define "mixture models" in which the latent classes are not ordered with regard to each other but "just" represent unobserved groups of people that have different observed preference distributions.

One might argue that some models presented in Chapter 6 actually imply something more about the ordered latent classes than just their order. Anton Formann and Thomas Kohlmann explicitly consider linear relationships between latent and observed variables and, with this, latent classes that form an interval level variable. They extend previous work on the logistic latent class model in Chapter 7. Logistic latent class models are linearly restricted models in the sense that the log-odds of choosing category a rather than category a' on indicator A increase or decrease linearly with the (fixed) values of the latent variable (as in logistic regression analysis). Logistic latent class models are very similar to standard psychometric item response models, such as the Rasch or the Birnbaum model. The authors generalize the logistic latent class model to a three-parameter model, in which besides the difficulty and discrimination parameters, guessing parameters are introduced. Identifiability of these models is discussed, and empirical data are used to show the substantive usefulness of the three-parameter model.

Because this volume emphasizes applications and focuses on real-world problems and data, the chapters in this volume cannot always be assigned unambiguously to one of the three main fields of application. Thus, important discussions of the latent class model as a measurement tool and as a means to correct for unreliability and invalidity of the measurements will also be found outside Chapters 3–7. First, there are overviews and introductions to measurement issues in Chapters 1 and 2. Then, the extensive examples offered in Chapter 8, especially the many variants of the basic Guttman scaling model and the applications to educational testing problems, provide practical insights into the potentialities of restricted latent class models as measurement models. Important measurement issues are also discussed in Chapter 9, where it is shown that the latent class model may consider not only random measurement errors, but also systematic measurement errors. Furthermore, with the use of the same data set, nominal and interval level latent class measurement models are systematically compared in Chapter 9. With regard to the interval level latent class model, two variants are presented: (1) the usual logistic latent class model in which the scores on all variables are fixed scores, and (2) a logistic latent class model in which the scores on the variables are treated as parameters to be estimated (in line with particular loglinear association models). Altogether, Chapters 3–7 plus the information in Chapters 1, 2, 8, and 9 provide a comprehensive overview of the latent class model as a measurement model.

Once a satisfactory latent typology or latent scale has been found, researchers often seek to investigate the causes and consequences of the latent variables, along with the way the several latent variables influence each other or develop over time. The Section on Causal Analysis and Dynamic Models, made up of Chapters 8–11, addresses these kinds of questions.

In Chapter 8, C. Mitchell Dayton and George Macready extend the basic latent class model by introducing independent variables that influence (or predict) the probabilities of belonging to particular latent classes. If the independent variables are categorical, the resulting latent class model is designated as a model with blocking variables. As the authors show, this "blocking" model is a straightforward extension of the multiple-group latent class model discussed in Chapter 2, and the same kinds of questions concerning the comparability (or homogeneity) of the measurement model apply. Formulated in a "causal" language, the question of homogeneity amounts to whether the associations between the categorical covariates and the indicators are completely mediated by the latent variable (the homogeneous case) or whether there remains a direct effect of the covariates on the indicators after the latent variable has been controlled for (the heterogeneous case, related to the psychometric concept of *item bias*). Along with categorical covariates, the authors show how to introduce continuous covariates into the standard latent class model. Because the dependent (latent) variable is still treated as categorical, the relationship between the continuous independent variables and the latent variable is modeled in the same way as in the logistic regression model. Insightful examples illustrate the practical applicability and usefulness of the approach.

The directed loglinear models, discussed by Jacques Hagenaars in Chapter 9, extend the contents of the previous chapter by including both independent categorical variables and categorical variables regarded as consequences of the latent variables. Analogous to the integration of factor analyses and path (regression) models into structural equation models with latent variables, latent class models and loglinear models can be integrated into one unifying framework for the causal analysis of categorical data. After an exposition of nominal and interval level latent class analyses, Hagenaars shows how the causes and consequences of the latent variables found can be integrated into a causal model with latent variables consisting of a series of loglinear (or logit) equations. Although an estimation of the parameters of these models is rather straightforward, model selection turns out to be a more difficult problem, and much attention is paid to this latter issue. Finally, the loglinear causal modeling approach is used to set up models for dealing with systematic measurement errors.

Within the context of causal modeling, longitudinal (panel) studies offer special possibilities because they enable the researcher to study individual changes over time. Linda Collins and Brian Flaherty describe in a very accessible and theoretically oriented way that latent class models can be used to model the nature of the "true" (latent) changes. In Chapter 10, they focus on latent transition analysis, a variant of the latent class model suited for analyzing stage-sequential developments in categorical characteristics. Their substantive analyses concern potential stages in development in drug use (alcohol, tobacco, and marijuana). Their analyses form exemplary applications of substantive latent class analyses, as they carefully consider possibly meaningful model restrictions and use a variety of tools for model selection.

Rolf Langeheine and Frank van der Pol present their work on latent Markov chains in Chapter 11. The Markov model and its variants form one of the most widely used models in the study of change in categorical characteristics. The authors discuss the essential properties of the ordinary "observed" Markov chain model. Then they show how to define a latent class model in the form of a Markov chain model at the latent level to account for (independent) classification errors in the observed data. They further extend the standard model by showing how to carry out multiple-group (latent) Markov analyses and how to deal with unobserved heterogeneity, that is, with the fact that there may be distinct, not directly observed groups in the population whose characteristics change over time according to a latent or observed Markov chain but with different parameters. The result is a very general Markov model that greatly extends the possibilities of the ordinary observed Markov model.

Further relevant information on causal analysis and dynamic models can be found in other chapters as well. As expected, given their nature, Chapters 1 and 2 are relevant. Furthermore, in Chapter 3, it is shown how to introduce exogenous variables into latent class cluster models to find out what kinds of people belong to the different clusters; similarly, the different background characteristics of people who use different latent budgets are covered in Chapter 4. In Chapter 13, independent variables are introduced to find the distinguishing characteristics of the indirectly observed groups in the population that have been characterized by means of latent class analysis and that behave differently regarding the pertinent research problem. Finally, in Chapter 14, additional observed variables are introduced to explain the "event histories" that take place, but especially to set up more realistic models to correct for the consequences of unobserved heterogeneity in event history analysis.

A common assumption in most data analyses is that the population is homogeneous, in the sense that all differences in the population that matter for the research question have been captured by the variables measured and that in all the other cases the population is homogeneous. In other words, it is assumed that there is no unobserved heterogeneity in the population that may influence the research outcomes. However, in reality, there will always be unobserved heterogeneity. As Chapter 1 and Chapters 12–15 in the section on Unobserved Heterogeneity and Nonresponse show, latent class analysis provides a very powerful tool to investigate the existence of unobserved heterogeneity and study its nature and possible consequences. This use of the latent class model is characterized as the employment of a form of finite mixture modeling. In Chapters 2–11, the latent class model is almost exclusively treated as a measurement model, that is, as a model that contains categorical latent variables that represent underlying theoretical concepts and observed variables that are assumed to be indicators of the underlying concepts. Occasionally, however, the same latent class model was used as representing distinct, latent groups of people who have different distributions on the observed variables, without the connotation that the latent classes and the latent variables these classes constitute represent theoretical constructs. From a purely formal statistical point of view, the latent class model as a measurement model is indistinguishable from the latent class model in the guise of a kind of finite mixture model; from a substantive, theoretical perspective, there may be a world of difference between these two.

In Chapter 12, Tamás Rudas uses the mixture approach to evaluate the fit of statistical models for categorical data. He assumes that a population can be divided into two latent classes. In the first latent class, the statistical model of interest holds (e.g., the independence model), whereas in the other latent class, an unrestricted, saturated model holds. The interest lies in finding out the possible maximum size of the first latent class and, with it, the largest fraction of the population for which the model of interest may be true. The larger this fraction, the better the model fits the population. Rudas shows how to obtain an estimate of this largest fraction and how this way of looking at model fit overcomes some problems inherent in the traditional chi-squared measures of fit. Examples illustrate the practical utility of this approach and show that it may lead to selecting models other than those based on traditional testing procedures.

A brief general introduction into (finite) mixture modeling is presented in Chapter 13. Michel Wedel and Wayne DeSarbo discuss the estimation, selection, and identification of mixture models. They then explain the mixture regression model and discuss applications of it. When researchers use regression models (either a standard linear regression model or any other member of the family of generalized linear models), it is assumed that the regression coefficients are constants and that all members of the population are described by the same regression equation. If it is assumed that some effects are different for some subgroups (e.g., women vs. men), interaction terms can be introduced to take care of such differential effects. However, there may exist still more, but unknown, groups whose regression equations are actually different from each other. The basic idea in the mixture regression model is to find these groups by combining latent class and regression analyses, that is, to postulate the existence of two or more such unobserved groups in the population and estimate a separate regression equation for each of these latent classes. Wedel and DeSarbo discuss the estimation, testing, and identification of mixture regression models and describe extensive examples. Finally, they discuss how extra concomitant variables can be introduced to help in the substantive interpretation of the latent classes.

Unobserved heterogeneity, caused by completely omitting relevant variables from the analyses or by including only incomplete or partial information (as in the case of censored data), may have a distorting effect on the outcomes of event history analyses. As Jeroen Vermunt shows in Chapter 14, this may be true even in situations in which the omission of variables would not create problems in regular regression analyses (e.g., in cases in which the omitted variable is not correlated with the other independent variables in the regression equation). Vermunt discusses several models and approaches used to investigate the consequences of unobserved heterogeneity in event history analysis. He mainly focuses on latent class analysis (finite mixture models). By integrating the event history model into a directed loglinear model with observed and unobserved covariates – where the unobserved variables form the mixing variables that cause the unobserved heterogeneity – he develops a very flexible and realistic general approach for dealing with unobserved heterogeneity in event history and many other kinds of models.

Unobserved heterogeneity may be seen as a form of missing data. Conversely, nonresponse can be considered a special form of unobserved heterogeneity: We know the pertinent characteristics of the respondents, but not of the nonrespondents. In Chapter 15, Christopher Winship, Robert Mare, and John Robert Warren show that (partial) nonresponse not only affects the precision of the parameter estimates, but may also lead to biased estimates and misleading substantive conclusions. They explain the various types of missing data, such as missing completely at random, missing at random, and nonignorable nonresponse. Using a latent class approach in the form of loglinear modeling with (partially) latent variables, they then show how the loglinear models may take care of the various types of (non)response. From their examples, it becomes clear that the assumption of a different response mechanism may lead to different substantive research results. Estimation and identification issues are discussed at the end of the chapter.

Several other chapters provide relevant information on unobserved heterogeneity. Besides the discussions in Chapter 1, those in Chapters 3 and 4 are relevant. The latent class cluster analysis dealt with in Chapter 3 and the latent budget model from Chapter 4 cover much the same ground as the mixture regression model in Chapter 13. The structure of these three models is very similar. The essential differences are mainly differences of interpretation. As the a priori bounds between indicators and concepts (latent variables) in measurement models are relaxed and the latent class analysis takes on a more open exploratory character and where classes are interpreted ex post facto, the latent clusters and latent budgets become "just" indirectly observed groups that behave differently with regard to their distributions of the joint observed variables. The measurement model then becomes a form of mixture model for unobserved heterogeneity. Or, formulated the other way around, the more the categories of the latent mixing variable(s) get an interpretation as being measured by particular observed variables acting as indicators, the more the mixture model becomes a measurement model. Still other chapters that provide relevant information on unobserved heterogeneity are Chapters 6, 9, and 11. Several models in Chapter 6 could be interpreted as representing unknown groups of people having different observed preference distributions. Also, a family of models in Chapter 9 for investigating systematic measurement error bias is actually a (causal) model for unobserved heterogeneity. Finally, explicit attention has been paid in Chapter 11 to mixture Markov models, in which the latent groups all follow a Markov model but with different parameters.

In summary, information on the use of the latent class model as a measurement model is provided in Chapters 3–7, but also in the first two chapters and in Chapters 8 and 9. Latent class analyses within the context of causal analyses and dynamic models are the main topic of Chapters 8–11, but relevant information on this issue is also found in Chapters 1, 2, 3, 4, 13, and 14. Finally, unobserved heterogeneity is mainly discussed in Chapters 12–15, but also in Chapters 1, 3, 4, 6, 9, and 11.

To enhance the practical utility of this book, several appendices have been added containing an exposition of the symbols and notation used in this book (Appendix A), a list of recommended literature (Appendix B), and information on the programs used (Appendix C). Moreover, a Webpage has been set up where the interested reader may find many data sets that have been analyzed in this book, as well as the program setups for the main models and several links to addresses where the programs might be obtained: at http://us.cambridge.org/titles/0521594510.html.

This volume is dedicated to the memory of Clifford C. Clogg. Next to the founding fathers of latent class analysis, that is, next to Lazarsfeld, Goodman, and Haberman, Clifford Clogg deserves consideration. He wrote two excellent reviews of the subject matter and developed one of the first computer programs for latent class analysis that was widely used. Most importantly, he did excellent methodological work. He showed how the latent class model might be used for purposes nobody had thought of previously, and he applied the model to problems that could not be solved otherwise. For the social and behavioral sciences, his contributions have been decisive. That even now his sudden death on May 7, 1995, is still vividly and shockingly remembered by his friends and colleagues demonstrates that he was much more than just a good scholar: He was first of all an excellent human being.

Clogg, C. C. (1981a). "New developments in latent structure analysis." In D. J. Jackson & E. F. Borgatta (eds.), *Factor Analysis and Measurement in Sociological Research*. Beverly Hills: Sage, pp. 215–246.

Clogg, C. C. (1995). "Latent class models." In G. Arminger, C. C. Clogg, & M. E. Sobel (eds.), *Handbook of Statistical Modeling for the Social and Behavioral Sciences*. New York: Plenum, pp. 311–360.

INTRODUCTION

ONE

Latent Class Analysis

The Empirical Study of Latent Types, Latent Variables, and Latent Structures

Leo A. Goodman

1. INTRODUCTION

I begin this introductory section on latent class analysis[1] by considering this subject in its simplest context; that is, in the analysis of the cross-classification of two dichotomous variables, say, variables A and B. In this context, we have the simple two-way 2×2 cross-classification table $\{A, B\}$, where the two rows of the 2×2 table correspond to the two classes of the dichotomous variable A, and the two columns of the 2×2 table correspond to the two classes of the dichotomous variable B. We let P_{ij} denote the probability that an observation will fall in the ith row ($i = 1, 2$) and jth column ($j = 1, 2$) of this 2×2 table. In other words, P_{ij} is the probability that an observation will be in the ith class ($i = 1, 2$) on variable A and in the jth class ($j = 1, 2$) on variable B. When variables A and B are statistically independent of each other, we have the simple relationship

$$P_{ij} = P_i^A P_j^B, \qquad (1)$$

where P_i^A is the probability that an observation will fall in the ith row of the 2×2 table, and P_j^B is the probability that an observation will fall in the jth column of the 2×2 table. In other words, P_i^A is the probability that an observation will be in the ith class on variable A, and P_j^B is the probability that an observation will be in the jth class on variable B; with

$$P_i^A = P_{i+} = \sum_j P_{ij}, \quad P_j^B = P_{+j} = \sum_i P_{ij}. \qquad (2)$$

When variables A and B are not statistically independent of each other, that is, when formula (1) does not hold true, which is often the case in many areas of empirical research (when both variables A and B are of substantive interest), the researcher analyzing the data in the 2×2 table will usually be interested in measuring the nonindependence between the two variables (A and B); and there are many different measures of this nonindependence. (Even for the simple 2×2 table, there are many such measures.) However, all of these measures of nonindependence (or almost all of them) are *deficient* in an important respect. Although these measures of nonindependence may help the researcher to determine the magnitude of the nonindependence between the two variables (A and B), they *cannot* help the researcher determine whether this nonindependence is *spurious*. In other words, none (or almost none) of the usual measures of the nonindependence between variables A and B can help the researcher to determine whether the observed relationship (the nonindependence) between variables A and B can be explained away by some other variable, say, variable X, where this variable X may be unobserved or unobservable, or latent. Is there a latent variable X that can explain away the observed (manifest) relationship between variables A and B, when we take into account the (unobserved) relationship that this latent variable X may have with variable A and the (unobserved) relationship that the latent variable may have with variable B? The use of latent class models can help the researcher to consider such questions.

The latent variable X introduced previously can be viewed as a possible explanatory variable. It can be used at times to explain away the observed relationship between variables A and B even when this observed relationship between the two observed variables (A and B) is statistically significant. At other times, the explanatory latent variable X can be used to help the researcher to explain more fully (rather than to explain away) the observed relationship between the two observed variables. With some sets of data, an appropriate latent class model might include several latent variables as explanatory variables; and these latent variables might be useful in helping the researcher to explain more fully (or to explain away) the observed relationships among the set of observed variables under consideration. Use of such latent class models can help the researcher in many ways, as we shall see later in this exposition on the use of latent class models and in the chapters that follow in this book on latent class analysis.

The problem of measuring the relationship (the nonindependence) between two (or more) observed dichotomous (or polytomous) variables has a long history. This problem has been considered by many researchers in many fields of inquiry at various times throughout the twentieth century, and it is a topic that was also considered by some eminent scholars in the nineteenth century. The use of latent class models as a tool to help researchers gain a deeper understanding of the observed relationships among the observed dichotomous (or polytomous) variables has, in contrast, a much shorter history in the twentieth century, but it might be worthwhile to note here that some mathematical models that were used earlier in some nineteenth-century work can now be viewed as special cases of latent class models or of other kinds of latent structures. With respect to these nineteenth-century models, we refer, in particular, to some work by C. S. Peirce, the great philosopher and logician, who was also an able scientist and mathematician. (In addition to the recognition he has received for some of his other work, he is also sometimes referred to as the "founder of pragmatism.") Peirce introduced such a model (i.e., a latent structure) in order to gain further insight into the relationship between two observed dichotomous variables in the context of measuring the success of predictions (Peirce, 1884; Goodman and Kruskal, 1959). We shall return to this example in a later section herein.

The main development of latent class models has taken place during the last half of the twentieth-century, and the practical application of these models by researchers in various fields of inquiry has become a realistic possibility only during the last quarter of the twentieth century (after more efficient and more usable statistical methods were developed and more general latent class models were introduced). Although the problem of measuring the relationship (the nonindependence) between two or more observed dichotomous (or polytomous) variables has arisen and has been considered in many fields of inquiry at various times throughout the nineteenth and twentieth centuries, we can expect that researchers in some of these fields of inquiry (and in other fields as well) will find that the introduction and application of latent class models can help them to gain further insight into the observed relationships among these observed variables of interest. The introduction of latent class models can insert a useful perspective into the study of the relationships among these variables.

Thus far in this introductory section on latent class analysis we have focused our attention on the possible use of a latent dichotomous or polytomous variable (or a set of such latent dichotomous or polytomous variables) as an explanatory variable (or as explanatory variables) in the study of the relationships among a set of observed (or manifest) dichotomous or polytomous variables. (In this case, our primary focus is on the set of observed variables and on possible explanations of the observed relationships among these variables.) We can also use the latent class models in those situations in which the observed dichotomous or polytomous variables may be viewed as indicators or markers for an unobserved latent variable X, where the unobserved variable is, in some sense, being measured (in an indirect way and with measurement error) by the observed variables. (In this case, our primary focus is on the unobserved latent variable; and the observed variables are, in some sense, ascriptive or attributive variables pertaining to the latent variable.) We can also use models of this kind in the study of the relationships among a set of unobserved (or latent) dichotomous or polytomous variables in the situation in which there are observed dichotomous or polytomous variables that can be viewed as indicators or markers for the unobserved (or latent) variables. (In this case, our primary focus is on the set of unobserved latent variables and on the unobserved relationships among these variables.)

2. THE LATENT CLASS MODEL

Now let us consider the latent class model in the situation in which variable A is an observed (or manifest) dichotomous or polytomous variable having I classes ($i = 1, 2, \ldots, I$), variable B is an observed (or manifest) dichotomous or polytomous variable having J classes ($j = 1, 2, \ldots, J$), and variable X is an unobserved (or latent) dichotomous or polytomous variable having T classes ($t = 1, 2, \ldots, T$). Let π_{ijt}^{ABX} denote the joint probability that an observation is in class i on variable A, in class j on variable B, and in class t on variable X; let $\pi_{it}^{\bar{A}X}$ denote the conditional probability that an observation is in class i on variable A, given that the observation is in class t on variable X; let $\pi_{jt}^{\bar{B}X}$ denote the conditional probability that an observation is in class j on variable B, given that the observation is in class t on variable X; and let π_t^X denote the probability that an observation is in class t on variable X. The latent class model in this situation can be expressed simply as follows:

$$\pi_{ijt}^{ABX} = \pi_t^X \pi_{it}^{\bar{A}X} \pi_{jt}^{\bar{B}X}, \quad \text{for } i = 1, \ldots, I;\ j = 1, \ldots, J;\ t = 1, \ldots, T. \tag{3}$$

This model states that variables A and B are conditionally independent of each other, given the class level on variable X. That is,

$$\pi_{ijt}^{\bar{A}\bar{B}X} = \pi_{ijt}^{ABX} / \pi_t^X = \pi_{it}^{\bar{A}X} \pi_{jt}^{\bar{B}X}, \tag{4}$$

where $\pi_{ijt}^{\bar{A}\bar{B}X} = \pi_{ijt}^{ABX} / \pi_t^X$ is the conditional probability that an observation is in class i on variable A and in class j on variable B, given that the observation is in class t on variable X.

We have presented the latent class model above for the situation in which there are only two observed (manifest) variables (say, A and B). This we do for expository purposes in order to consider this subject in its simplest context. However, it should be noted that some special problems arise when latent class models are considered in the situation in which there are only two observed variables that do not arise in the situation in which there are more than two observed variables. However, these problems need not deter us here. For illustrative purposes, next we shall consider some examples in which latent class models are applied in the analysis of cross-classified data in the situation in which there are two observed variables and also in the situation in which there are more than two observed variables.

3. FIRST EXAMPLE: THE ANALYSIS OF THE RELATIONSHIP BETWEEN TWO OBSERVED VARIABLES

To begin, let us consider the cross-classified data presented in Table 1. These data on the relationship between parental socioeconomic status and mental health status were analyzed earlier by various researchers using various methods of analysis. For the purposes of the present exposition, we note here that the data were used earlier to illustrate both the application of association models and the application of correlation models in measuring the observed relationship (the nonindependence) between the two polytomous variables in Table 1 (see, e.g., Goodman, 1979a, 1985; Gilula and Haberman, 1986). These data were also analyzed by using a latent class model (see, e.g., Goodman, 1987), but in the present

Table 1. Cross-Classification of 1660 Subjects

		Mental Health Status			
Parental Socioecon. Status		**Well 1**	**Mild Symptoms 2**	**Mod. Symptoms 3**	**Impaired 4**
High	1	64	94	58	46
	2	57	94	54	40
	3	57	105	65	60
	4	72	141	77	94
	5	36	97	54	78
Low	6	21	71	54	71

Note: This cross-classification of 1660 subjects is according to their parental socioeconomic status and their mental health status, as shown.

Source: Srole et al. (1962).

exposition we shall more fully examine how the analysis of latent classes in the present context can change in a dramatic way our view of the observed relationship between the two polytomous variables in Table 1 and in other such tables. We shall also include here some simplifications and improvements of some of the results presented in the earlier literature on the analysis of cross-classified data of the kind presented in Table 1 and of the kind presented in the other examples considered herein.

In Table 1, the row categories pertain to parental socioeconomic status (from high to low), and these categories have been numbered (six row categories numbered from 1 to 6); and the column categories pertain to mental health status (from well to impaired), and these categories have also been numbered (four column categories numbered from 1 to 4). These numbers have no special numerical meaning except to indicate which row is being referred to (and possibly where the row appears in the ordering of the rows, if the rows are considered to be ordered) and which column is being referred to (and possibly where the column appears in the ordering of the columns, if the columns are considered to be ordered). With the earlier analysis of Table 1 using correlation models, it was possible to find a set of meaningful numerical scores (different from the integers from 1 to 6) for the row categories and a set of meaningful numerical scores (different from the integers from 1 to 4) for the column categories in Table 1; and the correlation calculated between the meaningful scores for the row categories and the meaningful scores for the column categories for the cross-classified data in Table 1 was also meaningful. The correlation turned out to be small in magnitude, 0.16, but it was statistically significant. Also, with the earlier analysis of Table 1 using the association models, a somewhat similar kind of result was obtained; however, with these models, an index of intrinsic association (rather than an index of correlation) turns out to be meaningful, and the row scores and column scores that are obtained with these association models also turn out to be meaningful (but these scores differ in their meaning from the corresponding scores obtained with the correlation models).

The association models and the correlation models view the two-way 6×4 table in a symmetrical way; they consider the association (or the correlation) between the row variable and the column variable, treating the row and column variables symmetrically. However, these models can also be interpreted in an asymmetrical way in a situation in which we might be interested in the possible dependence of, say, the column variable on the row variable. Here we might be interested in, for example, the possible dependence of mental health status on parental socioeconomic status. With the application of the association models in this context, we can consider the odds of being, say, in mental health status 1 rather than 2, or the odds of being in mental health status 2 rather than 3, or the odds of being in mental health status 3 rather than 4; and we can use the association models to describe how these odds change in a systematic way as we consider these odds for those with different parental socioeconomic status – how these odds change as we move from considering those whose parental socioeconomic status is at the high level to those whose parental socioeconomic status is at a lower level. A somewhat similar kind of result can be obtained when the correlation models are applied to Table 1.

Both the association models and correlation models could be viewed as somewhat improved or somewhat more sophisticated forms of an ordinary regression analysis or an ordinary correlation analysis, or logit analysis or loglinear analysis. They describe, in one way or another, how the two observed variables, the row variable and the column variable, *appear* to be related to each other, or they describe how one of the variables, say, the column variable, *appears* to be related to the other variable or to be affected by the other variable. For the data in Table 1, the apparent relationship or the apparent effect is statistically significant – nevertheless, I continue to refer here to the apparent (or manifest) relationship or to the apparent (or manifest) effect. All of the methods just mentioned (regression analysis, correlation analysis, logit analysis, loglinear analysis, association model analysis, and correlation model analysis) are concerned with apparent (or manifest) relationships or apparent (or manifest) effects. With the introduction of latent class models, we can examine whether these statistically significant apparent relationships and apparent effects might actually be spurious.

Let us now suppose that there are, say, two kinds of families: One kind I shall call simply the "favorably endowed," and the other kind I shall call the "not favorably endowed." These can be viewed as latent categories or latent classes or latent types of families; and we can consider the latent variable E for "endowment" (favorable endowment or not favorable endowment) as a latent dichotomous variable. Further suppose that this latent variable E affects what parental socioeconomic status is attained, and also that it is this latent variable E that affects what the mental health status is of the individual. If this is the case, the relationships among variable S (parental socioeconomic status), M (mental health status), and E (endowment status) can be described by Figure 1(a), where variables S and M are conditionally independent of each other, given the level of variable E (i.e., given the endowment status of the family). In this

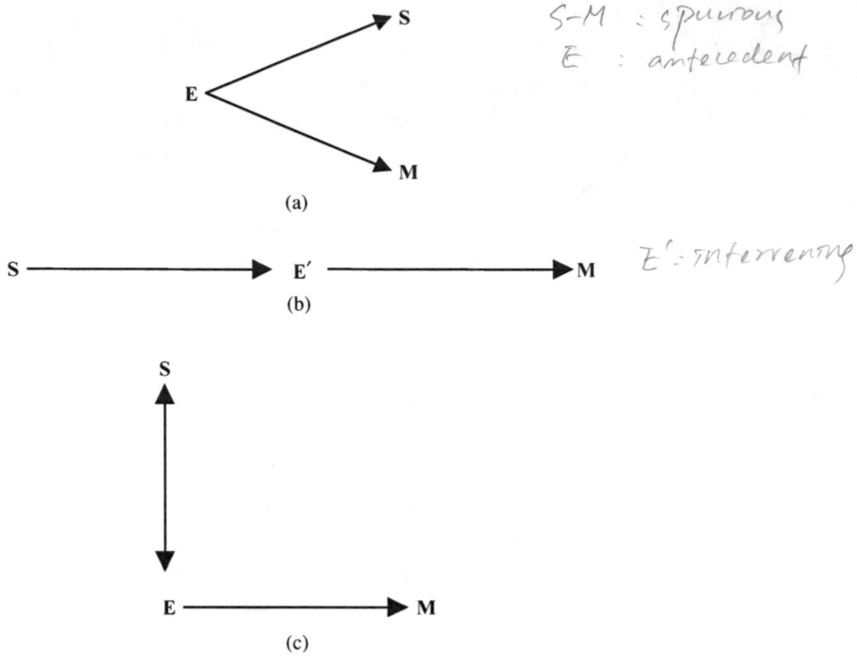

Figure 1. Three different views of the relationship between variables *S* and *M* and variable *E* or variable *E*′: (a) Explanatory latent variable *E* viewed as antecedent to variables *S* and *M*; (b) explanatory latent variable *E*′ viewed as intervening between variables *S* and *M*; and (c) explanatory latent variable *E* viewed as coincident or reciprocal with *S* and antecedent to *M*.

case, the statistically significant relationship observed between parental socioeconomic status and mental health status is spurious.

Next let us consider a somewhat different situation. Let us now suppose that there are, say, two kinds of individuals (rather than two kinds of families): One kind I shall call the "favorably endowed," and the other kind I shall call the "not favorably endowed." These can be viewed as latent categories or latent classes or latent types of individuals; and we can consider the latent variable *E*′ for "endowment" (favorable endowment or not favorable endowment) as a latent dichotomous variable. Further suppose that it is this latent variable *E*′ that affects what the mental health status is of the individual, and also that it is parental socioeconomic status that affects what the endowment status *E*′ is (favorable or not favorable) of the individual. If this is the case, the relationships among variables *S*, *M*, and *E*′ can then be described by Figure 1(b), with variable *S* affecting variable *E*′, and variable *E*′ affecting variable *M*. In this case too, variables *S* and *M* are again conditionally independent of each other, given the level of variable *E*′.

In Figure 1(a), the latent variable E is an antecedent variable, and in Figure 1(b), the latent variable E' is an intervening variable. In addition to Figures 1(a) and 1(b), we might also consider Figure 1(c). Here we again have a somewhat different situation, that is, different from the situations described by Figures 1(a) and 1(b). Figure 1(c) can be used to describe the situation in which variable S (parental socioeconomic status) and latent variable E (endowment status) reciprocally affect each other (or variables S and E are coincident with each other), and it is variable E that affects what the mental health status is of the individual. [In other contexts, where the column variable M might be viewed as prior to the row variable S, we could consider Figure 1(a), and we could also consider the corresponding figures obtained when the symbols S and M are interchanged in Figure 1(b) and also in Figure 1(c).] Each of these figures is congruent with the situation in which variables S and M are conditionally independent of each other, given the level of the latent variable (E or E').[2] Using latent class models, we can examine whether this conditional independence is congruent with the data in Table 1.

It may be worthwhile to note here that if variable E (or E') is viewed as antecedent to variables S and M, as in Figure 1(a), then we could conclude from Figure 1(a) that the observed relationship between S and M has been explained away by variable E (or E'); however, we could also conclude from Figure 1(a) that the observed relationship between S and M has been explained (rather than explained away) by variable E (or E') – that is, by the relationship between variable E (or E') and S and the relationship between variable E (or E') and M. If variable E (or E') is viewed as intervening between variables S and M, as in Figure 1(b), or if variable E (or E') is viewed as coincident or reciprocal with variable S and as antecedent to variable M, as in Figure 1(c), then we could also conclude from Figure 1(b) or Figure 1(c) that the observed relationship between S and M can be explained (rather than explained away) by variable E (or E').

For the cross-classified data in the 6 × 4 table (Table 1) considered here, we present in Table 2 the chi-square values and the corresponding

Table 2. Models Applied to the Cross-Classified Data in Table 1

			Chi-Square	
Model	**No. of Latent Classes**	**Degrees of Freedom**	**Goodness of Fit**	**Likelihood Ratio**
Independence H_0	1	15	45.99	47.42
Latent class H_1	2	8	2.74	2.75

degrees of freedom that are obtained when these data are analyzed by using the following two models: (a) the usual simple null model (H_0) of statistical independence between variables S and M, and (b) the latent class model (H_1) in which the latent variable is dichotomous. [Note that the usual null model H_0 can be described by the simple formula (1) presented earlier; and the latent class model H_1 can be described by formula (3) with $T = 2$, and with $I = 6$ and $J = 4$. We could also describe model H_0 by formula (3) with $T = 1$ and $\pi_1^X = 1$.] From the results presented in Table 2, we conclude that (a) the nonindependence between variables S and M is statistically significant, and (b) the latent class model H_1 fits the observed data extremely well.

From the results presented in Table 3, we can compare the 24 observed frequencies in the 6×4 table (Table 1) with the corresponding expected values estimated under model H_0 and under model H_1. From these results, we see clearly the dramatic improvement in fit that is obtained when the latent class model H_1 is used as a replacement for the usual null model H_0 of statistical independence between the row variable and the column variable in the two-way table. We also see that the goodness-of-fit chi-square value is reduced by 94% (from 45.99 to 2.74) when model H_0 is replaced by model H_1 (with the corresponding reduction in degrees of freedom from 15 to 8). Because model H_1 is the model described in Figures 1(a), 1(b), and 1(c), we find that the cross-classified data in Table 1 are congruent with those figures; thus, any of the corresponding interpretations presented earlier herein for these figures can be applied to these data.

The interpretations of the data obtained with Figures 1(a), 1(b), and 1(c) are very different in character from the kinds of interpretations obtained earlier when other statistical models (e.g., the association models, the correlation models) were used to analyze Table 1. With Figure 1(a), we could explain away the statistically significant relationship described earlier with the association and/or correlation models, or we could explain (rather than explain away) this statistically significant relationship. Also, with Figure 1(b) and/or Figure 1(c) we can explain the statistically significant relationship. With each of these three figures, we can provide a very different kind of explanation of the statistically significant relationship between parental socioeconomic status and mental health status than can be obtained by using the association and/or correlation models or any of the other more usual statistical methods of analysis (e.g., loglinear analysis, logit analysis). In the Appendix we further comment on the latent class analysis of Table 1.

Table 3. Observed and Estimated Expected Frequencies under the H_0 and H_1 Models

		Estimated Expected Frequency	
Cross-Classification	**Obs. Frequency**	**Model H_0**	**Model H_1**
(1,1)	64	48.45	62.22
(1,2)	94	95.01	98.18
(1,3)	58	57.13	56.26
(1,4)	46	61.40	45.34
(2,1)	57	45.31	59.21
(2,2)	94	88.85	92.04
(2,3)	54	53.43	52.55
(2,4)	40	57.41	41.21
(3,1)	57	53.08	58.21
(3,2)	105	104.08	105.26
(3,3)	65	62.59	62.26
(3,4)	60	67.25	61.27
(4,1)	72	71.02	70.03
(4,2)	141	139.26	139.03
(4,3)	77	83.74	83.80
(4,4)	94	89.99	91.14
(5,1)	36	49.01	36.08
(5,2)	97	96.10	93.13
(5,3)	54	57.79	58.61
(5,4)	78	62.10	77.18
(6,1)	21	40.13	21.26
(6,2)	71	78.70	74.36
(6,3)	54	47.32	48.52
(6,4)	71	50.85	72.86
Goodness-of-fit chi square		45.99	2.74
Percentage reduction		0	94
Degrees of freedom		15	8

Note: The H_0 model is the null model of independence; the H_1 model is the two-class latent class model; these models are applied to the cross-classified data in Table 1. The goodness-of-fit chi-square value obtained by applying each of these models to the cross-classified data is also included, and the corresponding percentage reduction in total chi-square value is also shown.

4. SECOND EXAMPLE: THE ANALYSIS OF THE RELATIONSHIPS AMONG FOUR OBSERVED VARIABLES

As our second example, let us next consider the analysis of the cross-classified data presented in Table 4, a four-way table with four observed (manifest) dichotomous variables (say, A, B, C, and D). This table describes the response patterns for respondents to four questionnaire items

Table 4. Cross-Classification of 216 Respondents

Response				
A	***B***	***C***	***D***	**Obs. Frequency**
+	+	+	+	42
+	+	+	−	23
+	+	−	+	6
+	+	−	−	25
+	−	+	+	6
+	−	+	−	24
+	−	−	+	7
+	−	−	−	38
−	+	+	+	1
−	+	+	−	4
−	+	−	+	1
−	+	−	−	6
−	−	+	+	2
−	−	+	−	9
−	−	−	+	2
−	−	−	−	20

Notes: Cross-classification is according to their response in four different situations of role conflict (situations A, B, C, and D). In each of the four different situations of role conflict, the response variable is dichotomous, and the respondent tends either toward *universalistic* values (+) or toward *particularistic* values (−) when responding to the situation with which he or she is confronted.

Source: Stouffer and Toby (1951).

in which four different situations of role conflict are considered. The respondents are cross-classified in Table 4 with respect to whether they tend toward "universalistic" values (+) or "particularistic" values (−) when confronted by each of the situations of role conflict.[3] The cross-classified data in Table 4 were analyzed earlier by Stouffer and Toby (1951) and Lazarsfeld and Henry (1968) by using a particular latent class model that had five latent classes, and by Goodman (1974b) by using much simpler latent class models that had two latent classes and three latent classes. In the present exposition, we shall more fully examine the interpretation of the two-class and three-class latent class models in the analysis of the data in Table 4 and in other such tables, and we shall see how the use of such latent class models in the present context can change in a dramatic way our view of the meaning and the character of the underlying latent variable on which our attention is focused. We shall also include here some simplifications and improvements of some of the results presented in the earlier literature on the analysis of these data.

For expository purposes, let us now present the formula for the latent class model in the situation in which there are four observed (or manifest) dichotomous or polytomous variables (say, A, B, C, and D) and one unobserved (or latent) dichotomous or polytomous variable (say, X):

$$\pi_{ijklt}^{ABCDX} = \pi_t^X \pi_{it}^{\bar{A}X} \pi_{jt}^{\bar{B}X} \pi_{kt}^{\bar{C}X} \pi_{lt}^{\bar{D}X}, \quad \text{for } i = 1, \ldots, I;$$
$$j = 1, \ldots, J;\ k = 1, \ldots, K;\ l = 1, \ldots, L;\ t = 1, \ldots, T, \qquad (5)$$

where π_{ijklt}^{ABCDX} is the joint probability that an observation is in class i on variable A, in class j on variable B, in class k on variable C, in class l on variable D, and in class t on variable X; and the other terms in formula (5) have the same kind of meaning as the corresponding terms in formula (3). Formula (5) states that variables A, B, C, and D are mutually independent of each other, given the class level on variable X. Next let us consider the following models applied to the cross-classified data in the four-way $2 \times 2 \times 2 \times 2$ table (Table 4): (a) the usual null model (M_0) in which variables A, B, C, and D are mutually independent of each other; (b) the latent class model (M_1) in which the latent variable is dichotomous; and (c) the latent class model (M_2) in which the latent variable is trichotomous. [Note that model M_0 can be described by formula (5) with $T = 1$ and $\pi_1^X = 1$; and the latent class models M_1 and M_2 can be described by formula (5) with $T = 2$ and $T = 3$, respectively. When these models are applied to the four-way table, Table 4, we also have $I = J = K = L = 2$ in formula (5).]

For the cross-classified data in Table 4, we present in Table 5a the chi-square values and the corresponding degrees of freedom that are obtained when these data are analyzed by using models M_0, M_1, and M_2 described previously. From the results presented in Table 5a, we see that (a) the usual null model M_0, which states that variables A, B, C, and D are mutually independent of each other, definitely does not fit the observed

Table 5a. Models Applied to the Cross-Classified Data in Table 4

			Chi-Square	
Model	**No. of Latent Classes**	**Degrees of Freedom**	**Goodness of Fit**	**Likelihood Ratio**
Independence M_0	1	11	104.11	81.08
Latent class M_1	2	6	2.72	2.72
Latent class M_2	3	2	0.42	0.39

Table 5b. More Parsimonious Models Applied to the Cross-Classified Data in Table 4

Model	No. of Latent Classes	Degrees of Freedom	Chi-Square: Goodness of Fit	Chi-Square: Likelihood Ratio
Latent class M_2	3	2	0.42	0.39
Latent class M_3	3	9	2.28	2.28
Latent class M_4	3	10	2.42	2.39
Latent class M_5	3	11	2.85	2.72

data in Table 4; and (b) the simple latent class model M_1 that has two latent classes fits the observed data very well indeed. When model M_0 is replaced by M_1, there is a dramatic reduction of 97% in the goodness-of-fit chi-square value (from 104.11 to 2.72), with the corresponding reduction in degrees of freedom from 11 to 6.

As we noted earlier, in addition to the results presented in Table 5a for the latent class model M_1, which has two latent classes, we also presented there the corresponding results obtained when the latent class model M_2, which has three latent classes, is applied to Table 4. With the three-class latent class model M_2, let us next consider in Table 5b several other latent class models (models M_3, M_4, and M_5) that also have three latent classes. Table 5b presents the chi-square values and the corresponding degrees of freedom that are obtained when these latent class models are applied to Table 4. Here again, we obtain some dramatic results.

Comparing the results presented in Table 5b for models M_3, M_4, and M_5 (each of these models having three latent classes) with the corresponding results presented in Table 5a for model M_1 (which has two latent classes), we find that models M_3, M_4, and M_5 are *more* parsimonious than model M_1, and that these more parsimonious models fit the observed data essentially as well as the less parsimonious model M_1. (From Tables 5a and 5b, we note that there are 6 degrees of freedom corresponding to model M_1, and 9, 10, and 11 degrees of freedom corresponding to M_3, M_4, and M_5, respectively; and the goodness-of-fit chi-square values range from 2.28 for M_3 to 2.85 for M_5, with a chi-square value of 2.72 for M_1.) Indeed, again comparing corresponding results in Tables 5a and 5b, we see that the latent class model M_5 in Table 5b turns out to be as parsimonious as the simple null model M_0 of mutual independence among the four observed variables (A, B, C, and D), with 11 degrees of freedom corresponding to model M_0 and also to model M_5; and the goodness-of-fit chi-square value

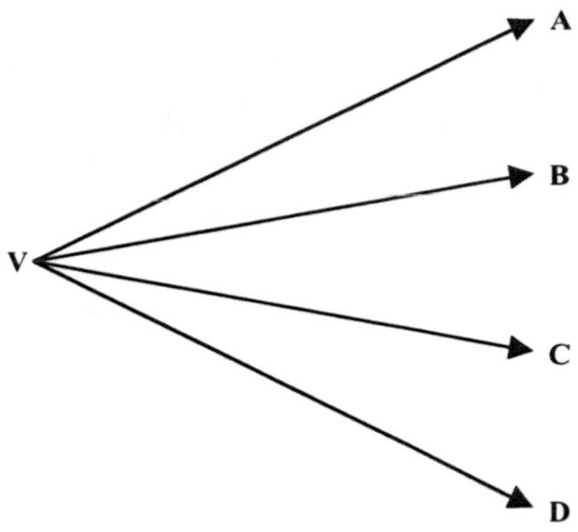

Figure 2. Explanatory latent variable V viewed as antecedent to response (manifest) variables A, B, C, and D.

obtained under M_5 is also 97% less than the corresponding chi-square value obtained under M_0.

Figure 2 can be used to describe any of the latent class models (M_1 to M_5) in Tables 5a and 5b, where the latent variable (dichotomous or trichotomous) is represented as variable V for "values" (universalistic values or particularistic values). In this figure, the latent variable V is viewed as an antecedent variable that can explain away the observed relationships among the observed (manifest) variables A, B, C, and D, or that can explain (rather than explain away) these observed relationships among the observed variables. We can also view the latent variable V as the variable of interest, and the observed variables A, B, C, and D as indicators or markers for the latent variable. Applying the latent class models to the observed data in Table 4, using the observed variables as indicators or markers, we are able to describe and measure the unobserved latent variable (i.e., the "true" latent types). We present the description of this latent variable next for models M_1 and M_3 in Tables 6 and 7.

In model M_1, there are two latent classes, a "universalistically inclined" latent class and a "particularistically inclined" latent class. From Table 6, we see that the probability of a universalistic response in situation A is 0.993 for those who are in the universalistically inclined latent class; and for situations B, C, and D, the corresponding probabilities are 0.940, 0.927, and 0.769, respectively. In addition, the probability of a universalistic response in situation A is 0.714 for those who are in the particularistically inclined latent class; and for situations B, C, and D, the corresponding probabilities are 0.330, 0.354, and 0.132, respectively. The modal response is universalistic in situations A, B, C, and D for those who are in the universalistically inclined latent class; and the modal response is particularistic in situations B, C, and D for those who are in the

Table 6. Distribution of Responses in Situations *A*, *B*, *C*, and *D*

Response	Universal. Incl.	Particular. Incl.
Situation *A*		
+	0.993	0.714
−	0.007	0.286
Situation *B*		
+	0.940	0.330
−	0.060	0.670
Situation *C*		
+	0.927	0.354
−	0.073	0.646
Situation *D*		
+	0.769	0.132
−	0.231	0.868

Notes: Responses are for those in the universalistically (+) inclined latent class and for those in the particularistically (−) inclined latent class, under the two-class latent class model M_1 applied to the cross-classified data in Table 4. In model M_1, 28% are estimated to be in the universalistically inclined latent class, and 72% in the particularistically inclined class.

Table 7. Distribution of Responses in Situations *A*, *B*, *C*, and *D*

Response	Strict. Universal.	Universal./Particular. Incl.	Strict. Particular.
Situation *A*			
+	1.000	0.796	0.000
−	0.000	0.204	1.000
Situation *B*			
+	1.000	0.420	0.000
−	0.000	0.580	1.000
Situation *C*			
+	1.000	0.437	0.000
−	0.000	0.563	1.000
Situation *D*			
+	1.000	0.175	0.000
−	0.000	0.825	1.000

Notes: Responses are for those in the strictly universalistic (+) latent class, for those in the mixed universalistically/particularistically (+/−) inclined latent class, and for those in the strictly particularistic (−) latent class, under the three-class latent class models M_3 applied to the cross-classified data in Table 4. In model M_3, 17% are estimated to be in the strictly universalistic latent class, 78% in the mixed universalistically/particularistically inclined class, and 5% in the strictly particularistic class.

particularistically inclined latent class. (In situation *A*, the modal response is universalistic both for those who are in the universalistically inclined latent class and for those who are in the particularistically inclined latent class; but the corresponding probability of a universalistic response is reduced in the latter latent class from 0.993 to 0.714.) If we think of the responses in items *A*, *B*, *C*, and *D* as indicators, measured with error, of the unobserved latent class (the "true" latent type), then the error rates for items *A*, *B*, *C*, and *D* are 0.007, 0.060, 0.073, and 0.231, respectively, for those who are in the universalistically inclined latent class; and the error rates are 0.714, 0.330, 0.354, and 0.132, respectively, for those who are in the particularistically inclined latent class. The proportion of individuals who are in the universalistically inclined latent class is estimated to be 0.28, and the proportion in the particularistically inclined latent class is estimated to be 0.72.

Now for model M_3 in Table 7: For this model, there are three different latent classes (rather than two); they are the "strictly universalistic," the "mixed universalistically/particularistically inclined," and the "strictly particularistic." The "strictly universalistic" always respond universalistically in situations *A*, *B*, *C*, and *D*; and the "strictly particularistic" always respond particularistically in situations *A*, *B*, *C*, and *D*. For those in the middle latent class, the "mixed universalistically/particularistically inclined," we see that the probability of a universalistic response varies from 0.796 in situation *A* to 0.175 in situation *D*. The proportion of individuals who are in each of the three latent classes is estimated to be 0.17, 0.78, and 0.05 for the strictly universalistic, the mixed universalistically/particularistically inclined, and the strictly particularistic, respectively.

Models M_1 and M_3 make different statements about the heterogeneity of the individuals whose responses in situations of role conflict are described in Table 4. Model M_1 states that there are two different types of individuals, and the model can be used to describe how these two types differ from each other; model M_3 states that there are three different types of individuals, and the model can be used to describe how these three types differ from each other. Each of the latent class models considered herein can be viewed as making a statement about the heterogeneity of the individuals under investigation.

From the information given in Tables 6 and 7, we can calculate, under model M_1 and under model M_3, the expected number of respondents who will be in each of the $2 \times 2 \times 2 \times 2 = 16$ possible cells of the four-way table who are from each of the latent classes; and, for each of the 16 cells in the four-way table, we can also calculate the chance that a respondent in that cell is from each of the latent classes. [Under model M_1, each of the two latent classes – the universalistically inclined class and the particularistically inclined class – contributes to the expected number of respondents who will be in each of the 16 cells of the four-way table; in contrast, under model M_3, only the mixed universalistically/particularistically inclined latent class contributes to the expected number of respondents who will be in each of the 16 cells of the four-way table. The strictly universalistic latent class contributes to the expected number of respondents only in the extreme (+,+,+,+) cell, and the strictly particularistic latent class contributes to the expected number of respondents only in the extreme (−,−,−,−) cell; see, e.g., Tables 4, 6, and 7.]

Models M_1 and M_3 are very different from each other. The corresponding latent variables (see Tables 6 and 7) are also very different from each other (with the universalistically inclined latent class and the particularistically inclined class in model M_1, and the strictly universalistic latent class, the strictly particularistic latent class, and the mixed universally/particularistically inclined class in model M_3). Whereas there are two latent classes in model M_1 and three latent classes in model M_3, we can view model M_3 as essentially consisting of only one latent class (the mixed universalistically/particularistically inclined latent class) contributing to each of the cells in the four-way table, with a special addition applied to the extreme (+,+,+,+) cell (contributed by the strictly universalistic latent class) and a special addition applied to the extreme (−,−,−,−) cell (contributed by the strictly particularistic latent class). As we noted earlier from Tables 5a and 5b, model M_3 is more parsimonious than model M_1, even though M_3 includes one more latent class than does M_1. We also noted from Tables 5a and 5b that both of these models fit the data very well indeed; and we also see from these tables that model M_3 fits the data slightly better than does M_1.

Models M_4 and M_5 in Table 5b are very similar to model M_3, with one added constraint (i.e., $\pi_{12}^{\bar{B}V} = \pi_{12}^{\bar{C}V}$) in model M_4, and two added constraints (i.e., $\pi_{12}^{\bar{B}V} = \pi_{12}^{\bar{C}V}$ and $\pi_{12}^{\bar{A}V} = \pi_{22}^{\bar{D}V}$) in model M_5. We present the results obtained with model M_5 in Table 8 (see also, e.g., Goodman, 1979b). As we noted earlier herein, model M_5 is more parsimonious than any of the other latent class models (the two-class and three-class latent class models) in Tables 5a and 5b, and it is as parsimonious as the simple null model M_0 of mutual independence (and, as we also noted earlier, model M_5 also fits the observed data very well indeed – with a 97% improvement in fit when the fit of this model is compared with that of the simple null model M_0). Compare the distribution of responses in Table 8 for

Table 8. Distribution of Responses in Situations *A*, *B*, *C*, and *D*

Response	Strict. Universal.	Universal./Particular. Incl.	Strict. Particular.
Situation *A*			
+	1.00	0.81	0.00
−	0.00	0.19	1.00
Situation *B*			
+	1.00	0.43	0.00
−	0.00	0.57	1.00
Situation *C*			
+	1.00	0.43	0.00
−	0.00	0.57	1.00
Situation *D*			
+	1.00	0.19	0.00
−	0.00	0.81	1.00

Notes: Responses are for those in the strictly universalistic (+) latent class, for those in the mixed universalistically/particularistically (+/−) inclined latent class, and for those in the strictly particularistic (−) latent class, under the three-class latent class models M_5 applied to the cross-classified data in Table 4. In model M_5, 17% are estimated to be in the strictly universalistic latent class, 77% in the mixed universalistically/particularistically inclined class, and 5% in the strictly particularistic class.

situations *B* and *C*, and also for situations *A* and *D*. We have here in Table 8 an explanation (or description) of the data in Table 4 that is very parsimonious indeed.

We have commented in this section on each of the models in Tables 5a and 5b, except for model M_2. This model will be considered later in the Appendix.

Before closing this section, it might be worthwhile to note here that the results presented in this section can serve to illustrate some different uses of latent class analysis: (1) When the observed (or manifest) variables (A, B, C, D) in Table 4 are viewed as indicators or markers (or as ascriptive or attributive variables) pertaining to the latent variable (or latent concept) V, the latent class model can serve as a "measurement model" in which the latent variable (or latent concept) V can, in some sense, be measured by the observed variables (see, e.g., models M_1, M_3, and M_5 in Tables 6, 7, and 8, respectively). (2) When the observed relationships among the responses on the observed variables (A, B, C, D) are thought to be a consequence of having respondents who differ among themselves in some unobserved manner, the latent class model can be used to investigate the unobserved heterogeneity among the respondents, with each latent class viewed as a homogeneous category, and with the different latent classes used to describe the unobserved heterogeneity among the respondents (see, e.g., how the responses of respondents in the first latent class in, say, model M_1 in Table 6 would differ from the responses of respondents in the second latent class in this model). (3) When the responses on the observed variables (A, B, C, D) are viewed as providing some information about the respondent's latent class membership (either when the latent class model serves as a "measurement model" or as a model that is used to investigate unobserved heterogeneity), the model can be used to classify (in a probabilistic manner) each respondent as being in one of the latent classes on the basis of the respondent's response on one or more of the observed variables (see, e.g., Goodman, 1974a).

5. ANOTHER EXAMPLE: THE SIMPLE 2 × 2 CROSS-CLASSIFICATION TABLE REVISITED: A NINETEENTH-CENTURY LATENT STRUCTURE AND SOME NEW ALTERNATIVES

Let us return now to the analysis of the relationship between two observed variables in the simple context of the 2×2 cross-classification table $\{A, B\}$, where the observed variables A and B are both dichotomous. In the Introduction, we began with this topic; and, in a comment in that section that pertained to nineteenth-century work on this topic, we included there a reference to the work of C. S. Peirce (1884) on the analysis of the 2×2 table in the context of measuring the success of predictions. We shall now show how Peirce's work can be interpreted as an example of the early use of a latent class model in the analysis of a 2×2 table; and we shall also show how the introduction of latent class models in this context can lead to some additional simple measures that are different from the one proposed by Peirce.

With respect to the prediction of, say, tornadoes, where the dichotomous row variable A pertains to tornado prediction (predict "tornado" or predict "no tornado"), and the dichotomous column variable B pertains to tornado occurrence (tornado occurs or no tornado occurs), Peirce pointed out that the observed results in the 2×2 cross-classification table could be viewed as having been obtained by using an infallible predictor a proportion θ of the time, and a completely ignorant predictor the remaining proportion $1 - \theta$ of the time. The infallible predictor predicts "tornado" if a tornado will occur, and he or she predicts "no tornado" if a tornado will not occur; for the ignorant predictor, the chance that he or she will predict "tornado" or "no tornado" is independent of whether the tornado will occur or will not occur. We are asked to contemplate a mixture of the following two 2 × 2 sets of probabilities, pertaining to the infallible predictor and to the ignorant predictor, respectively:

		B 1	B 2
A	1	P_1^B	0
A	2	0	P_2^B

		B 1	B 2
A	1	$\alpha_1 P_1^B$	$\alpha_1 P_2^B$
A	2	$\alpha_2 P_1^B$	$\alpha_2 P_2^B$

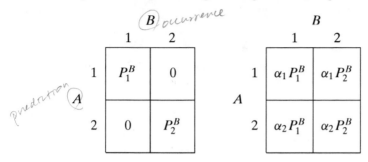

with weights θ and $1 - \theta$, respectively, where the meaning of the four cells is the same as in the 2 × 2 table $\{A, B\}$ for the dichotomous row variable A and the dichotomous column variable B defined earlier in this paragraph; and where P_1^B and P_2^B are the probability of tornado occurrence and the probability of no tornado occurrence, respectively; and where α_1 and α_2 are the probability of the "tornado" prediction and the probability of the "no tornado" prediction, respectively, for the ignorant predictor. Using this model, Peirce found that the proportion θ of the time that an infallible predictor is used can be determined by the following formula:

$$\theta = (P_{11}P_{22} - P_{12}P_{21})/(P_1^B P_2^B) = (P_{11}/P_1^B) - (P_{12}/P_2^B); \qquad (6)$$

see Peirce (1884) and Goodman and Kruskal (1959).

Peirce's model can be viewed as a latent structure in which the observed cross-classification in the 2 × 2 table $\{A, B\}$ is obtained from a mixture of the two 2 × 2 tables, which we described previously, pertaining to an unobserved (or latent) infallible predictor and an unobserved (or latent) ignorant predictor, respectively, where the proportion of the time that each kind of latent predictor is used is unknown. This latent structure is *not*, strictly speaking, a latent class model, because the first 2 × 2 table (i.e., the table pertaining to the infallible predictor) does *not* pertain to a latent class in which there is independence between the dichotomous row variable A and the dichotomous column variable B. We shall next introduce a latent structure that is more general than Peirce's model and that is a latent class model.

Let us consider a three-class latent class model for the 2 × 2 table $\{A, B\}$, in which the first two latent classes pertain to two different kinds of infallible predictors, and the third latent class pertains to an ignorant predictor, with

$$\pi_{11}^{\bar{A}X} = \pi_{11}^{\bar{B}X} = 1, \tag{7a}$$

$$\pi_{22}^{\bar{A}X} = \pi_{22}^{\bar{B}X} = 1. \tag{7b}$$

Peirce's model can be viewed as a special case of this three-class latent class model in which the following additional constraints are imposed on the parameters in the model:

$$\pi_{13}^{\bar{B}X} = P_1^B, \quad \pi_{23}^{\bar{B}X} = P_2^B, \tag{8a}$$

$$\pi_1^X = \theta P_1^B, \quad \pi_2^X = \theta P_2^B, \tag{8b}$$

where the value of θ is unknown. (Note that $\theta = \pi_1^X + \pi_2^X$ in this special case.) With the imposition of the additional constraints, Equations (8a) and (8b), in the previously shown three-class latent class model, we can then obtain formula (6) for θ.

Next let us briefly introduce two other special cases of the previously shown three-class latent class model [when Equations (7a) and (7b) hold true]. Instead of imposing the additional constraints of Equations (8a) and (8b) as in Peirce's model, let us consider, for example, the model obtained with the imposition of the simple constraint $\pi_2^X = 0$ in the three-class latent class model. With this model, there is just one kind of infallible predictor (rather than the two kinds) in the model, and the proportion θ' of the time that the infallible predictor is used can be determined simply by

$$\theta' = P_{11} - P_{12}P_{21}/P_{22}, \tag{9}$$

with $\theta' = \pi_1^X$. Similarly, with the imposition of the constraint $\pi_1^X = 0$ (rather than the constraint $\pi_2^X = 0$) in the three-class latent class model, it is the other kind of infallible predictor that is in the model, and the corresponding proportion θ'' of the time that this infallible predictor is used can be determined simply by

$$\theta'' = P_{22} - P_{12}P_{21}/P_{11}, \tag{10}$$

with $\theta'' = \pi_2^X$.

Previously, we have presented three simple but different formulas, that is, Equations (6), (9), and (10), for determining the proportion of time that use is made of the unobserved (latent) infallible predictor (or predictors). The three formulas (6), (9), and (10) define the corresponding three measures, θ, θ', and θ'', respectively; and the relationships among these three measures can be described simply as follows:

$$\theta = \theta' P_{22} / (P_1^B P_2^B) = \theta'' P_{11} / (P_1^B P_2^B), \tag{11a}$$

$$\theta' / \theta'' = P_{11} / P_{22}. \tag{11b}$$

In some substantive contexts, the three measures can yield very different results. The three latent class models that yield the three measures are quite different from each other [and the three sets of constraints pertaining to the three models – that is, Equations (8a) and (8b), and $\pi_2^X = 0$, and $\pi_1^X = 0$ – are quite different from each other]. Which of the three measures is appropriate will depend on the substantive context (i.e., on which of the three sets of constraints pertaining to the three latent class models is compatible with the substantive context).

The previously shown latent class models treat the cells on the main diagonal in the 2×2 cross-classification table in a way that is different from the way that the other cells in the table are treated (with the latent infallible predictor or predictors directly affecting the entries in the cells on the main diagonal and not directly affecting the entries in the other cells in the table). The need for special treatment for the cells on the main diagonal arises in various substantive contexts in the analysis of 2×2 tables and in the analysis of other square $I \times I$ cross-classification tables (i.e., $I \times J$ tables with $J = I$), where there is a one-to-one correspondence between the I row categories and the I column categories. (In the 2×2 table $\{A, B\}$ considered earlier in this section, the row variable A pertained to whether the prediction was that an event would occur or that it would not occur, and the column variable B pertained to whether the event then did occur or did not occur.) For the analysis of the 2×2 table $\{A, B\}$, we introduced in this section the three-class latent class model with the conditional probabilities pertaining to the first two latent classes ($\pi_{i1}^{\bar{A}X}, \pi_{i2}^{\bar{A}X}, \pi_{j1}^{\bar{B}X}$, and $\pi_{j2}^{\bar{B}X}$) restricted in a way such that these two latent classes would directly affect only the entries in the two cells on the main diagonal in the 2×2 table [see Equations (7a) and (7b)]; and for the analysis of the $I \times I$ table (with $I > 2$), we can consider the corresponding $(I + 1)$-class latent class model with corresponding restrictions imposed on the first I latent classes (so that these I latent classes would directly affect only the entries in the corresponding I cells on the main diagonal in the $I \times I$ table). The parameters $\pi_1^X, \pi_2^X, \ldots, \pi_I^X$ in the $(I + 1)$-class latent class model can then be estimated by using the following formula:

$$\pi_t^X = P_{tt} - \pi_{t,t,I+1}^{ABX}, \quad \text{for } t = 1, 2, \ldots, I, \tag{12}$$

with $\pi_t^X \geq 0$, and where P_{tt} is the probability that an observation will fall in the tth row and tth column of the $I \times I$ table (i.e., in the tth cell on the main diagonal, for $t = 1, 2, \ldots, I$), and $\pi_{t,t,I+1}^{ABX}$ is the corresponding probability, under the latent class model, that an observation will fall in the tth row and tth column of the $I \times I$ table *and* in the $(I+1)$th latent class of the latent class model. (Note that $P_{tt} \geq \pi_{t,t,I+1}^{ABX}$ in this model.) The formula for $\pi_{t,t,I+1}^{ABX}$ in Equation (12) is obtained from formula (3) for the latent class model, with

$$\pi_{t,t,I+1}^{ABX} = \pi_{I+1}^{X} \pi_{t,I+1}^{\bar{A}X} \pi_{t,I+1}^{\bar{B}X}. \tag{13}$$

[Compare formula (13) with formula (3); and also compare formula (12) with formulas (9) and (10).] The probability P_{tt} in formula (12) is estimated simply by the observed proportion p_{tt} of observations in the tth row and tth column of the $I \times I$ table; the estimate $\hat{\pi}_{t,t,I+1}^{ABX}$ of the corresponding $\pi_{t,t,I+1}^{ABX}$ in formula (12) can be obtained from formula (13) by using the model of "quasi-independence" applied to the observed cross-classification in the $I \times I$ table (when $I > 2$) with the entries deleted in the cells on the main diagonal in this table (see, e.g., Goodman, 1968, 1969a, 1969b).[4]

The preceding paragraph considered the analysis of the 2×2 table and the more general square $I \times I$ table (for $I \geq 2$) in the situation in which there is a need for special treatment for the I cells on the main diagonal of the table. [For this situation, we considered the $(I+1)$-class latent class model described previously, and the corresponding quasi-independence model applied with the entries deleted in the I cells on the main diagonal (when $I > 2$).] Similarly, we could also consider the two-way 2×2 table and the more general multiway table (i.e., the m-way table, for $m \geq 2$) in situations in which there is a need for special treatment for a specified subset of the cells in the table. Situations of this kind arise in various substantive contexts; and the method of analysis presented in the preceding paragraph can be directly extended to such situations. For example, in the situation in which there is a need for special treatment for, say, S specified cells in the m-way table, we could consider the $(S+1)$-class latent class model with appropriate restrictions (of the kind described previously) imposed on the first S latent classes [which correspond to the S specified cells in which there is a need for special treatment – see, e.g., Equations (7a) and (7b) for $S = 2$ and $m = 2$], and the corresponding m-way quasi-independence model applied with the entries deleted in the S specified cells of the m-way table. This general approach was applied, for example, in order to introduce a new latent class model for scaling response patterns in the analysis of, say, an m-way 2^m table (for $m > 2$) in which there is a need for special treatment for $m+1$ specified cells in the table – namely, the cells that correspond to the $m+1$ scalable types in the table (see, e.g., Goodman, 1975). In this special case, we have $S = m+1$, and the corresponding latent class model has $S+1 = m+2$ latent classes; and with the application of the corresponding m-way quasi-independence model, the entries are deleted in the cells that correspond to the $m+1$ scalable types.[5]

6. SOME NOTES ON THE HISTORY OF LATENT CLASS ANALYSIS

In the Introduction, we noted that in contrast to the problem of measuring the relationship (the nonindependence) between two (or more) observed dichotomous or polytomous variables, which many researchers have considered at various times throughout the twentieth century and also at times during the nineteenth century,[6] the main development of latent class models has taken place only during the last half of the twentieth century.[7] The literature on this topic that appeared in the 1950s and the 1960s was primarily limited to the situation in which all of the observed (manifest) variables were dichotomous (not polytomous). Lazarsfeld first introduced the term *latent structure models* in 1950; and the various models that he considered as latent structure models (including the latent class model) were concerned mainly with "dichotomous systems" of observed variables.

During the 1950s and 1960s, there were essentially five different methods that were proposed for estimating the parameters in the latent class model: (1) a method suggested by Green (1951) that resembled in some respects a traditional factor analysis; (2) a method suggested by Gibson (1951) that was quite different from the method suggested by Green (1951) but that also resembled factor analysis in some respects[8]; (3) a method that was based on the calculation of the solution of certain determinantal equations that was suggested by the work of Lazarsfeld and Dudman (1951) and Koopmans (1951) and that was developed by Anderson (1954a) and extended by Gibson (1955) and Madansky (1960); (4) a scoring method described by McHugh (1956) for obtaining maximum-likelihood estimates of the parameters in the model; and (5) a partitioning method developed by Madansky (1959) that is based on an examination of each of the possible assignments of the observations in the cross-classification table to the different latent classes.

It turned out that (1) the first method (Green's method), or a version of this method, can provide estimators of the parameters in the latent class model that are not consistent (see, e.g., Madansky, 1968)[9]; (2) the second method (Gibson's 1951 method) also can provide estimators of the parameters in the latent class model that are not consistent[10]; (3) the third method (the determinantal equations method) does provide (under certain conditions) consistent estimators, but these estimators are not efficient, and some of the estimates that are actually obtained by using this method are often not permissible (i.e., not admissible) as estimates of the corresponding parameters in the latent class model (see, e.g., Anderson, 1959; Anderson and Carleton, 1957)[11]; (4) the fourth method (the scoring method) can provide (under certain conditions) efficient estimators, but the estimates that are actually obtained with this method, as described in McHugh (1956, 1958), can have a similar kind of permissibility problem associated with it, as did the determinantal equations method, and an implementation of this procedure in the early 1960s was judged too costly for practical use (see, e.g., Madansky, 1968; Henry, 1983); and (5) the fifth method (the partitioning method) has certain merits (e.g., one version of this method can provide efficient estimators), but the shortcoming of this method is that it is too time consuming, even with fast computers and samples that are not very large, to enumerate and assess each of the possible assignments of the observations in the cross-classification table to the different latent classes (Madansky, 1968).

In assessing what the state of affairs was near the end of the 1960s with respect to the estimation of the parameters in the latent class model, Lazarsfeld and Henry (1968) comment as follows: "A great deal of imaginative thinking and sophisticated programming is still needed before latent class analysis can be routinely applied to a set of data." This state of affairs changed in the middle of the 1970s.

With the introduction then of more general latent class models and the introduction of a relatively simple method for obtaining the maximum-likelihood estimates of the parameters in these models, the practical application of these models by researchers in various fields of inquiry became a realistic possibility (see Goodman, 1974a, 1974b).[12] In contrast to the earlier latent class models that were limited to the analysis of only dichotomous (not polytomous) manifest variables and that included only one latent variable in the model, the new more general latent class models referred to previously have the following advantages: (1) They can be applied to both dichotomous and polytomous observed (manifest) variables; (2) they can include one or more than one unobserved (latent) variable or variables in the model; and (3) they can include a wide range of possible constraints that can be imposed (if desired) on the parameters in the model.

With the wide range of possible constraints that can be imposed (if desired) on the parameters in the more general latent class model, we can obtain a wide range of useful models. Models that can include properly specified constraints can enable the researcher to test a wide range of hypotheses about the structure of the data. With the more general latent class model and with the many possible models that can be obtained when different constraints are specified, we obtain models that can be applied in many different contexts. For example, the more general latent class model can be used to obtain various models for analyzing the scalability of response patterns (i.e., the scaling models; see, e.g., Goodman, 1975; Clogg and Sawyer, 1981), various models that can include error-rate parameters (measurement error, response error, or classification error parameters) to describe how each of the different latent classes (or some of these latent classes) in the model may respond (with error) in the corresponding observed manifest variables (i.e., the measurement models; see, e.g., Goodman, 1974a; Clogg and Sawyer, 1981), various models that allow for the special treatment of the entries in specified cells of the cross-classification table (i.e., the quasi-latent-structure models; see, e.g., Goodman, 1974a; Clogg, 1981a), various models that include both multiple observed indicator variables and multiple observed antecedent (exogenous or causal) variables (i.e., the multiple-indicator/multiple-cause, or MIMIC, models; see, e.g., Goodman, 1974a, Section 5.4; Clogg, 1981b), and various models that can be used for the simultaneous analysis and comparison of the latent structures pertaining to the cross-classified data in two or more multiway tables (i.e., the simultaneous latent structure models; see, e.g., Clogg and Goodman, 1984, 1985, 1986; Birkelund et al., 1996).[13]

With the estimation method introduced in Goodman (1974a, 1974b) for obtaining maximum-likelihood estimates for the more general latent class models, we obtained a method of estimation that is quite different from the five estimation methods (considered earlier in this section) that had been developed earlier specifically for latent class models. The relatively simple method introduced in Goodman (1974a, 1974b) can be viewed simply as a direct extension of (1) the iterative proportional fitting method (fitting one-way margins) used with the quasi-independence model in the analysis of a two-way table (or a multiway table) in which the usual independence model is of interest but some of the entries in the table are missing and/or there is a need for special treatment for some of the cells in the table (see, e.g., Goodman, 1968), and (2) the iterative proportional fitting method (e.g., fitting specified one-way marginals, specified two-way marginals) used with the usual hierarchical loglinear models in the analysis of a multiway table pertaining to the joint distribution of a set of manifest variables (see, e.g., Goodman, 1970). The iterative proportional fitting method for a quasi-independence model can be used to take into account the fact that there are missing (unobserved) entries in the two-way table (or the multiway table) pertaining to the joint distribution of two (or more) manifest variables; and the iterative proportional fitting method introduced in Goodman (1974a, 1974b) can be used to take into account the fact that there are missing (unobserved) variables (i.e., the latent variables) in the multiway table pertaining to the joint distribution of the manifest variables and the latent variables. With the iterative proportional fitting method for the analysis of a multiway table pertaining to the joint distribution of a set of manifest variables directly extended to obtain the iterative proportional fitting method for the analysis of a multiway table pertaining to the joint distribution of a set of manifest variables and latent variables, we obtained an estimation method that can be applied in many contexts in which a set of manifest variables and a related set of latent variables are of interest and also in many contexts in which a set of latent variables and a related set of manifest variables are of interest.

The analysis of a multiway table pertaining to the joint distribution of a set of manifest variables was extended earlier in order to obtain models for the simultaneous analysis and comparison of two or more such multiway tables (see, e.g., Goodman, 1973a), and it was also extended earlier in order to obtain models for the situation in which some of the manifest variables in the multiway table are posterior to other manifest variables in the table, where the manifest variables can be viewed as ordered from first to last, or where the set of manifest variables can be partitioned into mutually exclusive and exhaustive subsets that can then be viewed as ordered from first to last (see, e.g., Goodman, 1973b, 1973c, and the corresponding path diagrams presented in these articles). It may also be worth noting here that the analysis of the multiway table in Goodman (1970) listed all possible elementary loglinear models (for the m-way table, with $m = 2, 3, 4$) that can be described in terms of the concepts of independence and conditional independence; these models would include, for example, the Markov-type models.[14] The models referred to in the present paragraph for the analysis of a multiway table pertaining to the joint distribution of a set of manifest variables (or for the analysis of two or more such multiway tables) can now be extended to the analysis of the relationship between a set of manifest variables and a corresponding set of latent variables. See, for example, Goodman (1974a, Section 5.4), Hagenaars (1988, 1990, 1993), van de Pol and Langeheine (1990), and Vermunt (1997).

Reference should also be made, of course, to many other publications by others, in the 1970s and later, that also contributed to the further development of this subject. With respect to, say, the estimation procedures for latent class models, the reader is referred to, for example, Haberman (1976, 1977, 1988), Dempster et al. (1977), and Vermunt (1997, 1999).[15] In addition to the reference to Haberman's work (on estimation procedures for latent class models) cited previously, we also refer the interested reader to Haberman (1974, 1979) for related material. With respect to, say, the various reviews of the latent class models literature, which covered work done in the late 1970s and later, we cite here, for example, Clogg (1981b, 1988, 1995), Clogg and Sawyer (1981), Andersen (1982), Bergan (1983), McCutcheon (1987), and Langeheine (1988). The reader is referred to these reviews for additional references. Many other contributions to this field could also be cited; however, this would be beyond the scope of this exposition on latent class analysis and these brief notes on its history. We are now pleased to be able to refer the reader to the chapters that follow in this book on latent class analysis.

APPENDIX

Here let us comment further on the latent class analysis presented earlier of the cross-classified data in the two-way 6×4 table (Table 1) and the cross-classified data in the four-way $2 \times 2 \times 2 \times 2$ table (Table 4). For the analysis of the data in the 6×4 table, the latent class model that was used had two latent classes (i.e., the latent variable was dichotomous). With the formula for this model [see formula (3)], we can determine how many basic parameters are included in this model. Replacing the symbols A, B, and X in formula (3) by the corresponding S (parental socioeconomic status), M (mental health status), and E (endowment status), we see that the basic parameters in the model are π_1^E (with $\pi_2^E = 1 - \pi_1^E$), and

$$\pi_{1t}^{\bar{S}E}, \pi_{2t}^{\bar{S}E}, \ldots, \pi_{5t}^{\bar{S}E} \left(\text{with } \pi_{6t}^{\bar{S}E} = 1 - \sum_{i=1}^{5} \pi_{it}^{\bar{S}E} \right), \quad \text{for } t = 1, 2,$$

$$\pi_{1t}^{\bar{M}E}, \pi_{2t}^{\bar{M}E}, \pi_{3t}^{\bar{M}E} \left(\text{with } \pi_{4t}^{\bar{M}E} = 1 - \sum_{j=1}^{3} \pi_{jt}^{\bar{M}E} \right), \quad \text{for } t = 1, 2.$$

Thus, we see that there are $1 + 5(2) + 3(2) = 17$ basic parameters in the model. It is possible to estimate the expected frequencies under this model by using the cross-classified data in the 6×4 table (see the corresponding estimates in Table 3) without actually estimating the basic parameters in the model (see, e.g., Goodman, 1974b); but if these parameters are of substantive interest, then we must note that they become identifiable only after two of these parameters are specified. Any 2 of these parameters can be specified (in any way that does not constrain the expected frequencies estimated under the model) and the other 15 parameters then become identifiable, and they can be estimated by using efficient statistical methods.

Table A1 presents two different sets of numerical values that are obtained as estimates for the parameters in the latent class model after two of the parameters in the model are specified in different ways. The two different sets of estimates in Table A1 yield the same estimated expected frequencies presented in Table 3 for the latent class model H_1. The first set of estimates in Table A1 describes a model (say, model H_1') in which there is a stringent threshold for the favorably endowed (with $\pi_{61}^{\bar{S}E} = 0$ and $\pi_{41}^{\bar{M}E} = 0$; i.e., where the conditional probability is zero of being in the lowest parental socioeconomic class for those who are in the favorably endowed latent class, and the conditional probability is zero of being in the worst mental health class for those who are in the favorably endowed class); and similarly, the second set of estimates in Table A1 describes a model (say, model H_1'') in which there is a stringent threshold for the not favorably endowed (with $\pi_{12}^{\bar{M}E} = 0$ and $\pi_{22}^{\bar{S}E} = 0$; i.e., where the conditional probability is zero of being in the best mental health class for those who are in the not favorably endowed latent class, and the conditional probability is zero of being in the second highest parental socioeconomic class for those who are in the not favorably endowed class). In the first model (H_1'), 30% are estimated to be in the favorably endowed latent class; and in the second model (H_1''), 23% are estimated to be in the not favorably endowed latent class. As we have already noted, these two models (H_1' and H_1'') are equivalent in the sense that they yield the same estimated expected frequencies (see Table 3). The two equivalent models can be viewed as different versions of model H_1. The estimate of the proportion favorably endowed (i.e., π_1^E) under model H_1 cannot be less than 0.30, and the estimate of the proportion not favorably endowed (i.e., π_2^E) under model H_1 cannot be less than 0.23.

Table A2 presents another two different sets of numerical values that are obtained as estimates for the parameters in the latent class model after

Table A1. Distribution of Parental Socioeconomic Status and Distribution of Mental Health Status

	Endowed	
Model	**Favorably**	**Not Favorably**
H_1'		
Parental socioeconomic status		
1	0.25	0.12
2	0.24	0.11
3	0.21	0.16
4	0.22	0.23
5	0.07	0.20
6	0.00	0.19
Mental health status		
1	0.39	0.10
2	0.41	0.34
3	0.21	0.22
4	0.00	0.34
H_1''		
Parental socioeconomic status		
1	0.20	0.01
2	0.19	0.00
3	0.19	0.12
4	0.23	0.24
5	0.12	0.30
6	0.07	0.33
Mental health status		
1	0.24	0.00
2	0.38	0.32
3	0.21	0.23
4	0.17	0.45

Notes: Distributions are for those in the favorably endowed latent class and for those in the not favorably endowed latent class, under two different but equivalent latent class models (models H_1' and H_1'') applied to the cross-classified data in Table 1. In the first model (H_1'), there is a stringent threshold for the favorably endowed; and in the second model (H_1''), there is a stringent threshold for the not favorably endowed. In model H_1', 30% are estimated to be in the favorably endowed latent class, and 70% in the not favorably endowed class; and in model H_1'', 23% are estimated to be in the not favorably endowed latent class, and 77% in the favorably endowed class.

two of the parameters in the model are again specified in different ways. Again, the two different sets of estimates in Table A2 yield the same estimated expected frequencies presented in Table 3 for the latent class model H_1. The first set of estimates in Table A2 describes a model (say, model

Table A2. Distribution of Parental Socioeconomic Status and Distribution of Mental Health Status

Model	Endowed	
	Favorably	Not Favorably
H_1'''		
Parental socioeconomic status		
1	0.25	0.01
2	0.24	0.00
3	0.21	0.12
4	0.22	0.24
5	0.07	0.30
6	0.00	0.33
Mental health status		
1	0.24	0.10
2	0.38	0.34
3	0.21	0.22
4	0.17	0.34
H_1''''		
Parental socioeconomic status		
1	0.20	0.12
2	0.19	0.11
3	0.19	0.16
4	0.23	0.23
5	0.12	0.20
6	0.07	0.19
Mental health status		
1	0.39	0.00
2	0.41	0.32
3	0.21	0.23
4	0.00	0.45

Notes: Distributions are for those in the favorably endowed latent class and for those in the not favorably endowed latent class, under two additional different but equivalent latent class models (models H_1''' and H_1'''') applied to the cross-classified data in Table 1. In model H_1''', the difference is maximized between the distribution of parental socioeconomic status for the favorably endowed and the corresponding distribution of parental socioeconomic status for the not favorably endowed; and in model H_1'''', the difference is maximized between the distribution of mental health status for the favorably endowed and the corresponding distribution of mental health status for the not favorably endowed. In model H_1''', 61% are estimated to be in the favorably endowed latent class, and 39% in the not favorably endowed class; in model H_1'''', 48% are estimated to be in the favorably endowed latent class, and 52% in the not favorably endowed class.

H_1''') in which the difference is maximized between the conditional distribution of parental socioeconomic status for the favorably endowed and the corresponding conditional distribution for the not favorably endowed (with $\pi_{61}^{\bar{S}E} = 0$ and $\pi_{22}^{\bar{S}E} = 0$); and similarly, the second set of estimates in Table A2 describes a model (say, model H_1'''') in which the difference is maximized between the conditional distribution of mental health status for the favorably endowed and the corresponding conditional distribution for the not favorably endowed (with $\pi_{41}^{\bar{M}E} = 0$ and $\pi_{12}^{\bar{M}E} = 0$). In the first model (H_1''') in Table A2, 61% are estimated to be in the favorably endowed latent class and 39% in the not favorably endowed class; and in the second model (H_1''''), 48% are estimated to be in the favorably endowed latent class and 52% in the not favorably endowed class.

Each latent class model in Tables A1 and A2 is described in terms of the estimates for the four conditional distributions ($\pi_{i1}^{\bar{S}E}$, $\pi_{i2}^{\bar{S}E}$, $\pi_{j1}^{\bar{M}E}$, and $\pi_{j2}^{\bar{M}E}$) and the corresponding estimate of the distribution (π_1^E and π_2^E) of the latent classes. Comparing the estimates for models H_1' and H_1'' in Table A1, we see the extent to which the conditional distributions can change as the distribution of the latent classes goes from one extreme (when there is a stringent threshold for the favorably endowed) to the other extreme (when there is a stringent threshold for the not favorably endowed); and comparing the estimates for models H_1''' and H_1'''' in Table A2, we see the extent to which the conditional distributions can change as the differences in the conditional distributions ($|\pi_{i1}^{\bar{S}E} - \pi_{i2}^{\bar{S}E}|$ and $|\pi_{j1}^{\bar{M}E} - \pi_{j2}^{\bar{M}E}|$) go from one extreme (when $|\pi_{i1}^{\bar{S}E} - \pi_{i2}^{\bar{S}E}|$ is maximized) to the other extreme (when $|\pi_{j1}^{\bar{M}E} - \pi_{j2}^{\bar{M}E}|$ is maximized).

As we noted earlier in this Appendix, the 17 basic parameters in the latent class model for the 6×4 table [see formula (3)], become identifiable and can be estimated after 2 of these parameters are specified. Tables A1 and A2 present four different examples of how two of the parameters in each example can be specified. The basic parameters can also be estimated by specifying the numerical value of the parameter π_1^E and the numerical value of a second parameter, which I call the σ multiplier. The parameter σ can be introduced into the latent class model in a way that can increase or decrease the difference ($\pi_{i1}^{\bar{S}E} - \pi_{i2}^{\bar{S}E}$) and can thus decrease or increase the corresponding difference ($\pi_{j1}^{\bar{M}E} - \pi_{j2}^{\bar{M}E}$). As noted earlier in this Appendix, for the data in Table 1 the π_1^E parameter has to be within the closed range from 0.30 to 0.77, and the σ parameter is also limited in its range. [For further details, see, e.g., formulas (A9) and (A10) and footnote 23.]

Before this part of the Appendix on the estimation of the parameters in the latent class model for the data in Table 1 is closed, it is interesting to note from Tables A1 and A2 that the estimates of $\pi_{i1}^{\bar{S}E}$ are the same under H_1' and H_1''' and under H_1'' and H_1'''' (and this is also true for the estimates of $\pi_{j2}^{\bar{M}E}$); and the estimates of $\pi_{j1}^{\bar{M}E}$ are the same under H_1' and H_1'''' and under H_1'' and H_1''' (and this is also true for the estimates of $\pi_{i2}^{\bar{S}E}$). The parameter $\pi_{61}^{\bar{S}E}$ was specified in the same way under H_1' and H_1''' (but it was not specified under H_1'' and H_1''''); the parameter $\pi_{41}^{\bar{M}E}$ was specified in the same way under H_1' and H_1'''' (but it was not specified under H_1'' and H_1'''); the parameter $\pi_{22}^{\bar{S}E}$ was specified in the same way under H_1'' and H_1''' (but it was not specified under H_1' and H_1''''); and the parameter $\pi_{12}^{\bar{M}E}$ was specified in the same way under H_1'' and H_1'''' (but it was not specified under H_1' and H_1''').

The description of the two-class latent class models in Tables A1 and A2 (models H_1', H_1'', H_1''', and H_1'''') applied to the cross-classified data in the 6×4 table (Table 1) viewed the two-way table in a symmetrical way, in the sense that the models could be described by Figure 1(a) and by formula (3) presented earlier, with *A*, *B*, and *X* in formula (3) replaced by *S*, *M*, and *E*, respectively, in Figure 1(a).[16] These tables presented the distribution of parental socioeconomic status and the distribution of mental health status for the favorably endowed latent class and the not favorably endowed latent class. As we noted earlier, in addition to the symmetrical view of the two-way table presented in Figure 1(a), we also can view the two-way table in an asymmetric way with Figures 1(b) and 1(c). With the asymmetric view presented in Figure 1(b), our interest is focused on the distribution of the favorably endowed and the not favorably endowed (i.e., the endowment distribution) for those in each parental socioeconomic status category (rather than on the distribution of parental socioeconomic status for those in the favorably endowed latent class and for those in the not favorably endowed class); see footnote 2 earlier herein. We next consider Table A3, which presents the endowment distribution for those in each parental socioeconomic status category, and also the corresponding endowment distribution for those in each mental health status category,[17] under the latent class models H' and H'' (which we considered earlier in Table A1).[18]

As we noted earlier from Table A1, for the favorably endowed latent class, the estimate of π_1^E, under models H' and H'', was 0.30 and 0.77, respectively; thus, for the not favorably endowed latent class, the estimate of π_2^E under H' and H'' was 0.70 and 0.23, respectively. (Under model H', there was a stringent threshold for the favorably endowed; under model

Table A3. Distribution of the Favorably Endowed and the Not Favorably Endowed

	Endowed	
Model	**Favorably**	**Not Favorably**
$H_1{}'$		
Parental socioeconomic status		
1	0.48	0.52
2	0.50	0.50
3	0.36	0.64
4	0.29	0.71
5	0.13	0.87
6	0.00	1.00
Mental health status		
1	0.63	0.37
2	0.34	0.66
3	0.28	0.72
4	0.00	1.00
$H_1{}''$		
Parental socioeconomic status		
1	0.98	0.02
2	1.00	0.00
3	0.84	0.16
4	0.75	0.25
5	0.56	0.44
6	0.41	0.59
Mental health status		
1	1.00	0.00
2	0.79	0.21
3	0.75	0.25
4	0.55	0.45

Notes: Distribution is for those in each parental socioeconomic status category and for those in each mental health status category, under two different but equivalent latent class models (models $H_1{}'$ and $H_1{}''$) applied to the cross-classified data in Table 1. With respect to models $H_1{}'$ and $H_1{}''$, see corresponding note below Table A1.

H'', there was a stringent threshold for the not favorably endowed.) The difference between the endowment distributions under model H' in Table A3 and the corresponding endowment distributions under model H'' in Table A3 reflects, to a large extent, the difference between the estimate of π_1^E under H' and the corresponding estimate of π_1^E under H'' (and thus it also reflects the difference between the estimate of π_2^E under H' and the corresponding estimate of π_2^E under H''). This comment about the difference between the endowment distributions under model H' and under model H'' will be clarified further with formulas given in Equation (A11).

Returning now to our basic formula (3) for the latent class model for manifest variables A and B and latent variable X, and replacing the symbols S, M, and E used in the present section and in Figure 1 by the corresponding symbols A, B, and X used in formula (3), we noted in footnote 2 that the latent class model described by formula (3) views the row variable A and the column variable B in a symmetric way [as was the case with Figure 1(a), viewing row variable S and column variable M], whereas the equivalent latent class model, which was obtained simply by replacing $\pi_t^X \pi_{it}^{\bar{A}X}$ in formula (3) by the equivalent $\pi_i^A \pi_{ti}^{\bar{X}A}$, views these variables asymmetrically [as was the case with Figure 1(b)]. (Recall that π_i^A here is the probability that an observation is in class i on variable A, and $\pi_{ti}^{\bar{X}A}$ is the conditional probability that an observation is in class t on variable X, given that the observation is in class i on variable A.) Similarly, another equivalent latent class model can be obtained simply by replacing $\pi_t^X \pi_{jt}^{\bar{B}X}$ in formula (3) by the equivalent $\pi_j^B \pi_{tj}^{\bar{X}B}$, where π_j^B is the probability that an observation is in class j on variable B, and $\pi_{tj}^{\bar{X}B}$ is the conditional probability that an observation is in class t on variable X, given that the observation is in class j on variable B.[19] The conditional probabilities $\pi_{ti}^{\bar{X}A}$ and $\pi_{tj}^{\bar{X}B}$ can be expressed simply as

$$\pi_{ti}^{\bar{X}A} = \pi_t^X \pi_{it}^{\bar{A}X} / \pi_i^A, \qquad \pi_{tj}^{\bar{X}B} = \pi_t^X \pi_{jt}^{\bar{B}X} / \pi_j^B. \tag{A1}$$

Note that the symbols π_i^A and π_j^B here have the same meaning as the symbols P_i^A and P_j^B in formulas (1) and (2) given earlier. The endowment distributions in Table A3 present the estimates of $\pi_{ti}^{\bar{X}A}$ and $\pi_{tj}^{\bar{X}B}$ (i.e., the corresponding estimates of $\pi_{ti}^{\bar{E}S}$ and $\pi_{tj}^{\bar{E}M}$), under models H' and H'' (for $t = 1, 2$; $i = 1, \ldots, 6$; and $j = 1, \ldots, 4$).

Having returned now to our basic formula (3) for the latent class model for manifest variables A and B and latent variable X, and replacing the symbols S, M, and E used earlier by the corresponding symbols A, B, and X used in formula (3), we see that Tables A1 and A2 present the estimated $\pi_{it}^{\bar{A}X}$, $\pi_{jt}^{\bar{B}X}$, and π_t^X, under the equivalent latent class models H', H'', H''', and H''''; and Table A3 presents the corresponding estimated $\pi_{ti}^{\bar{X}A}$ and $\pi_{tj}^{\bar{X}B}$, under the equivalent models H' and H''. We shall now present some relatively simple *explicit* formulas that describe the relationship between the estimates of $\pi_{it}^{\bar{A}X}$ under two equivalent models (e.g., models H' and H''), between the estimates of $\pi_{jt}^{\bar{B}X}$ under the two models, between the estimates of π_t^X under the two models, between the estimates of $\pi_{ti}^{\bar{X}A}$ under the two models, and between the estimates of $\pi_{tj}^{\bar{X}B}$ under the two models.

With the latent class model described by formula (3), in the case in which the latent variable is dichotomous we find that

$$\pi_{ij}^{AB} = \sum_{t=1}^{2} \pi_t^X \pi_{it}^{\bar{A}X} \pi_{jt}^{\bar{B}X}, \tag{A2}$$

and

$$\pi_i^A = \sum_{t=1}^{2} \pi_t^X \pi_{it}^{\bar{A}X}, \qquad \pi_j^B = \sum_{t=1}^{2} \pi_t^X \pi_{jt}^{\bar{B}X}, \tag{A3}$$

where π_{ij}^{AB} is the joint probability that an observation is in class i on variable A and in class j on variable B, where π_i^A is the probability that an observation is in class i on variable A, and where π_j^B is the probability that an observation is in class j on variable B. From Equations (A2) and (A3), we find that

$$\pi_{ij}^{AB} - \pi_i^A \pi_j^B = \pi_1^X \pi_2^X \delta_i^A \delta_j^B, \tag{A4}$$

where

$$\delta_i^A = \pi_{i1}^{\bar{A}X} - \pi_{i2}^{\bar{A}X}, \qquad \delta_j^B = \pi_{j1}^{\bar{B}X} - \pi_{j2}^{\bar{B}X}. \tag{A5}$$

This set of formulas, namely, Equations (A2)–(A5), pertains to the estimates of the parameters (π_t^X, $\pi_{it}^{\bar{A}X}$, and $\pi_{jt}^{\bar{B}X}$) under the latent class model H (e.g., under model H'), and a corresponding set of formulas pertains to the estimates of the parameters (say, $\tilde{\pi}_t^X$, $\tilde{\pi}_{it}^{\bar{A}X}$, and $\tilde{\pi}_{jt}^{\bar{B}X}$) under the equivalent latent class model $\tilde{H}$ (e.g., under model H''). Because the models H and $\tilde{H}$ are equivalent, we see from Equation (A4) that

$$\gamma \delta_i^A \delta_j^B = \tilde{\gamma} \tilde{\delta}_i^A \tilde{\delta}_j^B, \tag{A6}$$

where $\gamma = \pi_1^X \pi_2^X$ and $\tilde{\gamma} = \tilde{\pi}_1^X \tilde{\pi}_2^X$, and where $\tilde{\delta}_i^A$ and $\tilde{\delta}_j^B$ are defined by using $\tilde{\pi}_{it}^{\bar{A}X}$ and $\tilde{\pi}_{jt}^{\bar{B}X}$ in the same way that δ_i^A and δ_j^B were defined in Equation (A5) by using $\pi_{it}^{\bar{A}X}$ and $\pi_{jt}^{\bar{B}X}$. We now define σ as the ratio $\tilde{\delta}_i^A / \delta_i^A$; and from Equation (A6) we see that

$$\sigma = \tilde{\delta}_i^A / \delta_i^A = (\delta_j^B / \tilde{\delta}_j^B)(\gamma / \tilde{\gamma}). \tag{A7}$$

Thus,

$$\tilde{\delta}_i^A = \delta_i^A \sigma, \quad \tilde{\delta}_j^B = \delta_j^B (\gamma / \tilde{\gamma}) / \sigma. \tag{A8}$$

From formulas in (A8) we see that the quantity σ can be viewed as a parameter that transforms the numerical value of δ_i^A (from δ_i^A to $\tilde{\delta}_i^A$) in direct proportion to the magnitude of σ, and that transforms the numerical value of δ_j^B (from δ_j^B to $\tilde{\delta}_j^B$) in inverse proportion to the magnitude of σ.[20] From these formulas we can see why I have called σ the *multiplier parameter*.

With the definition of σ introduced previously, we can also obtain the following general formulas[21]:

$$\begin{aligned}(\tilde{\pi}_{it}^{\bar{A}X} - P_i^A) &= (\pi_{it}^{\bar{A}X} - P_i^A)[(\pi_t^X/\tilde{\pi}_t^X)/(\gamma/\tilde{\gamma})]\sigma,\\ (\tilde{\pi}_{jt}^{\bar{B}X} - P_j^B) &= (\pi_{jt}^{\bar{B}X} - P_j^B)(\pi_t^X/\tilde{\pi}_t^X)/\sigma.\end{aligned} \tag{A9}$$

Note also that the formulas in Equation (A9) can be rewritten and simplified as follows when[22] $\tilde{\pi}_t^X = \pi_t^X$:

$$\begin{aligned}\tilde{\pi}_{it}^{\bar{A}X} &= P_i^A + (\pi_{it}^{\bar{A}X} - P_i^A)\sigma,\\ \tilde{\pi}_{jt}^{\bar{B}X} &= P_j^B + (\pi_{jt}^{\bar{B}X} - P_j^B)/\sigma.\end{aligned} \tag{A10}$$

When $\tilde{\pi}_t^X = \pi_t^X$, we see from Equation (A10) that $(\pi_{it}^{\bar{A}X} - P_i^A)$ is transformed directly into $(\tilde{\pi}_{it}^{\bar{A}X} - P_i^A)$ by using the multiplier parameter σ, and that $(\pi_{jt}^{\bar{B}X} - P_j^B)$ is transformed directly into $(\tilde{\pi}_{jt}^{\bar{B}X} - P_j^B)$ by using $1/\sigma$ as the multiplier. We also see from Equation (A10) that when $\tilde{\pi}_t^X = \pi_t^X$, the $\tilde{\pi}_{it}^{\bar{A}X}$ can be expressed simply as a linear function of the multiplier parameter σ, and the $\tilde{\pi}_{jt}^{\bar{B}X}$ can be expressed similarly as a linear function of $1/\sigma$.[23]

Returning now for a moment to the more general formulas in Equation (A9), we see that $(\pi_{it}^{\bar{A}X} - P_i^A)\pi_t^X/\gamma$ is transformed there directly into $(\tilde{\pi}_{it}^{\bar{A}X} - P_i^A)\tilde{\pi}_t^X/\tilde{\gamma}$ by using the multiplier parameter σ, and that $(\pi_{jt}^{\bar{B}X} - P_j^B)\pi_t^X$ is transformed there directly into $(\tilde{\pi}_{jt}^{\bar{B}X} - P_j^B)\tilde{\pi}_t^X$ by using $1/\sigma$ as the multiplier. Note that $(\pi_{it}^{\bar{A}X} - P_i^A)\pi_t^X/\gamma$ and $(\pi_{jt}^{\bar{B}X} - P_j^B)\pi_t^X$ can be expressed as $(\pi_{it}^{AX} - P_i^A\pi_t^X)/\gamma$ and $(\pi_{jt}^{BX} - P_j^B\pi_t^X)$, respectively, where π_{it}^{AX} denotes the joint probability that an observation will be in class i on variable A and in class t on variable X under the latent class model H, and where π_{jt}^{BX} denotes the corresponding joint probability pertaining to class j on variable B and class t on variable X under model H. We can thus rewrite the formulas in Equations (A9) and (A10) to see how $(\pi_{it}^{AX} - P_i^A\pi_t^X)/\gamma$ and $(\pi_{jt}^{BX} - P_j^B\pi_t^X)$ are transformed into the corresponding $(\tilde{\pi}_{it}^{AX} - P_i^A\tilde{\pi}_t^X)/\tilde{\gamma}$ and $(\tilde{\pi}_{jt}^{BX} - P_j^B\tilde{\pi}_t^X)$ under the equivalent latent class model $\tilde{H}$. We also note that $(\pi_{it}^{AX} - P_i^A\pi_t^X)/\gamma$ and $(\pi_{jt}^{BX} - P_j^B\pi_t^X)$ can be expressed as $(\pi_{ti}^{\bar{X}A} - \pi_t^X)P_i^A/\gamma$

and $(\pi_{tj}^{\bar{X}B} - \pi_t^X)P_j^B$, respectively, where $\pi_{ti}^{\bar{X}A}$ and $\pi_{tj}^{\bar{X}B}$ denote the conditional probabilities considered in Equation (A1). Thus we can also rewrite the formulas in Equations (A9) and (A10) to see how $(\pi_{ti}^{\bar{X}A} - \pi_t^X)/\gamma$ and $(\pi_{tj}^{\bar{X}B} - \pi_t^X)$ are transformed into the corresponding $(\tilde{\pi}_{ti}^{\bar{X}A} - \tilde{\pi}_t^X)/\tilde{\gamma}$ and $(\tilde{\pi}_{tj}^{\bar{X}B} - \tilde{\pi}_t^X)$ under the equivalent latent class model $\tilde{H}$. With this rewriting of the formulas in Equations (A9) and (A10), we thus obtain the following formulas corresponding to Equation (A9):

$$\begin{aligned} \tilde{\pi}_{ti}^{\bar{X}A} &= \tilde{\pi}_t^X + \left[\pi_{ti}^{\bar{X}A} - \pi_t^X\right](\tilde{\gamma}/\gamma)\sigma, \\ \tilde{\pi}_{tj}^{\bar{X}B} &= \tilde{\pi}_t^X + \left[\pi_{tj}^{\bar{X}B} - \pi_t^X\right]/\sigma; \end{aligned} \tag{A11}$$

and we obtain the following formulas corresponding to Equation (A10) when $\tilde{\pi}_t^X = \pi_t^X$:

$$\begin{aligned} \tilde{\pi}_{ti}^{\bar{X}A} &= \pi_t^X + \left(\pi_{ti}^{\bar{X}A} - \pi_t^X\right)\sigma, \\ \tilde{\pi}_{tj}^{\bar{X}B} &= \pi_t^X + \left(\pi_{tj}^{\bar{X}B} - \pi_t^X\right)/\sigma. \end{aligned} \tag{A12}$$

From formulas in Equations (A10) and (A12), we see what the effects are of the multiplier parameter σ when $\tilde{\pi}_t^X = \pi_t^X$; and from formulas in Equations (A9) and (A11), we see what the effects are of σ and of $\tilde{\pi}_t^X$ when $\tilde{\pi}_t^X$ may differ from π_t^X.[24]

As we have seen here, in the analysis of the data in the two-way table (the 6×4 table; Table 1), we can estimate the parameters in the latent class model H_1 (a two-class latent class model; see Table 2) after two of the parameters in the model have been specified; and it is possible to specify the two parameters in the model in a way that is meaningful in the substantive context pertaining to the data. Next, we return to the analysis of the data in the four-way table (the $2 \times 2 \times 2 \times 2$ table; Table 4). For these data, we shall now see that we can estimate the parameters in the latent class model M_2 (a three-class latent class model; see Tables 5a and 5b) after one of the parameters in the model has been specified; and it is possible to specify this one parameter in the model in a way that is meaningful in the substantive context pertaining to these data.[25]

Model M_2 in Tables 5a and 5b is very different from latent class model M_1 (a two-class latent class model) in Table 5a and in Table 6; it is also very different from the other three-class latent class models in Table 5b, namely, models M_3, M_4, and M_5. In model M_3, we see from Table 7 that the parameters that were specified were

$$\begin{aligned} \pi_{11}^{\bar{A}V} &= \pi_{11}^{\bar{B}V} = \pi_{11}^{\bar{C}V} = \pi_{11}^{\bar{D}V} = 1, \\ \pi_{13}^{\bar{A}V} &= \pi_{13}^{\bar{B}V} = \pi_{13}^{\bar{C}V} = \pi_{13}^{\bar{D}V} = 0; \end{aligned}$$

and this was also the case in models M_4 and M_5, with one more constraint ($\pi_{12}^{\bar{B}V} = \pi_{12}^{\bar{C}V}$) introduced in model M_4, and two more constraints ($\pi_{12}^{\bar{B}V} = \pi_{12}^{\bar{C}V}$ and $\pi_{12}^{\bar{A}V} = \pi_{22}^{\bar{D}V}$) introduced in model M_5.

Two different but equivalent versions of model M_2 are presented in Table A4; these are models M_2' and M_2''. The expected frequencies

Table A4. Distribution of Responses in Situations *A*, *B*, *C*, and *D*

Response	Universal. Incl.	Universal./Particular. Incl.	Particular. Incl.
Model M_2'			
Situation A			
+	0.995	0.806	0.288
−	0.005	0.194	0.712
Situation B			
+	0.968	0.428	0.000
−	0.032	0.572	1.000
Situation C			
+	0.976	0.407	0.241
−	0.024	0.593	0.759
Situation D			
+	0.863	0.170	0.057
−	0.137	0.830	0.943
Model M_2''			
Situation A			
+	0.998	0.845	0.483
−	0.002	0.155	0.517
Situation B			
+	0.980	0.500	0.096
−	0.020	0.500	0.904
Situation C			
+	1.000	0.448	0.269
−	0.000	0.552	0.731
Situation D			
+	0.913	0.202	0.075
−	0.087	0.798	0.925

Notes: Responses are for those in the universalistically (+) inclined latent class, for those in the mixed universalistically/particularistically (+/−) inclined class, and for those in the particularistically (−) inclined latent class, under two different but equivalent latent class models (the three-class latent class models M_2' and M_2'') applied to the cross-classified data in Table 4. In model M_2', there is a more stringent threshold for the particularistically inclined; in model M_2'', there is a more stringent threshold for the universalistically inclined. In model M_2', 22% are estimated to be in the universalistically inclined latent class, 67% in the mixed universalistically/particularistically inclined class, and 11% in the particularistically inclined class. In model M_2'', 19% are estimated to be in the universalistically inclined class, 58% in the mixed universalistically/particularistically inclined class, and 23% in the particularistically inclined class.

estimated under model M_2' for the cells in the four-way table will be equal to the corresponding expected frequencies estimated under model M_2''. In model M_2', there is a more stringent threshold specified for the particularistically inclined, with $\pi_{13}^{\bar{B}V}$ specified as zero; and in model M_2'', there is a more stringent threshold specified for the universalistically inclined, with $\pi_{11}^{\bar{C}V}$ specified as one. From Table A4, we see that under model M_2', the estimated values for $\pi_{13}^{\bar{A}V}$, $\pi_{13}^{\bar{B}V}$, $\pi_{13}^{\bar{C}V}$, and $\pi_{13}^{\bar{D}V}$ are 0.288, 0.000, 0.241, and 0.057, respectively, with the more stringent threshold for the particularistically inclined; and under model M_2'', the corresponding estimated values for $\pi_{21}^{\bar{A}V}$, $\pi_{21}^{\bar{B}V}$, $\pi_{21}^{\bar{C}V}$, and $\pi_{21}^{\bar{D}V}$ are 0.002, 0.020, 0.000, and 0.087, respectively, with the more stringent threshold for the universalistically inclined. If we think of the responses to items A, B, C, and D as indicators or markers for the unobserved latent classes (i.e., the "true" latent types), where the unobserved variable is, in some sense, being measured (in an indirect way and with measurement error) by the observed variables, then the error rates for items A, B, C, and D are as noted previously for the particularistically inclined under M_2'; and the corresponding error rates are also as noted previously for the universalistically inclined under M_2''. The proportion of individuals in each of the three latent classes is estimated as $\pi_1^V = 0.22$, $\pi_2^V = 0.67$, and $\pi_3^V = 0.11$ under model M_2'; and the corresponding proportion is estimated as $\pi_1^V = 0.19$, $\pi_2^V = 0.58$, and $\pi_3^V = 0.23$ under model M_2''. As the proportion π_1^V decreases from 0.22 to 0.19 (i.e., from model M_2' to M_2''), the corresponding π_2^V also decreases, the corresponding π_3^V therefore increases, and all of the $\pi_{1t}^{\bar{A}V}$, $\pi_{1t}^{\bar{B}V}$, $\pi_{1t}^{\bar{C}V}$, and $\pi_{1t}^{\bar{D}V}$ increase.

There is a one-dimensional continuum of models ranging between models M_2' and M_2'' (as π_1^V ranges between 0.22 and 0.19) that will yield the same estimated expected frequencies as obtained under models M_2' and M_2''. Thus, even though the parameters in model M_2 cannot be estimated until one of the parameters in the model is specified, by comparing the corresponding parameter values in models M_2' and M_2'' we can see what the possible range of the parameter values is; and we find that this range is relatively small here for most of the parameters in the model.

As we noted previously, when the three-class unrestricted latent class model M_2 is applied to the cross-classified data in Table 4 (the four-way $2 \times 2 \times 2 \times 2$ table), there is a one-dimensional continuum of models, ranging between models M_2' and M_2'', that will yield the same estimated expected frequencies as obtained under both models M_2' and M_2''; and when the two-class unrestricted latent class model H_1 is applied to the cross-classified data in Table 1 (the two-way 6×4 table), there is a two-dimensional continuum of models (contained within a subset of two-dimensional space), ranging between models H_1' and H_1'' and between models H_1''' and H_1'''', that will yield the same estimated expected frequencies as obtained under models H_1', H_1'', H_1''', and H_1''''. With respect to the possible range of the parameter values in the model under consideration, for some sets of data the range will be relatively small (as is the case for most of the parameters when model M_2 is applied to the data in Table 4) and for other sets of data this will not be the case.

As we have seen with the two examples considered in this Appendix, when the expected frequencies estimated under a particular model (say, model M) are congruent with the cross-classified data of interest, and this model is of the kind in which the parameters in the model can be estimated from the data only after some of the parameters are specified, it will often be possible to specify the appropriate parameters in a way that is meaningful in the substantive context pertaining to the data. Also, as we have seen with these two examples, it will often be possible to specify the appropriate parameters in different ways that yield the same estimated expected frequencies but provide opposite extremes with respect to some meaningful dimensions, thus obtaining, say, models M' and M'' when only one parameter must be specified, and models M', M'', M''', and M'''' when two parameters must be specified, with M' and M'' providing opposite extremes with respect to one dimension, and M''' and M'''' providing opposite extremes with respect to a second dimension. Comparisons of the corresponding parameter has in models M' and M'' when only one parameter must be specified, or in models M', M'', M''', and M'''' when two parameters must be specified, will shed additional light on the meaning of model M and on the cross-classified data of interest.

REFERENCES

Andersen, E. B. (1982). "Latent structure analysis: a survey," *Scandinavian Journal of Statistics*, **9**, 1–12.

Anderson, T. W. (1954a). "On estimation of parameters in latent structure analysis," *Psychometrika*, **19**, 1–10.

Anderson, T. W. (1954b). "Probability models for analyzing time changes in attitudes." In P. F. Lazarsfeld (ed.), *Mathematical Thinking in the Social Sciences*. Glencoe, IL: Free Press, pp. 17–66.

Anderson, T. W. (1959). "Some scaling models and estimation procedures in the latent class model." In U. Grenander (ed.), *Probability and Statistics*. New York: Wiley, pp. 9–38.

Anderson, T. W., & Carleton, R. O. (1957). "Sampling theory and sampling experiments in latent structure analysis," *Journal of the American Statistical Association*, **52**, 363.

Anderson, T. W., & Goodman, L. A. (1957). "Statistical inference about Markov chains," *Annals of Mathematical Statistics*, **28**, 89–110.

Benini, R. (1928). "Gruppi chiusi e gruppi aperti in alcuni fatti collettivi di combinazioni," *Bulletin de l'Institut International de Statistique*, **23**, 362–83.

Bergan, J. R. (1983). "Latent-class models in educational research." In E. W. Gordon (ed.), *Review of Research in Education*, Vol. 10. Washington, DC: American Educational Research Association, pp. 305–60.

Birkelund, G. E., Goodman, L. A., & Rose, D. (1996). "The latent structure of job characteristics of men and women," *American Journal of Sociology*, **102**, 80–113.

Clogg, C. C. (1977). "Unrestricted and restricted maximum likelihood latent structure analysis: a manual for users," Working Paper 1997-09, Population Issues Research Office, The Pennsylvania State University, University Park, PA.

Clogg, C. C. (1979). "Some latent structure models for the analysis of Likert-type data," *Social Science Research*, **8**, 287–301.

Clogg, C. C. (1980). "Characterizing the class organization of labor market opportunity. A modified latent structure approach," *Sociological Methods and Research*, **8**, 243–72.

Clogg, C. C. (1981a). "Latent structure models of mobility," *American Journal of Sociology*, **86**, 836–68.

Clogg, C. C. (1981b). "New developments in latent structure analysis." In D. M. Jackson & E. F. Borgatta (eds.), *Factor Analysis and Measurement*. Beverly Hills, CA: Sage, pp. 215–46.

Clogg, C. C. (1988). "Latent class models for measuring." In R. Langeheine & J. Rost (eds.), *Latent Trait and Latent Class Models*. New York: Plenum, pp. 173–206.

Clogg, C. C. (1995). "Latent class models." In G. M. Arminger, C. C. Clogg, & M. E. Sobel (eds.), *Handbook of Statistical Modeling for the Social and Behavior Sciences*. New York: Plenum, pp. 311–60.

Clogg, C. C., & Goodman, L. A. (1984). "Latent structure analysis of a set of multidimensional contingency tables," *Journal of the American Statistical Association*, **79**, 762–71.

Clogg, C. C., & Goodman, L. A. (1985). "Simultaneous latent structure analysis in several groups." In N. B. Tuma (ed.), *Sociological Methodology, 1985*. San Francisco: Jossey–Bass, pp. 81–110.

Clogg, C. C., & Goodman, L. A. (1986). "On scaling models applied to data from several groups," *Psychometrika*, **51**, 123–35.

Clogg, C. C., Rudas, T., & Xi, L. (1995). "A new index of structure for the analysis of mobility tables and other cross-classifications." In P. V. Marsden (ed.), *Sociological Methodology, 1995*. Oxford, UK: Basil Blackwell, pp. 197–222.

Clogg, C. C., & Sawyer, D. O. (1981). "A comparison of alternative models for analyzing the scalability of response patterns." In S. Leinhardt (ed.), *Sociological Methodology, 1981*. San Francisco: Jossey–Bass, pp. 240–80.

Cournot, A. A. (1838). "Mémoire sur les applications du calcul des chances á la statistique judiciaire," *Journal de Mathématique Pures et Appliquées*, **3**, 257–334.

de Leeuw, J., & van der Heijden, P. G. M. (1988). "The analysis of time-budgets with a latent time-budget model." In E. Diday (ed.), *Data Analysis and Informatics, V*. Amsterdam: North-Holland, pp. 159–66.

de Meo, G. (1934). "Su di alcuni indici atti a misurare l'attrazione matrimoniale in classificazioni dicotome," *Rendiconto dell'Accademia delle Scienze Fisiche e Matematiche, Napoli*, **73**, 62–77.

Dempster, A. P., Laird, N. M., & Rubin, D. B. (1977). "Maximum likelihood from incomplete data via the EM algorithm," with discussion, *Journal of the Royal Statistical Society, Series B*, **39**, 1–38.

Gibson, W. A. (1951). "Applications of the mathematics of multiple-factor analysis to problems of latent structure analysis," Ph.D. dissertation, Department of Psychology, University of Chicago.

Gibson, W. A. (1955). "An extension of Anderson's solution for the latent structure equations," *Psychometrika*, **20**, 69–73.

Gilula, Z. (1983). "Latent conditional independence in two-way contingency tables: a diagnostic approach," *British Journal of Mathematical and Statistical Psychology*, **36**, 114–22.

Gilula, Z., & Haberman, S. (1986). "Canonical analysis of contingency tables by maximum likelihood," *Journal of the American Statistical Association*, **81**, 780–8.

Goodman, L. A. (1962). "Statistical methods for analyzing processes of change," *American Journal of Sociology*, **68**, 57–78.

Goodman, L. A. (1968). "The analysis of cross-classified data: independence, quasi-independence, and interactions in contingency tables with or without missing entries," *Journal of the American Statistical Association*, **63**, 1091–1131.

Goodman, L. A. (1969a). "How to ransack social mobility tables and other kinds of cross-classification tables," *American Journal of Sociology*, **75**, 1–40.

Goodman, L. A. (1969b). "On the measurement of social mobility: an index of status persistence," *American Sociological Review*, **34**, 832–50.

Goodman, L. A. (1970). "The multivariate analysis of qualitative data: interactions among multiple classifications," *Journal of the American Statistical Association*, **65**, 226–56.

Goodman, L. A. (1971). "Partitioning of chi-square, analysis of marginal contingency tables, and estimation of expected frequencies in multidimensional contingency tables," *Journal of the American Statistical Association*, **66**, 339–44.

Goodman, L. A. (1973a). "Guided and unguided methods for the selection of models for a set of multidimensional contingency tables," *Journal of the American Statistical Association*, **68**, 165–75.

Goodman, L. A. (1973b). "The analysis of multidimensional contingency tables when some variables are posterior to others: a modified path analysis approach," *Biometrika*, **60**, 179–92.

Goodman, L. A. (1973c). "Causal analysis of data from panel studies and other kinds of surveys," *American Journal of Sociology*, **78**, 1135–91.

Goodman, L. A. (1974a). "The analysis of systems of qualitative variables when some of the variables are unobservable. Part I: a modified latent structure approach," *American Journal of Sociology*, **79**, 1179–1259.

Goodman, L. A. (1974b). "Exploratory latent structure analysis using both identifiable and unidentifiable models," *Biometrika*, **61**, 215–31.

Goodman, L. A. (1975). "A new model for scaling response patterns: an application of the quasi-independence concept," *Journal of the American Statistical Association*, **70**, 755–68.

Goodman, L. A. (1978). *Analyzing Qualitative/Categorical Data: Loglinear Models and Latent Structure Analysis.* Cambridge, MA: Abt Books; Lanham, MD: University Press of America.

Goodman, L. A. (1979a). "Simple models for the analysis of association in cross-classifications having ordered categories," *Journal of the American Statistical Association*, **74**, 537–52.

Goodman, L. A. (1979b). "The analysis of qualitative variables using more parsimonious quasi-independence models, scaling models, and latent structures." In R. K. Merton, J. S. Coleman, & P. H. Rossi (eds.), *Qualitative and Quantitative Social Research: Papers in Honor of Paul F. Lazarsfeld*. New York: Free Press, pp. 119–37.

Goodman, L. A. (1979c). "On the estimation of parameters in latent structure analysis," *Psychometrika*, **44**, 123–8.

Goodman, L. A. (1984). *The Analysis of Cross-Classified Data Having Ordered Categories*. Cambridge, MA: Harvard University Press.

Goodman, L. A. (1985). "The analysis of cross-classified data having ordered and/or unordered categories: association models, correlation models, and asymmetry models for contingency tables with or without missing entries," *Annals of Statistics*, **13**, 10–69.

Goodman, L. A. (1987). "New methods for analyzing the intrinsic character of qualitative variables using cross-classified data," *American Journal of Sociology*, **93**, 529–83.

Goodman, L. A. (2000). "The analysis of cross-classified data: notes on a century of progress in contingency table analysis, and some comments on its prehistory and its future." In C. R. Rao & G. Szekely (eds.), *Statistics for the 21st Century*. New York: Marcel Dekker, pp. 189–231.

Goodman, L. A., & Clogg, C. C. (1992). "New methods for the analysis of occupational mobility tables and other kinds of cross-classifications." Symposium on the American Occupational Structure: Reflections after Twenty-Five Years, *Contemporary Sociology*, **21**, 609–22.

Goodman, L. A., & Kruskal, W. H. (1959). "Measures of association for cross-classifications. II: further discussion and references," *Journal of the American Statistical Association*, **54**, 123–63.

Goodman, L. A., & Kruskal, W. H. (1979). *Measures of Association for Cross-Classifications*. New York: Springer-Verlag.

Green, B. F., Jr. (1951). "A general solution for the latent class model of latent structure analysis," *Psychometrika*, **16**, 151–66.

Haberman, S. J. (1974). "Loglinear models for frequency tables derived by indirect observation. Maximum-likelihood equations," *Annals of Statistics*, **2**, 911–24.

Haberman, S. J. (1976). "Iterative scaling procedures for loglinear models for frequency data derived by indirect observation," *Proceedings of the Statistical Computing Section, American Statistical Association, 1975*, 45–50.

Haberman, S. J. (1977). "Product models for frequency tables involving indirect observation," *Annals of Statistics*, **5**, 1124–47.

Haberman, S. J. (1979). *Analysis of Qualitative Data*, Vol. 2. New York: Academic Press.

Haberman, S. J. (1988). "A stabilized Newton–Raphson algorithm for loglinear models for frequency tables derived by indirect observations." In C. C. Clogg (ed.)., *Sociological Methodology, 1988*. San Francisco: Jossey–Bass, pp. 193–211.

Hagenaars, J. A. (1988). "Latent structure models with direct effects between indicators: local dependence models," *Sociological Methods of Research*, **16**, 379–405.

Hagenaars, J. A. (1990). *Categorical Longitudinal Data; Loglinear Panel, Trend and Cohort Analysis*. Newbury Park, CA: Sage.

Hagenaars, J. A. (1993). *Loglinear Models with Latent Variables*. Newbury Park, CA: Sage.

Hagenaars, J. A., & Halman, L. C. (1989). "Searching for ideal types: the potentialities of latent class analysis," *European Sociological Review*, **5**, 81–96.

Henry, N. W. (1983). "Latent structure analysis." In S. Kotz & N. L. Johnson (eds.), *Encyclopedia of Statistical Sciences*, Vol. 4. New York: Wiley, pp. 497–504 .

Koopmans, T. C. (1951). "Identification problems in latent structure analysis," Cowles Commission Discussion Paper: Statistics No. 360, Cowles Commission, University of Chicago.

Langeheine, R. (1988). "New developments in latent class theory." In R. Langeheine & J. Rost (eds.), *Latent Traits and Latent Class Models*. New York: Plenum, pp. 77–108 .

Lazarsfeld, P. F. (1950). "The logical and mathematical foundations of latent structure analysis." In S. S. Stouffer et al. (eds.), *Measurement and Prediction*. Princeton, NJ: Princeton University Press, pp. 362–412.

Lazarsfeld, P. F., & Dudman, J. (1951). "The general solution of the latent class case." Paper No. 5 in *The Use of Mathematical Models in the Measurement of Attitudes*, RAND Research Memorandum No. 455, Part II, edited by P. F. Lazarsfeld et al. RAND Corp., Santa Monica, CA. See also the Introduction to Paper No. 6 in RAND Research Memorandum No. 455, Part II.

Lazarsfeld, P. F., & Henry, N. W. (1968). *Latent Structure Analysis*. Boston, MA: Houghton Mifflin.

Lindsay, B. G., Clogg, C. C., & Greco, J. M. (1991). "Semi-parametric estimation in the Rasch model and related exponential response models, including a simple latent class model for item analysis," *Journal of the American Statistical Association*, **86**, 96–107.

Madansky, A. (1959). "Partitioning methods in latent class analysis," RAND Paper P-1644, RAND Corp., Santa Monica, CA.

Madansky, A. (1960). "Determinantal methods in latent class analysis," *Psychometrika*, **25**, 183–98.

Madansky, A. (1968). "Latent structures." In D. L. Sills (ed.), *International Encyclopedia of the Social Sciences*, Vol. 9. New York: Macmillan and Free Press, pp. 33–8.

McCutcheon, A. L. (1987). *Latent Class Analysis*. Newbury Park, CA: Sage.

McHugh, R. B. (1956). "Efficient estimation and local identification in latent class analysis," *Psychometrika*, **21**, 331–47.

McHugh, R. B. (1958). "Note on 'Efficient estimation and local identification in latent class analysis,'" *Psychometrika*, **23**, 273–4.

Mooijaart, A., & van der Heijden, P. G. M. (1992). "The EM algorithm for latent class analysis with equality constraints," *Psychometrika*, **57**, 261–9.

Peirce, C. S. (1884). "The numerical measure of the success of predictions," *Science*, **4**, 453–4.

Rudas, T., Clogg, C. C., & Lindsay, B. G. (1994). "A new index of fit based on mixture methods for the analysis of contingency tables," *Journal of the Royal Statistical Society, Series B*, **56**, 623–39.

Srole, L., Langer, T. S., Michael, S. T., Opler, M. K., & Rennie, T. A. C. (1962). *Mental Health in the Metropolis: The Midtown Manhattan Study*. New York: McGraw–Hill.

Stigler, S. M. (1986). *The History of Statistics: The Measurement of Uncertainty Before 1900*. Cambridge, MA: Harvard University Press.

Stouffer, S. A., & Toby, J. (1951). "Role conflict and personality," *American Journal of Sociology*, **56**, 395–406. Reprinted in S. A. Stouffer (1962). *Social Research to Test Ideas*. New York: Free Press. Reprinted in part in M. W. Riley (1963). *Sociological Research: A Case Approach*, New York: Harcourt, Brace & World.

van de Pol, F., & Langeheine, R. (1990). "Mixed Markov latent class models." In C. C. Clogg (ed.), *Sociological Methodology, 1990*. Oxford, UK: Basil Blackwell, pp. 213–48 .

van der Ark, L. A. (1999). *Contributions to Latent Budget Analysis: A Tool for the Analysis of Compositional Data*. Leiden, Netherlands: DSWO Press.

van der Ark, L. A., & van der Heijden, P. G. M. (1998). "Graphical display of latent budget analysis and latent class analysis, with special reference to correspondence analysis." In J. Blasius & M. Greenacre (eds.), *Visualization of Categorical Data*. San Diego: Academic Press, pp. 489–508.

van der Heijden, P. G. M., Gilula, Z., & van der Ark, L. A. (1999). "An extended study into the relationship between correspondence analysis and latent class analysis." In M. E. Sobel & M. P. Becker (eds.), *Sociological Methodology, 1999*. Oxford, UK: Basil Blackwell, pp. 147–86.

Vermunt, J. K. (1997). *Loglinear Models for Event Histories*. Thousand Oaks, CA: Sage.

Vermunt, J. K. (1999). "*lEM*: A general program for the analysis of categorical data," Department of Methodology, Tilburg University, Tilburg, The Netherlands.

Weinberg, W. (1902). "Beiträge zur physiologie und pathologie der mehrlingsgeburten beim menschen," *Pflügers Archiv für die Gesamte Physiologie des Menschen und der Terre*, **88**, 346–430.

NOTES

1. The material presented in this exposition is a further development of some of the material presented by the author in the Clifford C. Clogg Memorial Lecture at the International Social Science Methodology Conference, held at the University of Essex in Colchester, England, in July, 1996; it is also a further development of some of the material presented by the author in the

Keynote Address at the Conference on Social Science and Statistics, held in honor of the late Clifford C. Clogg at The Pennsylvania State University in September 1996.

2. Figure 1(a) can be used to describe the latent class model expressed by formula (3) in the preceding section [with A, B, and X in formula (3) now replaced by S, M, and E, respectively]; similarly, Figure 1(b) can be used to describe the equivalent latent class model in which $\pi_t^X \pi_{it}^{\bar{A}X}$ in formula (3) is replaced simply by the equivalent $\pi_i^A \pi_{ti}^{\bar{X}A}$, where here π_t^A is the probability that an observation is in class i on variable A, and $\pi_{ti}^{\bar{X}A}$ is the conditional probability that an observation is in class t on variable X, given that the observation is in class i on variable A; and Figure 1(c) can be used to describe the equivalent latent class model in which $\pi_t^X \pi_{it}^{\bar{A}X}$ in formula (3) is replaced simply by the equivalent π_{it}^{AX}, where π_{it}^{AX} is the joint probability that an observation is in class i on variable A and in class t on variable X. Note that the symbol π_i^A has the same meaning as the symbol P_i^A in formulas (1) and (2) earlier herein. Figure 1(a) views the row variable S and the column variable M in a symmetrical way, whereas Figures 1(b) and 1(c) view these variables asymmetrically. [With Figure 1(b), we presented an asymmetrical view of the latent class model that is sometimes nowadays referred to as the latent budget model; see, e.g., Goodman, 1974a; Clogg, 1981a; de Leeuw and van der Heijden, 1988; van der Ark and van der Heijden, 1998; and van der Ark, 1999.]
3. The four questionnaire items pertain to the respondent's possible expectation (A) that a friend, who was in the respondent's car when the respondent (driving at least at 35 mph in a 20 mph zone) hit a pedestrian, would "testify under oath that the car speed was only 20 mph" (which, according to the respondent's lawyer, "might save the respondent from serious consequences"); and (B) that a doctor friend, who works for an insurance company and examines the respondent when the respondent needs insurance and finds that he has some doubts on some minor points that are difficult to diagnose (in his examination of the respondent), would "shade the doubts in favor of the respondent"; and (C) that a drama-critic friend, who is reviewing a play that he (the drama critic) really thinks is no good, a play in which the respondent has sunk all his savings, would "go easy in his review of the respondent's play"; and (D) that a friend, who is a member of the board of directors of a company and has secret company information that could be financially ruinous to the respondent (unless the respondent gains this information), would "tip off the respondent."
4. Formula (12) can be used to estimate π_t^X by replacing P_{tt} and $\pi_{t,t,I+1}^{ABX}$ in this formula by the corresponding estimates described previously (p_{tt} and $\hat{\pi}_{t,t,I+1}^{ABX}$) when $p_{tt} \geq \hat{\pi}_{t,t,I+1}^{ABX}$, for $t = 1, 2, \ldots, I$. When this inequality is not satisfied for some values of t (say, for $t = t_0$), then the corresponding estimates of P_{tt} (for $t = t_0$) and $\pi_{t,t,I+1}^{ABX}$ (for $t = 1, 2, \ldots, I$) have to be modified in a straightforward way. The estimates of P_{tt} and $\pi_{t,t,I+1}^{ABX}$ in formula (12) can also be obtained directly from the observed cross-classified data in the $I \times I$ table by using the $(I+1)$-class latent class model with appropriate

restrictions (of the kind described previously) imposed on the first I latent classes [see, e.g., restrictions given by Equations (7a) and (7b) for $I = 2$].

5. For the m-way table (with $m \geq 2$), we could also consider more general latent class models in the situation in which there is a need for special treatment for, say, S specified cells in the table. Instead of the $(S+1)$-class latent class model with appropriate restrictions imposed on the first S latent classes (which correspond to the S specified cells in which there is a need for special treatment), we could consider, say, the more general $(S+2)$-class latent class model with appropriate restrictions imposed on the first S latent classes; and, instead of the corresponding m-way quasi-independence model, we would then have the more general m-way quasi-latent-structure model (with two latent classes rather than with one such latent class) applied to the cross-classified data in the m-way table with the entries deleted in the S specified cells of the table (see, e.g., Goodman, 1974a; Clogg, 1981a).
6. For comments on the history and the prehistory (in the twentieth and the nineteenth centuries, respectively) of work on the problem of measuring the relationship (the nonindependence) between two or more observed dichotomous or polytomous variables, see, e.g., Goodman and Kruskal (1979), Stigler (1986), and Goodman (2000).
7. However, it might also be worthwhile to note here (as we noted in the Introduction) that some mathematical models that were used earlier (in some nineteenth-century and early twentieth-century work) can now be viewed as special cases of latent class models or of other kinds of latent structures. The work by C. S. Peirce (1884) was considered, for example, in the preceding section; and, for other early references, see, for example, Cournot (1838), Weinberg (1902), Benini (1928), and De Meo (1934).
8. The methods suggested by Green (1951) and by Gibson (1951) were described also in Anderson (1959); and the method suggested by Gibson (1951) was also described and applied in Lazarsfeld and Henry (1968).
9. Madansky (1968) showed that estimators that are not consistent can be obtained with a version of Green's method in a special case of the two-class latent class model applied to the three-way $2 \times 2 \times 2$ table; but the version of Green's method that was used and reported in Madansky (1968) considered the result that is obtained when the initial estimates suggested by Green are used without iteration (rather than the result that is obtained after the method is used in an iterative manner, as described in Green, 1951). However, Madansky has informed the author that he has now applied Green's method, with the iterative procedure described by Green, to the three-way table example used in Madansky (1968), and the iterative procedure does not converge in this case. A systematic investigation of the conditions under which convergence is obtained with the iterative procedure, and of the conditions under which consistent estimators are obtained, has not been carried out. In addition, with respect to the application of Green's method to the m-way 2^m table, it should be noted that a particular difficulty inherent in this method (the need to estimate the "elements with recurring subscripts" described in Green, 1951) will be more pronounced, in a certain sense, when

$m = 3, 4$, and 5, and it will become less pronounced as m increases for $m \geq 5$. (The ratio of the number of elements with recurring subscripts to the number of elements with nonrecurring subscripts decreases as m increases for $m \geq 3$; and the difference between the number of elements with nonrecurring subscripts and the number of elements with recurring subscripts increases as m increases for $m \geq 4$; and this difference for $m = 3$ is equal to the difference for $m = 5$.) In commenting on Green's method, Lazarsfeld and Henry (1968, p. XXX) state that "... we do not think that it is a workable estimation method in practice."

10. For example, when Gibson's (1951) method is applied by using a two-class latent class model in analyzing a four-way $2 \times 2 \times 2 \times 2$ table, the estimates that are actually obtained (for the parameters in the model) will depend on the particular angle of rotation that is selected with this method (see, e.g., the application of Gibson's method in Lazarsfeld and Henry, 1968, Sections 5.1 and 5.2). However, because the parameters in the two-class latent class model are identifiable (under certain conditions) when applied to the four-way $2 \times 2 \times 2 \times 2$ table (see, e.g., model M_1 in Table 5a and Table 6 earlier herein), the fact that Gibson's method would yield a range of estimates (depending on the angle of rotation that is used) in this case indicates that Gibson's estimators are not consistent in this case. With respect to Gibson's (1951) method, Lazarsfeld and Henry (1968, p. XXX) comment as follows: "As far as we know, there do not exist any computer programs which will carry out the rotations of factors required by Gibson's method." With respect to the rotations required by Gibson's method, Anderson (1959, p. XXX) comments as follows: "The rotation is laborious; this is an art that factor analysts have developed. The final result depends on the centering [i.e., on some "centering principles" suggested by Gibson]; and it is impossible to give any mathematical results about this."

11. Because all of the parameters in the latent class model are probabilities and conditional probabilities (the π_t^X, and the $\pi_{it}^{\bar{A}X}$, $\pi_{jt}^{\bar{B}X}$, ...), all of these quantities must lie in the closed interval from zero to one; however, it is often the case that some of the estimates that are actually obtained by using the determinantal equations method will lie outside this interval (i.e., these estimates will be less than zero or more than one), and are thus not permissible.

    With respect to the "efficiency" of an estimator, this statistical term is used here to mean "asymptotic efficiency." Whenever the determinantal equations method is applicable (i.e., under certain conditions when all of the observed variables in the m-way cross-classification table are dichotomous), this method will always yield estimators that are not efficient *except* in one special case; namely, in the special case of the two-class latent class model applied to the three-way table in which, under certain conditions, the method could yield efficient estimators (see, e.g., Goodman, 1974b). Even in this special case, when the method could yield efficient estimators, some of the estimates that are actually obtained by using this method may not be permissible as estimates of the corresponding parameters in the latent class model (so the permissibility problem, which is present when the determinantal equations method is applied more generally, is also present even in this special case).

12. In commenting on the estimation method introduced in Goodman (1974a, 1974b) for obtaining the maximum-likelihood estimates of the parameters in the more general latent class models, McCutcheon (1987, p. XXX) writes as follows: "Goodman's procedure is simpler and more general [than earlier procedures]. . .Goodman's estimators provide a crucial breakthrough beyond the earlier approaches for the estimation of latent class parameters." McCutcheon (1987, p. XXX) also notes that "Goodman's estimation procedure has been implemented in a readily accessible computer program MLLSA (Maximum Likelihood Latent Structure Analysis)," and he refers the reader to Clogg (1977).
13. These simultaneous latent structure models can be viewed as a direct extension of the corresponding simultaneous loglinear models that were introduced earlier for the simultaneous analysis and comparison of the loglinear models pertaining to the cross-classified data in two or more multiway tables; see, for example, Goodman (1973a).
14. See, for example, Model 6 in Table 4 [with the corresponding formula (4.1), where the order of the variables in the model is *C, A, B, D*], and the related more general material on pp. 240–241, in Goodman (1970). In addition, Markov-type models were considered more explicitly and in more detail in Goodman (1971); and various generalizations of the Markov-type models were introduced in Goodman (1973b). For related material, see also, for example, Anderson (1954b), Anderson and Goodman (1957), and Goodman (1962).
15. The iterative proportional fitting procedure in Haberman (1976), when it is used with the general unconstrained latent class model, is equivalent to the estimation method introduced in Goodman (1974a, 1974b); and the EM algorithm in Dempster et al. (1977), when it is used with the unconstrained latent class model, is also equivalent to the estimation method introduced in Goodman (1974a, 1974b). With respect to constrained latent class models and the particular kinds of equality constraints and fixed-value constraints that were described in Goodman (1974a, Section 5.2) for constraining the parameters in the model, the relatively simple procedures introduced in that article (for estimating the parameters in the corresponding constrained model) will provide maximum-likelihood estimates, as was noted in that article. (The particular kinds of constraints referred to here were also described in Goodman, 1974b, Section 4.) With these particular kinds of constraints, we can obtain a wide range of useful models, as was demonstrated in Goodman (1974a, 1974b). These kinds of constraints can also be generalized still further in a straightforward manner, and an even wider range of useful models is thereby obtained; but for this generalized set of constraints, the relatively simple iterative method introduced in Goodman (1974a, 1974b) would have to be modified somewhat. Mooijaart and van der Heijden (1992) present such a modified version; a somewhat simpler modified version can also be obtained by using, as a part of the iterative method, an appropriate version of the unidimensional Newton algorithm introduced in Goodman (1979a); see Vermunt (1997, 1999).
16. To simplify notation, we shall replace the notation for models H_1', H_1'', H_1''', and H_1'''' by H', H'', H''', and H'''', respectively.

17. With the asymmetric view presented in Figure 1(b), the endowment distribution for those in each parental socioeconomic status category is directly relevant (and the distribution of mental health status for those in the favorably endowed latent class and for those in the not favorably endowed latent class is also directly relevant), whereas the corresponding endowment distribution for those in each mental health status category is not directly relevant. Nevertheless, for the sake of completeness, in Table A3 we have included both the endowment distribution for those in each parental socioeconomic status category and the corresponding endowment distribution for those in each mental health status category.
18. In addition to the endowment distributions presented here under the latent class models H' and H'', we could also have presented, for example, the corresponding endowment distributions under the latent class models H''' and H'''', which we considered earlier in Table A2. This we leave for the interested reader.
19. The $\pi_t^X \pi_{jt}^{\bar{B}X}$ referred to earlier is obtained by rewriting the $\pi_t^X \pi_{it}^{\bar{A}X} \pi_{jt}^{\bar{B}X}$ in formula (3) simply as $\pi_t^X \pi_{jt}^{\bar{B}X} \pi_{it}^{\bar{A}X}$.
20. From the second formula in Equation (A8), we also see that $\tilde{\delta}_j^B / \delta_j^B$ is directly proportional to the magnitude of $\gamma / \tilde{\gamma}$ and is inversely proportional to the magnitude of σ.
21. The formulas in Equations (A9) can be obtained by using the formulas in Equation (A3) pertaining to the estimates of the parameters $(\pi_t^X, \pi_{it}^{\bar{A}X}, \pi_{jt}^{\bar{B}X})$ under the latent class model H, and the corresponding formulas pertaining to the estimates of the parameters $(\tilde{\pi}_t^X, \tilde{\pi}_{it}^{\bar{A}X}, \tilde{\pi}_{jt}^{\bar{B}X})$ under the equivalent latent class model $\tilde{H}$. From the usual formulas for calculating the corresponding maximum-likelihood estimates (see, e.g., Goodman, 1974a, 1974b), we find that the estimates of π_i^A and π_j^B in the formulas in Equation (A3) are equal to the corresponding estimates of P_i^A and P_j^B in the formulas in Equations (1) and (2), namely the corresponding observed proportion in class i on variable A and the corresponding observed proportion in class j on variable B.
22. The formulas in Equation (A10) and the more general formulas in Equation (A9) are, in some respects, simpler than some related (but different) formulas introduced in Goodman (1987); the formulas presented here can help to shed additional light on the earlier results. For some other related (but quite different) kinds of results, see, for example, Gilula (1983), van der Ark (1999), and van der Heijden et al. (1999).
23. Because the probabilities $\tilde{\pi}_{it}^{\bar{A}X}$ and $\tilde{\pi}_{jt}^{\bar{B}X}$ in the formulas in Equation (A10), and in the more general formulas in Equation (A9), have to satisfy the usual inequality constraints for probabilities (that $0 \leq \tilde{\pi}_{it}^{\bar{A}X} \leq 1$ and $0 \leq \tilde{\pi}_{jt}^{\bar{B}X} \leq 1$), the σ parameter is limited in its range (as we noted earlier herein). A similar result can also be obtained by using the corresponding probabilities $\tilde{\pi}_{ti}^{\bar{X}A}$ and $\tilde{\pi}_{tj}^{\bar{X}B}$; see the corresponding formulas presented later herein in Equations (A12) and (A11). With respect to the linear function of the multiplier parameter σ and the linear function of $1/\sigma$ described by the formulas in Equation (A10), see also the corresponding linear functions in Equation (A12).

24. The formulas in Equation (A8) can be applied when $\tilde{\pi}_t^X = \pi_t^X$ and also when $\tilde{\pi}_t^X \neq \pi_t^X$.
25. With respect to the identifiability of the parameters in an unrestricted three-class latent class model applied to a $2 \times 2 \times 2 \times 2$ table, Lazarsfeld and Henry (1968, p. 65) present what they claim is a proof that the parameters in this case are unidentifiable; and Clogg (1981b, p. 243), citing Lazarsfeld and Henry (1968), repeats this claim. However, there is an error in the Lazarsfeld–Henry proof. (Their proof crucially depends on there being 15 basic parameters in the latent class model, whereas there actually are only $2 + (3 \times 4) = 14$ such, parameters in this model.) For the cross-classified data in the particular $2 \times 2 \times 2 \times 2$ table, Table 4, Goodman (1974b) states that the parameters in the three-class latent class model are not "locally identifiable," and he explains why this is the case. (The reader should note that "identifiability" and "local identifiability" are different but related concepts. For further details, see, e.g., Goodman, 1974b.)

TWO

Basic Concepts and Procedures in Single- and Multiple-Group Latent Class Analysis

Allan L. McCutcheon

Latent class analysis is frequently used when the researcher has a set of categorically scored observed measures that are highly interrelated. The latent class model (LCM) – which is often characterized as the categorical data analog to factor analysis – is most appropriately used when the observed indicator variables are associated because of some underlying unobserved factor rather than being causally related.[1] For example, the correctness (incorrectness) of answers to questions on an exam may be highly interrelated as a result of mastery; those who have mastered the material will tend to answer items correctly, and those who have yet to master the material will tend to answer them incorrectly. Thus, in a sufficiently large sampling of student exams, we would anticipate that those who correctly answered question 1 would also be more likely to have correctly answered questions 2, 3, and so forth, yielding a clear association among the "variables" (exam questions). Frequent instances of this kind of association can be found in the social and behavioral sciences (e.g., self-esteem, religiosity, partisan identification, consumer loyalty).

Since the early 1990s, the LCM has emerged as a powerful new method for the analysis of categorically scored data. As is clear from the range of applications in this volume, the range of topics to which the LCM can be fruitfully applied is quite broad. A major reason for the utility of the LCM is that we can use two quite different, although highly interrelated and completely equivalent, parameterizations for this model: probabilistic and loglinear. In this chapter, these two parameterizations will be considered. In addition to exploring the differences between them, we will examine the isomorphisms of the two parameterizations and show how the selection of one may focus the analyst's attention on certain aspects of the model, whereas selection of the other may focus attention on a somewhat different aspect of the model.

Although the alternative parameterizations of the LCM are clearly a strength of the LCM and make the model highly flexible and useful across a wide range of research applications, they also present a set of special challenges. In particular, restrictions that can be readily imposed on one LCM parameterization may be difficult, or even impossible, to replicate in the other parameterization. Thus, although the basic, unrestricted latent class model is identical across the two parameterizations and yields identical "fits" to observed data, restricted models with one parameterization may not be readily translated into the alternative parameterization. Consequently, the researcher may need to consider both parameterizations in order to determine which of the two – the probabilistic or the loglinear – is most appropriate to the research problem.

In this chapter, we first examine the basic unrestricted latent class model in its probabilistic and loglinear parameterizations. This section focuses on model equivalence and interpretation across the two parameterizations. In the second section, model estimation issues are discussed. There are two main approaches to maximum-likelihood estimation (MLE) for LCMs: the expectation–maximization (EM) algorithm and the Newton–Raphson (NR) algorithm. These two iterative algorithms will be briefly introduced, along with some associated issues such as identification and problematic solutions. In the third section, model evaluation criteria are considered. These criteria are critical for guiding our decisions about the appropriateness of accepting a specific model as adequately characterizing the associations observed in the data. In the fourth section, we focus on the model restrictions for each of the parameterizations. Model restrictions are critical for establishing the equivalence of latent class structures when we wish to examine an identical set of categorically scored measures in two or more samples. We will discuss cases in which restrictions that are complicated under one parameterization are readily managed under the other. In the final section, we briefly examine the simultaneous latent class models (SLCMs) in which LCMs are compared in two or more populations. These SLCMs have proven highly useful in comparative research, as well as in the examination of cross-time trends.

1. PARAMETERIZATIONS OF THE BASIC LATENT CLASS MODEL

In this section, we examine the two parameterizations of the basic LCM and demonstrate the equivalence of these two parameterizations. We examine each of the parameterizations in some detail, showing how the common properties of the two parameterizations (e.g., local independence) are manifest in each.

A. Probabilistic Parameterization

Perhaps the most widely used and most intuitively grasped parameterization of the latent class model is the probabilistic parameterization. As Goodman (this volume) has discussed in detail, the probabilistic parameterization of the basic unrestricted LCM is characterized by two types of categorical variables – observed (manifest) indicator variables and unobserved (latent) variables – and two types of parameters: latent class and conditional probabilities. The LCM postulates that the relationship between any two manifest variables is accounted for by the latent variable; this is typically referred to as the *axiom of local independence*. Thus, for an LCM with a single latent variable (X) and four manifest variables (A, B, C, and D), we can formally express the basic LCM as the product of the latent class probabilities and conditional probabilities:

$$\pi_{ijklt}^{ABCDX} = \pi_t^X \pi_{it}^{A|X} \pi_{jt}^{B|X} \pi_{kt}^{C|X} \pi_{lt}^{D|X}, \tag{1}$$

where the latent class probability (π_t^X) is the probability[2] that a randomly selected observation in the sample is located in latent class t, and the conditional probabilities (e.g., $\pi_{it}^{A|X}$) are the probabilities that a member of latent class t will be at a specified level of an observed indicator variable. For example, if our latent variable (X_t) is a measure of religiosity ($t = 1$, religious; $t = 2$, not religious), the first indicator variable (A_i) might be a self-report of church attendance in the previous week ($i = 1$, yes; $i = 2$, no). Thus, the conditional probability $\pi_{11}^{A|X}$ is the probability that a randomly selected religious (i.e., latent class 1) respondent would report having attended church in the previous week.

Within the LCM, hypotheses are tested by imposing restrictions and determining how these restrictions affect the fit of the model to the data. Goodman (1974a, 1974b) has shown that the LCM can be made identifiable by imposing a set of logical constraints (restrictions) on the basic LCM. Thus, for the basic LCM with a single latent variable (X_t) and four observed indicator variables (A_i, B_j, C_k, and D_l), we can express the restrictions as

$$\sum_t \pi_t^X = \sum_i \pi_{it}^{A|X} = \sum_j \pi_{jt}^{B|X} = \sum_k \pi_{kt}^{C|X} = \sum_l \pi_{lt}^{D|X} = 1.0. \tag{2}$$

The restriction that $\sum_t \pi_t^X = 1.0$ requires that the latent classes sum to

1.0 – that there is a latent class for each of the possible response patterns observed in the data. The remaining restrictions require that each of the indicator variables sum to one within each of the T classes.

Consider, for example, the data collected by Stouffer and Toby (1951) in their study of role conflict. As noted elsewhere in this volume (see Goodman), in February 1950 these researchers asked a group of Harvard and Radcliffe students about a set of situations "involving conflict between obligations to a friend and more general social obligations" (1951, p. 396). Students were presented with four scenarios in which either they or their friend confronted each of four role conflicts. For students responding to the friend's role conflict, each scenario was of the following type (item A):

Your close friend is riding in a car that you are driving, and you hit a pedestrian. He knows that you were going at least 35 miles an hour in a 20-mile-an-hour zone. There are no other witnesses. Your lawyer says that if your friend testifies under oath that the speed was only 20 miles an hour, it may save you from serious consequences. What right do you have to expect him to protect you?

The other scenarios involve similar role conflicts for the respondent's close friend who is a drama critic (item B), an insurance doctor (item C), and a member of a board of directors (item D). An equal number of students were given scenarios with the same four situations, but modified so that the students themselves were in the role conflict position. The distributions of the responses for the four scenarios in these two situations are reported in Table 1.

Table 1. Responses to Four Role Conflict Scenarios for Ego and Ego's Close Friend

			Dilemma			
Items			Ego Faces		Ego's Friend Faces	
A	B	C	Item D (+)	Item D (−)	Item D (+)	Item D (−)
+	+	+	20	2	20	3
+	+	−	6	1	4	3
+	−	+	9	2	23	3
+	−	−	4	1	4	2
−	+	+	38	7	25	6
−	+	−	25	6	15	6
−	−	+	24	6	29	5
−	−	−	23	42	31	37

Source: Stouffer and Toby (1951).

As the data in Table 1 indicate, there are 16 (2^4) response patterns for each of the two dilemma situations (Ego and Ego's Friend). In general, if there are k dichotomous items, there are 2^k possible response patterns. Each of the 16 response patterns was selected by one or more of the student respondents for each of the dilemma situations. In this section, we focus only on the responses to the situation confronting Ego. We consider both situations later, when we take up the issue of multisample latent class analysis.

One rather obvious question that emerges from Table 1 is whether we must consider the 16 response patterns for each scenario as representing 16 distinct types, or whether some lesser number of patterns might account for the observed distribution of responses. If we allow for measurement error in each of the four indicator variables, we might view some of the responses as the result of misclassification caused by the measurement error, rather than true response types.

Setting aside, for a moment, the issue of how we determine if a particular latent class model is an adequate representation of the observed data, we can present the two-class LCM for the data in the 16 cells of the leftmost two columns in Table 1. The conditional probabilities (e.g., $\pi_{it}^{A|X}$) for respondents saying that their friend had a right to expect them to violate their role responsibilities (i.e., the positive or *particularistic* response), along with the latent class probabilities (π_t^X), for the probabilistic parameterization of the unrestricted basic LCM are reported in Table 2.

As the data in Table 2 indicate, approximately three-quarters (0.7206) of the sample are Class 1–type respondents and, by the restriction that $\sum_t \pi_t^X = 1.0$ specified in Equation (2), about one-quarter ($1.0 - 0.7206 = 0.2794$) of the sample are Class 2–type respondents. We can use the conditional probabilities to characterize the two latent classes in much the same way that factor loadings are used to characterize the factors in factor

Table 2. Latent Class and Conditional Probabilities for Ego's Dilemma

Indicator	Class 1	Class 2
A. Passenger Friend	0.286	0.007
B. Drama Critic Friend	0.646	0.074
C. Insurance Doctor Friend	0.670	0.060
D. Board of Directors Friend	0.868	0.231
Latent Class Probabilities	0.7206	0.2794

analysis.[3] Thus, we can see that Class 1–type respondents appear to have a consistently higher probability of giving the particularistic response – that is, of saying their friend had a right to ask them to violate their role obligation – than do Class 2–type respondents. Thus, we may wish to regard Class 1 respondents as the "particularistic" respondents and the Class 1 as the "universalistic" respondents (Stouffer and Toby, 1951, pp. 395–396; Parsons, 1949, Chapter 8).[4]

Before we turn to the loglinear representation of the LCM, it is worth noting that the restrictions on the conditional probabilities of the LCM expressed in Equation (2) mean that it is not necessary to report both the probability that the respondent would choose "my friend has a right" and the probability that he or she would choose the response "my friend does not have a right," because these two probabilities must sum to 1.0. Thus, for the dichotomous response options reported here (i.e., $I = J = K = L = 2$), we need report only the probability of a single response (e.g., the particularistic response), because the probability of giving the universalistic response is equal to the difference between 1.0 and the probability of giving the particularistic response (e.g., $\pi_{21}^{A|X} = 1.0 - 0.286 = 0.714$).

B. Loglinear Parameterization

It is also possible to represent the LCM as a loglinear model (Goodman, 1974a; Haberman, 1979, especially Chapter 10). The loglinear parameterization of the unrestricted basic LCM, however, differs from the usual loglinear model in two important ways. First, the loglinear model for the LCM includes an unobserved, latent variable (X_t). Second, only the two-variable parameters between the latent variable and each indicator variable are included – all higher-order terms involving combinations among the indicator variables of the loglinear LCM are set to zero. Thus, the loglinear LCM also exhibits local independence among the indicator variables. This basic model is expressed as

$$\ln\left(f_{ijklt}^{ABCDX}\right) = \lambda + \lambda_t^X + \lambda_i^A + \lambda_j^B + \lambda_k^C + \lambda_l^D + \lambda_{it}^{AX} + \lambda_{jt}^{BX} + \lambda_{kt}^{CX} + \lambda_{lt}^{DX}. \tag{3}$$

As Equation (3) indicates, the loglinear LCM includes only the single-variable lambda parameters along with the two-variable association parameters between each of the indicator variables and the latent variable.

Table 3. Loglinear Parameters for Ego's Dilemma Latent Class Model

	Parameter	
Indicator	**Single Variable**	**Two Variable**
A. Passenger Friend	−1.472	1.016
B. Drama Critic Friend	−0.483	0.784
C. Insurance Doctor Friend	−0.509	0.864
D. Board of Directors Friend	0.169	0.771
X. Latent Class Variable	0.474	—

As with the probabilistic parameterization (and with ordinary loglinear models), it is necessary to impose a set of identifying restrictions.

$$\begin{aligned}\sum_t \lambda_t^X &= \sum_i \lambda_i^A = \sum_j \lambda_j^B = \sum_k \lambda_k^C = \sum_l \lambda_l^D = \sum_i \lambda_{it}^{AX} \\ &= \sum_t \lambda_{it}^{AX} = \sum_j \lambda_{jt}^{BX} = \sum_t \lambda_{jt}^{BX} = \sum_k \lambda_{kt}^{CX} \\ &= \sum_t \lambda_{kt}^{CX} = \sum_l \lambda_{lt}^{DX} = \sum_t \lambda_{lt}^{DX} = 0. \end{aligned} \tag{4}$$

These restrictions to the loglinear model require that the product of odds and odds ratios are 1 (i.e., that the natural logarithm is 0). This, for example, means that if you have twice the likelihood of being at level 1 relative to level 2, you have the reciprocal likelihood (0.5) of being at level 2 relative to level 1.

The estimates of the loglinear parameters for the Stouffer and Toby study are reported in Table 3. Typically, of greatest interest are the two-variable association parameters that relate each of the indicator variables to the latent variable. Although not necessarily intuitive, the lambda parameters range from negative to positive infinity, with zero indicating complete independence. Thus, the positive values for each of the four two-variable parameters indicate that the student respondents' particularistic responses are positively associated with their location as Class 1–type respondents.

An alternative and somewhat more intuitive approach to interpreting the two-variable loglinear LCM parameters is to convert them to odds ratios. For example, we might note that the estimate of 1.016 for latent variable item A (passenger friend) corresponds to an estimated log cross-product ratio of

$$\tau^{AX} = 4\lambda^{AX} = 4.064$$

with an estimated cross-product ratio of

$$e^{4.064} = 58.21.$$

Thus, we would conclude from the estimates reported in Table 3 that the odds that a respondent with a Class 1–type of latent attitude will give a particularistic response to the passenger friend indicator are 58 times as great as the odds if he or she is a Class 2–type of respondent. The other two-variable lambda coefficients may be similarly interpreted.

Finally, it is important to note that the unrestricted, basic LCMs expressed in Equations (1) and (3) are essentially equivalent, requiring the estimation of an identical number of parameters and yielding identical expected values. This equivalence can be illustrated, in part, through the equivalence between the conditional probabilities of the probabilistic parameterization and the loglinear parameters (Haberman, 1979, p. 551):

$$\pi_{it}^{A|X} = \exp\left(\lambda_i^A + \lambda_{it}^{AX}\right) / \sum_i \exp\left(\lambda_i^A + \lambda_{it}^{AX}\right). \tag{5}$$

We can use the lambda coefficients reported in Table 3 to calculate the conditional probabilities reported in Table 2. For example, we can calculate the probability that a particularistic (latent Class 1)-type of respondent would give the particularistic response to the passenger friend indicator as

$$\begin{aligned}\pi_{11}^{A|X} &= \frac{\exp(-1.472 + 1.016)}{\exp(-1.472 + 1.016) + \exp(1.472 - 1.016)}\\ &= \frac{0.633560}{0.633560 + 1.578382} = 0.286427.\end{aligned}$$

Each of the other conditional probabilities in Table 2 can be derived in an analogous manner from the loglinear lambda coefficients presented in Table 3.

The probabilistic and loglinear parameterizations each permit the researcher to test a number of interesting hypotheses the researcher might wish to test by imposing restrictions on the model parameters. However, the two parameterizations lend themselves to somewhat different hypotheses. Before we consider restricted models, a brief discussion of model evaluation will be useful.

2. MODEL ESTIMATION

There are two main alternatives to estimating the parameters of the LCM – the expectation–maximization and the Newton–Raphson algorithms – both of which are iterative, maximum-likelihood estimation approaches. As iterative approaches, both algorithms begin with a set of "start values" and proceed with a series of steps of parameter estimation and reestimation iterations until some designated criterion is reached. Usually, the "stop" criteria focuses on convergence – each additional iteration in the parameter reestimation procedure finally approaches some predesignated "small" change, and the procedure stops. As a number of excellent expositions of the EM and NR algorithms and their variants now exist (see, e.g., Everitt, 1987; McLachlan and Krishnan, 1997; see also Vermunt, 1997a, and Wedel and Kamakura, 1998), only a brief discussion of these approaches is considered here.

The EM algorithm has become one of the most widely used approaches to LCM estimation. Certainly among the contributing factors to the popularity of the EM algorithm are its robustness with respect to the initial (start) values – these can be quite distant from the final estimates and will still reach at least a local maximum – and its relative ease to program. Two of the more frequently mentioned disadvantages of the EM algorithm are that this approach may require a large number of iterations to reach a final solution and that the EM algorithm does not directly provide estimates of the standard errors. The rapid increase in computational speed has reduced the first of these problems.

The EM algorithm consists of two steps. In the first step – the expectation, or *E*, step – the expected value of the log of the likelihood function is computed, conditional on the observed data and the initial parameter estimates. In the second step – the maximization, or *M*, step – the function is maximized in order to give updated values of the parameter estimates. These new estimates of the parameters replace the initial estimates and the algorithm returns to the *E* step. The algorithm continues in this iterative manner until the changes in either the parameter estimates or the changes in the likelihood function (or its logarithm) reach some predefined level of precision, at which point the iterative process halts.[5]

The NR method also uses an iterative approach to produce maximize likelihood estimates of the LCM parameters. The NR method has the advantage of being relatively fast and producing the standard errors of the parameter estimates as a by-product of parameter estimation. Among the disadvantages of this method are the need to invert the Hessian matrix at every iteration and the requirement that the initial estimates must be close to the final estimates or the matrix may become negative definite and thus cannot be inverted. Consequently, when the start values used with this method are too far from the final solution, the NR algorithm may not converge on a set of final estimates.

The NR method begins with a set of initial parameter values (θ) and improves on these values by modifying them by the product of the inverse of the Hessian matrix (H) and the gradient vector (g) of derivatives of the log-likelihood function:

$$\theta_{i+1} = \theta_i - H_i^{-1} g_i . \tag{6}$$

The parameter estimates are updated at the end of each iteration. As noted earlier, the convergence of this method is quite fast when the initial parameter estimates (θ) are near the ML estimates.

An important problem with likelihood approaches is that they can have several local optima. That is, both the EM and NR estimation methods may converge to local optima and not the true (global) maximum of the likelihood function. To determine whether this has occurred, the researcher can repeat the procedure by using different start values. If the results of the repeated estimation nets different parameter estimates and a higher likelihood (i.e., a lower value for the test statistic G^2; see the next section), then the result with the highest likelihood (the lowest G^2 value) is the MLE.

Figure 1 is a hypothetical illustration of the local optima issue for a parameter space from the probabilistic parameterization. Because probabilities are bounded by zero and one, no parameter estimate may lie

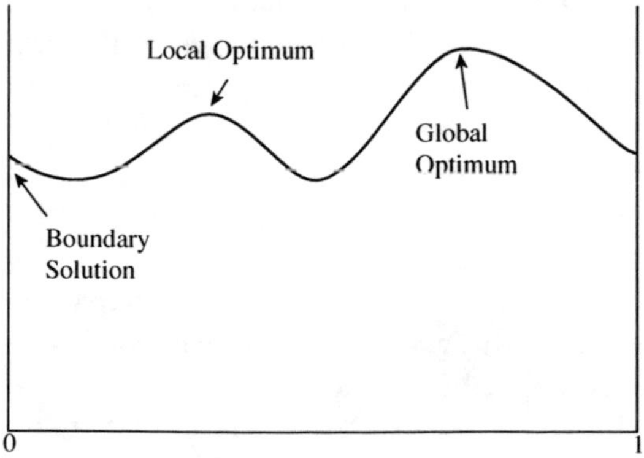

Figure 1. Maxima and boundaries.

"outside" of those values. In Figure 1 we see that there are three maxima – one at 0, one at about 0.4 and one at about 0.8 – although only one of these is the global maximum (at about 0.8). Thus, if the start value for the parameter estimate is too close to the boundary 0, the iterative process will move this particular parameter estimate to the local maximum (0), but it will not provide the true (global) maximum estimate. Thus, it is important to note that a consideration of the start values and the final estimates – especially if one or more of these estimated parameter values lie on the boundary of the parameter space – may be of critical importance, and should be given considerable attention.[6]

A final cautionary note should be made with respect to model estimation. As will be discussed in the next section, the issue of model identification – whether there is sufficient information in the observed cross-tabulation to estimate the parameters of the proposed model – is of crucial importance in latent class analysis. One practical approach to exploring whether the model is identified is to begin with quite different start values for the each of the model parameters and estimate the same model several times. If the final estimates of the model parameters are quite different for the several analyses but the estimated frequencies and the chi squares are the same for each of the analyses, it is a sure sign that the specified model is not identified. A more complete discussion of model identification issues is presented by Goodman (1974a, 1974b) and Clogg (1981; see also McCutcheon, 1987).

3. MODEL EVALUATION

Several model evaluation criteria have become more or less standard in the evaluation of LCMs. All of these criteria are evaluations of how well the expected cell counts under the model hypothesis replicate the originally observed cell counts. Four in particular – the Pearson chi square (X^2), the likelihood ratio chi square (L^2), the Akaike information criteria (AIC), and the Baysian information criteria (BIC) – have become widely used and appear throughout this volume. In this section, we briefly consider the basics of these four criteria.

Each of the four evaluation criteria relies upon a comparison between the expected cell frequency count (f_{ijkl}) given by the estimated LCM parameters and the actual (observed) cell frequency count (F_{ijkl}) found in the sample data. LCMs that lead to expected cell frequencies that are too far from the observed cell frequencies are deemed unacceptable or implausible, whereas models that yield expected cell counts that are similar to what has actually been observed are believed to be more plausible or acceptable. Models with more parameters (e.g., a model with more latent classes) usually provide a better "fit" with the data; that is, the expected cell frequencies are typically closer to the observed cell frequencies for models with more parameters. More parsimonious models tend to have a somewhat poorer fit. Thus, the usual task is to find the most parsimonious model – one with the fewest parameters[7] – that has an acceptable fit to the observed data.

As is clear from the two LCM parameterizations expressed in Equations (1) and (3), however, neither directly yields expected frequency counts for the cells of the observed contingency table. Instead, each of the outcomes on the left side of Equations (1) and (2) include the latent variable (X). By summing over the T classes of the latent variable, however, we can obtain the LCM's expected frequency counts for the observed table. For example, for Equation (1), we see that

$$\pi_{ijkl}^{ABCD} = \sum_t \pi_t^X \pi_{it}^{A|X} \pi_{jt}^{B|X} \pi_{kt}^{C|X} \pi_{lt}^{D|X}. \tag{7}$$

By multiplying the expected joint probability of being at level *ijkl* in the *ABCD* contingency table (π^{ABCD}) by the sample size (here, $N = 216$), we obtain the expected cell count (f_{ijkl}).

These expected cell count values can be compared with the observed cell counts (F_{ijkl}) in order to evaluate the model fit. For example, the expected values can be used with the Pearson chi-square statistic to evaluate how well the expected cell counts from the specified latent class model compare to the observed distribution.

$$X^2 = \sum_{ijkl} \frac{(F_{ijkl} - f_{ijkl})^2}{f_{ijkl}}. \tag{8}$$

The degrees of freedom (df) for the chi-square statistic of the unrestricted LCM are typically calculated as

$$\text{df} = (IJKL - 1) - [T(I + J + K + L + d) - 1], \tag{9}$$

where I, J, K, and L represent the number of levels for the respective indicator variables. In the case of four dichotomies, $I = J = K = L = 2$, T indicates the number of latent classes in the LCM, and d is the number of indicator variables minus 1. In the current example, we obtain $15 - [2(5) - 1] = 6$ df for the model test. Although Equation (9) works in nearly all instances, a necessary and sufficient condition for determining the local identifiability of an LCM involves determining the rank of the

Hessian matrix of second-order partial derivatives. As mentioned earlier, more complete discussions of model identification issues are presented by Goodman (1974a, 1974b) and Clogg (1981; see also McCutcheon, 1987).

The likelihood ratio chi-square test (G^2) provides a general-purpose test for evaluating models and, as Goodman (1968, 1970) and Agresti (1990) note, is of special importance for comparing alternative models. This latter point will be of considerable interest as we take up the issues of hypothesis testing and model restrictions in the following sections. The likelihood ratio chi-square statistic is a function of the ratio of the observed to expected cell counts:

$$G^2 = 2\sum_{ijkl} F_{ijkl} \ln(F_{ijkl}/f_{ijkl}) \tag{10}$$

As with the Pearson chi-square statistic, the likelihood ratio chi-square statistic has asymptotic chi-square distributions with respect to the degrees of freedom, and thus the probability of acceptance of the alternative hypothesis – the LCM model – can be determined. Unlike with the Pearson chi-square statistic, however, when one of two nested models is true, the difference between the likelihood ratio chi-square statistics for the two models can expressed explicitly as the conditional likelihood ratio chi-square statistic with degrees of freedom equal to the difference between the degrees of freedom for the two models, thus allowing statistical comparisons of successive models. The conditional likelihood (G^2) test statistic follows a chi-square distribution when the "baseline" (i.e., less restricted) model is acceptable and when the more restricted model is nested within the less restricted model. This *partitioning* of the likelihood ratio chi square has contributed to its widespread adoption in statistical modeling.

Although the Pearson chi-square and the likelihood ratio chi-square statistics are used throughout this volume and throughout the latent class analysis literature, alternative model evaluation criteria for LCMs have been identified in recent years – notably, information criteria. These alternative criteria avoid some limitations of the traditional X^2. First, the chi-square statistics tend to be conservative when sample sizes are large; that is, it is difficult to reject the significance of even quite modest parameters when the sample size is large. Second, LCMs can require the estimation of a rather large number of parameters even for models of modest size.

Information criteria approaches penalize the likelihood for the increased number of parameters required to estimate more complex (i.e., less parsimonious) models. Because more parameters (i.e., more complex models) yield a greater likelihood, each of the information criteria penalizes the likelihood by reducing it by a function of the increased number of estimated parameters. The two most widely used of the information evaluation criteria are the AIC (Akaike, 1974) and the BIC (see Rafftery, 1995):

$$\text{AIC} = G^2 - 2\text{df}$$
$$\text{BIC} = G^2 - \text{df} * [\ln(\text{N})],$$

where df is the number of degrees of freedom and N is the sample size. Thus, we see that the AIC penalizes the G^2 by the total number of parameters required for model estimation (by subtracting two times the number of degrees of freedom), and the BIC penalizes the G^2 by both the total number of parameters required for model fit and the total sample size (by subtracting the natural log of the sample N times the number of degrees of freedom). Consequently, models with lower AICs and BICs are preferred to those with higher values for these criteria.

In Table 4, these four evaluation criteria are presented for the LCM in the current example. As these data indicate, the independence model – that is, the model in which the four indicator variables are independent of one another – is clearly an unacceptable hypothesis; all of the evaluation criteria are unequivocal with respect to this conclusion. The unrestricted two-class LCM – that is, the model in which these four indicator variables are independent of one another within the two classes of the latent variable – is clearly an acceptable solution; all of the evaluation criteria support this conclusion.

As nearly all of the advances to the basic LCM presented in this volume demonstrate, there are a wide range of interesting variations on the basic LCM that make this modeling approach a powerful tool for data analysis and research. Among the many modifications are restrictions on the parameters of the LCM. In the next section, we examine some of the basics of these restrictions.

Table 4. LCM Evaluation Criteria for Ego's Dilemma Data

Model	X^2	G^2	AIC	BIC	df
Independence	104.11	81.08	59.08	21.96	11
Two-Class LCM	2.72	2.72	−9.28	−29.53	6

4. RESTRICTED LATENT CLASS MODELS

A variety of hypotheses may be tested by restricting the parameters of the LCM. Because the two parameterizations are essentially equivalent, a number of parameter restrictions are equivalent across the two parameterizations. However, there are also several somewhat substantively different parameter restrictions in the loglinear parameterization compared with the probabilistic parameterization. Thus, we will examine the two parameterizations separately, beginning with the probabilistic parameterization.

There are two basic types of restrictions that we can impose on the parameters of the probabilistic parameterization of the LCM: deterministic and equality restrictions. When these restrictions are applied to the conditional probabilities across classes and across indicators, as well as to the latent class probabilities across classes, these two general types of restrictions yield a number of different hypotheses we might wish to test. Moreover, as we will note later, imposing restrictions on parameters "frees up" additional model parameters, which may allow us to fit LCMs with more latent classes than are permitted according to Equation (9). We must proceed with caution when imposing model restrictions because by imposing certain restrictions we may actually turn an identified model into an unidentified model.

Consider the use of equality restrictions in the case of Ego's dilemma presented earlier. We may wish to test the *parallel* indicators hypothesis (Goodman, 1974a, 1974b; Hagenaars, 1990), in which we hypothesize that two or more of the indicator variables have identical error rates with respect to each of the latent classes. As with all of the parameter restrictions, we should suggest our hypotheses prior to reviewing the outcome of prior model estimation because post-hoc hypothesizing capitalizes on chance findings. For purposes of illustration, however, let us assume that we had earlier hypothesized that indicators *B* and *C* (Drama Critic Friend and Insurance Doctor Friend) were parallel indicators – that is, that these two indicators have identical error rates with respect to the two latent classes. We can formally test this hypothesis by imposing the following restrictions:

$$\pi_{11}^{B|X} = \pi_{11}^{C|X}, \qquad \pi_{12}^{B|X} = \pi_{12}^{C|X}. \tag{11}$$

By imposing these two restrictions, we reduce by two the number of parameters that must be estimated for the model. Thus, this model will yield two additional degrees of freedom relative to the unrestricted two-class model. The evaluation criteria for the estimated LCM with the parallel indicators restriction are reported in Table 5.

Table 5. LCM Evaluation Criteria for Ego's Dilemma Data

Model	X^2	G^2	AIC	BIC	df
H_1: two-class LCM	2.72	2.72	−9.28	−29.53	6
H_2: $H_1 + B$ & C parallel indicators	2.84	2.89	−13.11	−40.12	8
H_3: $H_2 + D$ equal error rate	3.60	3.65	−14.35	−44.73	9
H_4: $H_3 + A$ as perfect indicator for class 2	3.61	3.66	−16.34	−50.09	10

The data in Table 5 indicate that we can accept the hypothesis that indicator variables B and C are parallel indicators (H_2); that is, that latent Class 1– and latent Class 2–type of respondents are equally likely to err in their responses with respect to the Drama Critic Friend and Insurance Doctor Friend indicators in their respective classes. Both the Pearson (2.84) and the likelihood ratio (2.89) chi squares (with 8 df) indicate that H_2 is a plausible model. Moreover, because H_2 is nested within H_1, we can use the conditional likelihood ratio X-square test to examine whether the newly imposed restrictions are acceptable, or whether they result in an unacceptable erosion of fit to the observed data. In the current instance, the conditional difference between the two model G^2s ($2.89 - 2.72 = 0.17$, with $8 - 6 = 2$ df) clearly indicates that the newly hypothesized model (H_2) produces only a very modest erosion of fit and is, consequently, preferred to the less parsimonious, unrestricted model represented by H_1. In addition to these X-square tests, the greater negative values for both the AIC and BIC indicate that model H_2 is preferred to H_1. Thus, we accept H_2 as our new "baseline" model.

Another equality model restriction that we might wish to examine with the probabilistic parameterization of the LCM is the *equal error rate* hypothesis (Goodman, 1974a, 1974b; Hagenaars, 1990). The equal error rate hypothesis suggests that we can impose equality constraints on the conditional probabilities to test whether an indicator variable has the same error rate across the two classes. For example, we might wish to test that indicator variable D (Board of Directors Friend) has an equal error rate for Classes 1 and 2.[8] This hypothesis can be formally stated as

$$\pi_{21}^{D|X} = \pi_{12}^{D|X}. \tag{12}$$

This hypothesis indicates that the likelihood of a particularistic-type (Class 1) respondent giving a universalistic ($l = 2$) response is equal to a universalistic-type (Class 2) respondent giving a particularistic ($l = 1$) response.

Once again, the data in Table 5 support the acceptance of this hypothesis. All of the evaluation criteria for H_3 indicate acceptance. With

9 df for this model, both the Pearson (3.60) and likelihood ratio chi squares (3.65) indicate acceptance. Also, the conditional G^2 (0.76, 1 df) indicates that the additional restriction is acceptable. Finally, the two information criteria values (−14.35 and −44.73) also indicate that H_3 is preferred over H_2. Thus, we accept the hypothesis that indicator D has equal error rates for Classes 1 and 2.

A second type of hypothesis that we can consider with respect to the LCM's conditional probabilities is that of a *deterministic model restriction*. A deterministic restriction on a conditional probability tests the hypothesis that the conditional probability equals a specific value – usually, this value is either 1.0 or 0. For example, we might wish to test the hypothesis that indicator A (Passenger Friend) is a perfect indicator of Class 2; that universalistic-type (Class 2) respondents will have a zero probability of giving a particularistic response to question A. This restriction can be formally stated as

$$\pi_{12}^{A|X} = 0. \tag{13}$$

Because the model with this restriction requires the estimation of one fewer parameter, the model will have an additional degree of freedom.

As the data in Table 5 indicate, model H_4 provides an acceptable fit to the data. The Pearson (3.61) and likelihood ratio (3.66) chi squares (10 df) are clearly acceptable. Also, the conditional G^2 (0.01 with 1 df) clearly indicates acceptance of this model. The AIC (−16.34) and the BIC (−50.09) indicate that this model is a preferable model to all of those presented in Table 5. The parameter estimates for the probabilistic LCMs of H_4 are presented in Table 6.

The parameter estimates in Table 6 reflect the three types of model restrictions on the conditional probabilities. We see, for example, that for Class 1 and Class 2 the conditional probabilities are identical for indicator

Table 6. Latent Class and Conditional Probabilities for the Restricted Ego's Dilemma LCM

Indicator	Class 1	Class 2
A. Passenger Friend	0.275	0.000^b
B. Drama Critic Friend	0.636^a	0.046^a
C. Insurance Doctor Friend	0.636^a	0.046^a
D. Board of Directors Friend	0.852^a	0.148^a
Latent Class Probabilities	0.7574	0.2426

[a] Equality restriction.
[b] Deterministic restriction.

variables B and C; thus, we can conclude that these two are parallel indicators for our model. Also, we see that indicator D has equal error rates for both classes; the probability of giving a level 2 response to D among Class 1 respondents is $(1.0 - 0.852) = 0.148$, which is identical to the probability of giving a level 1 response among Class 2 respondents. Finally, we see that Class 2–type respondents have a zero probability of responding that they have a right to expect their friend to falsely report that the respondent was driving within the speed limit.

A third type of restriction that may be of interest to researchers is the *inequality restriction* on conditional probabilities (see Croon, this volume). For example, inequality restrictions could be useful when a three-class solution is found and the researcher wishes to test whether the three classes lie on a continuum, with one of the classes being intermediate to the other two "end" classes. If this were the case using the current data, for example, we may wish to test the hypothesis that

$$\pi_{11}^{A|X} \geq \pi_{12}^{A|X} \geq \pi_{13}^{A|X}, \tag{14}$$

which indicates that Class 1 respondents are the most likely to respond at level 1 of A; Class 2 respondents are more likely than Class 3 but less likely than Class 1 to respond at level 1 of A; and Class 3 are the least likely to respond at level 1 of A. As Croon notes, however, there are problems associated with the determination of the degrees of freedom for these model tests.

In principle, each of these types of probabilistic parameter restrictions – equality, deterministic and inequality – may also be imposed on the latent class probabilities. In practice, however, deterministic restrictions are rarely used; unless there exists some a priori hypothesis regarding the exact size of a class, there would be no practical use for a deterministic restriction on a latent class probability. In our current example, an equality restriction on the two latent class probabilities – although possible – appears implausible.

A. Restricted Loglinear Models

A brief consideration of Equation (5) indicates that we can replicate in the loglinear parameterization each of the equality restrictions we imposed on the conditional probabilities of the LCM. For example, the parallel indicator restriction imposed on the conditional probabilities of indicators B and C can be replicated by imposing the restrictions

$$\lambda_1^{B} = \lambda_1^{C}, \qquad \lambda_{11}^{BX} = \lambda_{11}^{CX}. \tag{15}$$

The equal error rate hypothesis imposed on the conditional probabilities of indicator D can be replicated by imposing the restriction

$$\lambda_1^D = 0. \tag{16}$$

By imposing these three restrictions on the loglinear parameterization, we obtain the model specified as H_3 in Table 5.

Imposing the deterministic restriction that the Class 2 conditional probabilities of indicator item A equals 0 and 1, however, is problematic in the loglinear parameterization. For Equation (5) to yield a conditional probability of 0 requires that the numerator on the right side of the equation also equal 0. This, of course, means that the sum of the two lambdas in the numerator would have to equal negative infinity. Thus, to test this particular deterministic restriction the loglinear parameterization, our hypothesis must include one or more *structural zeros*.

In contrast to the probabilistic parameterization, the loglinear parameterization tends to focus attention on the association between the indicator variables and the latent variable. For example, with the loglinear restriction, we can impose an equal association restriction to test the hypothesis that two or more indicator items have an identical pattern of association with the latent variable. When either the indicator items or the latent variable has more than two levels, the loglinear parameterization permits linear restrictions on the two-variable association parameters.

As Clogg noted (1988, 1995; see also McCutcheon, 1996; Heinen, 1996), the linear restrictions that were first applied to the usual loglinear models in which all variables are observed (Haberman, 1974; Goodman, 1979; Clogg and Shihadeh, 1994) can also be applied to the loglinear parameterization of the LCM. For an example of this approach, we consider a set of four dichotomous items about the approval of social reasons for abortion taken from the 1982 General Social Survey (Davis and Smith, 1982). In this survey, a sample of American respondents were asked about their approval of legal abortions for a woman if she is single and does not wish to marry the man (A), if the woman is married and too poor to have any more children (B), if the woman wants an abortion for any reason (C), and if the woman is married and does not want any more children (D). The observed cell counts for the responses to these items are reported in Table 7.

We begin by fitting an unrestricted two-class latent model to these data, as this is the simplest and most parsimonious model. As the model evaluation criteria reported in Table 8 indicate, however, the unrestricted two-class LCM does not provide an acceptable fit to the observed data.

Table 7. Responses to Four Social Reasons for Abortion Items (1982 General Social Survey)

Items			Approval of Legal Abortion	
A	*B*	*C*	Item *D*(+)	Item *D*(−)
+	+	+	567	11
+	+	−	62	32
+	−	+	17	10
+	−	−	22	38
−	+	+	18	13
−	+	−	42	62
−	−	+	9	11
−	−	−	28	719

Thus, we must reject this model as implausible. The next model we might wish to test is the unrestricted three-class model. As Goodman (1974a) notes, however, this model is not identified, even though Equation (9) indicates that this model would be identified with a single degree of freedom for the model test.

An alternative to the unrestricted three-class model is a three-class model in which the three-classes are linearly ordered from more approving to less approving. We can estimate this model by modifying Equation (3) to include a set of linear restrictions on the two-variable indicator-by-latent variable parameters:

$$\ln\left(f_{ijklt}^{ABCDX}\right) = \lambda + \lambda_t^X + \lambda_i^A + \lambda_j^B + \lambda_k^C + \lambda_l^D + g_t\lambda_{i*}^{AX} + g_t\lambda_{j*}^{BX} + g_t\lambda_{k*}^{CX} + g_t\lambda_{l*}^{DX}. \qquad (17)$$

The single variable parameters of this loglinear parameterization are subject to the restrictions expressed in Equation (4). The two-variable parameters in Equation (16) are now restricted only with respect to summation

Table 8. LCM Evaluation Criteria for Abortion Approval Data

Model	X^2	G^2	AIC	BIC	df
H_1: two-class LCM	216.45	182.13	170.13	137.64	6
H_2: three-class model with linear restrictions	2.76	2.75	−7.25	−34.32	5
H_3: $H_2 + A, B, C$ restricted to equal association	4.52	4.57	−9.43	−47.32	7

over the levels of the indicator variable

$$\sum_i g_t \lambda_{i*}^{AX} = \sum_j g_t \lambda_{j*}^{BX} = \sum_k g_t \lambda_{k*}^{CX} = \sum_l g_t \lambda_{l*}^{DX} = 0, \tag{18}$$

because there is only a single estimate of each two-variable lambda for all T levels of the latent variable. The linear coefficient, g_t, takes on the values -1, 0, and 1 for the three levels of the latent variable.[9]

As the evaluation criteria in Table 8 indicate, the three-class linearly restricted model provides a good fit to the observed data. Both the Pearson (2.76) and likelihood ratio (2.75) chi squares (5 df) indicate a good fit to the data, as do the AIC (-7.25) and the BIC (-34.32). Because H_1 is a two-class model and H_2 is a three-class model, these are not nested; thus, no conditional G^2 test can be made. Given the four evaluation criteria listed in Table 8, we can conclude that the loglinear parameters for the four indicator variables can be linearly restricted across the three classes of the latent variable.

Although we have linearly restricted the dichotomously scored variables in the abortion attitudes example across the three classes of the latent variable, it is important to note that linear restrictions may also be imposed on the indicator variables when they have three or more ordered categories. Also, linear-by-linear restrictions can be imposed when both the indicator and latent variables have three or more ordered levels. McCutcheon (1996) has also shown that such restrictions are also possible on parameters related to a grouping variable when it has three or more ordered levels.

Our final hypothesis for these data tests the equality of linear association across the three indicator variables A, B, and C. As the evaluation criteria in Table 8 indicate, this hypothesis (H_3) is clearly acceptable, with a X^2 test statistic of 4.52 and a G^2 test statistic of 4.57 (7 df). Because H_3 is nested in H_2, we can use the conditional G^2 to evaluate the improvement of fit (1.82, 2 df) and find that, like the AIC (-9.43) and the BIC (-47.32), the conditional G^2 recommends acceptance of the equal linear association hypothesis for indicator variables A, B, and C.

As the parameters in Table 9 indicate, the indicator variables are very highly related to the three levels of the latent abortion attitudes variable in this 1982 sample of American adults. Moreover, the fourth indicator variable – whether married women who want to have no more children should be able to obtain a legal abortion – appears to be the indicator most highly related to the unobserved latent classes.

Table 9. Linearly Restricted Loglinear Parameters for Abortion Attitudes LCM (H_3)

Indicator	Parameter	
	Single Variable	Two Variables
A. Single woman	−0.095	4.151[a]
B. Poor married woman	0.124	4.151[a]
C. Any reason	−0.629	4.151[a]
D. Wants no more children	−0.084	6.917
X. Latent class variable		
($t = 1$)	0.141	—
($t = 2$)	−0.441	—

[a] Equality restrictions imposed.

5. MULTI-SAMPLE LATENT CLASS MODELS

Often researchers are confronted with a set of responses to identical indicator items in sample data from two or more populations. These samples may be from different social, cultural, or economic groups (McCutcheon, 1987); different regions, states, or nations (see, e.g., McCutcheon and Nawojczk, 1995; McCutcheon and Hagenaars, 1997); or the samples may be from the same group at two or more points in time (see, e.g., McCutcheon, 1986, 1996). Indeed, the samples may be from any mutually exclusive groups. When we are in such a situation, we can use a multi-sample, or simultaneous, LCM (SLCM) to compare the latent structures in the samples (Clogg and Goodman, 1984, 1985, 1986).

We begin our consideration of the multisample LCM by noting that our original parameterizations, shown in Equations (1) and (3), must be modified to reflect the addition of a grouping variable that reflects the populations from which we have samples. We designate that variable as G with S samples. Thus, if we have samples from four nations (i.e., $S = 4$), for example, we would have G_1, G_2, G_3, and G_4. The formal multisample probabilistic parameterization of the LCM can be expressed as

$$\pi_{ijklts}^{ABCDXG} = \pi_{s}^{G}\pi_{ts}^{X|G}\pi_{its}^{A|XG}\pi_{jts}^{B|XG}\pi_{kts}^{C|XG}\pi_{lts}^{D|XG}. \tag{19}$$

We must also note that an inclusion of a grouping variable in this parameterization results in changes in the model restrictions:

$$\sum_t \pi_{ts}^{X|G} = \sum_i \pi_{its}^{A|XG} = \sum_j \pi_{jts}^{B|XG} = \sum_k \pi_{kts}^{C|XG} = \sum_l \pi_{lts}^{D|XG} = 1.0. \tag{20}$$

We must also add similar modifications to the loglinear parameterizations of Equations (3):

$$\ln\left(f_{ijklts}^{ABCDXG}\right) = \lambda + \lambda_s^G + \lambda_t^X + \lambda_i^A + \lambda_j^B + \lambda_k^C + \lambda_l^D + \lambda_{it}^{AX} + \lambda_{jt}^{BX} + \lambda_{kt}^{CX} + \lambda_{lt}^{DX} + \lambda_{ts}^{XG} + \lambda_{is}^{AG} + \lambda_{js}^{BG} + \lambda_{ks}^{CG} + \lambda_{ls}^{DG} + \lambda_{its}^{AXG} + \lambda_{jts}^{BXG} + \lambda_{kts}^{CXG} + \lambda_{lts}^{DXG}. \tag{21}$$

Similarly, we must modify the model restrictions for Equation (21). In addition to the restrictions noted in Equation (4), we must add

$$\begin{aligned}
\sum_s \lambda_s^G &= \sum_i \lambda_{is}^{AG} = \sum_j \lambda_{js}^{BG} = \sum_k \lambda_{ks}^{CG} = \sum_l \lambda_{ls}^{DG} = \sum_s \lambda_{is}^{AG} \\
&= \sum_s \lambda_{js}^{BG} = \sum_s \lambda_{ks}^{CG} = \sum_s \lambda_{ls}^{DG} = \sum_i \lambda_{its}^{AXG} = \sum_t \lambda_{its}^{AXG} \\
&= \sum_s \lambda_{its}^{AXG} = \sum_j \lambda_{jts}^{BXG} = \sum_t \lambda_{jts}^{BXG} = \sum_s \lambda_{jts}^{BXG} \\
&= \sum_k \lambda_{kts}^{CXG} = \sum_t \lambda_{kts}^{CXG} = \sum_s \lambda_{kts}^{CXG} = \sum_l \lambda_{lts}^{DXG} \\
&= \sum_t \lambda_{lts}^{DXG} = \sum_s \lambda_{lts}^{DXG} = 0.
\end{aligned} \tag{22}$$

As with the single-sample LCM, the two parameterizations of the SLCM are equivalent. Thus, for convenience, we will adopt the usage of referring to the specific SLCM by the marginals that correspond to the highest-order interaction terms because in hierarchical loglinear models these terms represent the minimal sufficient statistics (Goodman, 1978; Agresti, 1990; Hagenaars, 1990). Because of the axiom of local independence, the SLCM is one of the few instances when we hypothesize a model with three-variable interactions. Consequently, we can refer to the model specified in Equations (19) and (21) as the {*AXG*, *BXG*, *CXG*, *DXG*} SLCM.

Typically, the researcher's first interest in SLCM is to establish *structural equivalence* – that is, to establish that the indicator variables are independent of the grouping variable. This means that the preferred instance is one in which the model specified in Equations (19) and (21) can be reduced to {*XG*, *AX*, *BX*, *CX*, *DX*}. In this instance, we can represent the latent variable as *structurally homogeneous* in the two or more samples, meaning that the associations between the latent variable and each of the indicator variables are identical across all of the samples. In this instance, we are confident that we are measuring the identical latent variable in each of the samples.

The less our data conform to this structural equivalence ideal, the less our certainty that we have truly measured the identical phenomenon in each of the samples. Divergences from complete structural equivalence in the several samples must be evaluated by the researcher. In some instances, divergences may be readily explained by factors such as differing historical, political, cultural, social, and economic circumstances (see, e.g., McCutcheon and Nawojczk, 1995).

Consider the data presented in Table 1. You will recall that in this example, students were asked about four scenarios involving role conflict. One group of 216 students was asked about the situations in which they personally were confronted with the dilemmas, and another group of 216 was asked about the situations in which their close friend was confronted by the dilemmas. Recall that a two-class model fit the Ego's Dilemma data well. Thus, the first model we consider is the two-class model for each of these two samples. The evaluation criteria presented in Table 10 suggest that the unrestricted two-class model (H_1) fits the data well.

As we see from these evaluation criteria, the two-class per sample model appears to provide an acceptable characterization of these two samples of students. Both the X^2 (9.06) and the G^2 (8.25) with 12 df support this conclusion. Moreover, both of the information criteria statistics also support the characterization of the two samples as each having two classes (H_1).

The next hypothesis we test (H_2) is the two-class *structural equivalence* model. This model tests the hypothesis that the two latent classes for the Ego's Dilemma sample are the same two classes as in the Friend's Dilemma sample. As the evaluation criteria indicate, this too is an acceptable hypothesis: the X^2 (24.78) and L^2 (23.47) on 20 df support this conclusion. Because H_2 is nested within H_1, we can also examine the conditional G^2 test (15.22 with 8 df), which also supports the structural equivalence hypothesis at the 0.05 alpha level ($p = .0550$). Finally, both of the information criteria statistics – the AIC (−16.53) and the BIC (−97.90) – indicate a preference for H_2 over H_1.

Table 10. Simultaneous LCM Evaluation Criteria for Ego's Dilemma and Friend's Dilemma Data: Two-Latent-Class Solution ($T = 2$)

Model	X^2	G^2	AIC	BIC	df
H_1: $\{AXG, BXG, CXG, DXG\}$	9.06	8.25	−15.75	−64.57	12
H_2: $\{XG, AX, BX, CX, DX\}$	24.78	23.47	−16.53	−97.90	20
H_3: $\{G, AX, BX, CX, DX\}$	24.82	23.48	−18.52	103.96	21

The final hypothesis we might wish to examine is the *complete homogeneity* hypothesis (see H_3). The complete homogeneity hypothesis tests whether the latent structure is identical in both (all) samples *and* that the distribution of the latent variable is identical in both (all) samples. Because these two samples were both from the same population and the questionnaires with Ego's and Friend's Dilemmas were randomly assigned, in this particular instance the complete homogeneity hypothesis allows us to assess whether problem framing (Ego vs. Friend) appears to have a significant influence on the outcome of the observed response patterns (beyond what might occur by chance variation).

As the data in the last row of Table 10 indicate, the complete homogeneity hypothesis is clearly preferred to the other two hypotheses. Both of the chi-square statistics support this conclusion – the X^2 (24.82) and L^2 (23.48) on 21 df. Because H_3 is nested within H_2, we can examine the conditional G^2 test (0.01 with 1 df), which indicates virtually no increase in the likelihood ratio chi square from imposing this constraint. Finally, the two information criteria statistics – the AIC (−18.52) and the BIC (−103.96) – clearly indicate a preference for H_3 over H_2. Thus, we might conclude from these data that the manner in which one frames these scenarios – whether as Ego's Dilemma or as Friend's Dilemma – has no consequences for the pattern of responses; at least, not among the population of Harvard and Radcliffe social science students of 1950.

As the conditional probabilities and latent class probabilities reported in Table 11 indicate, there are no differences between the models for the two question framings. In every instance, the conditional probabilities for group 1 (Ego's Dilemma) equals the conditional probabilities for group 2 (Friend's Dilemma). Moreover, the distribution of the two classes of respondents is also identical.

Table 11. Latent Class and Conditional Probabilities for the SLCM for Ego's and Friend's Dilemmas: Complete Homogeneity (H_3)

	Ego's Dilemma		Friend's Dilemma	
Indicator	Class 1	Class 2	Class 1	Class 2
A. Passenger Friend	0.354	0.010	0.354	0.010
B. Drama Critic Friend	0.567	0.108	0.567	0.108
C. Insurance Doctor Friend	0.717	0.021	0.717	0.021
D. Board of Directors Friend	0.849	0.319	0.849	0.319
Latent Class Probabilities	0.7083	0.2917	0.7083	0.2917

Table 12. Loglinear Parameters for the SLCM for Ego's and Friend's Dilemmas: Complete Homogeneity (H_3)

Indicator	Single-Var. Parameters	Two-Var. Parameters (With *X*)	Two-Var. Parameters (With *G*)	Three-Var. Parameters (With *XG*)
A. Passenger Friend	−1.296	0.976	0.000	0.000
B. Drama Critic Friend	−0.461	0.596	0.000	0.000
C. Insurance Doctor Friend	−0.731	1.195	0.000	0.000
D. Board of Directors Friend	0.242	0.622	0.000	0.000
X. Latent Class Variable	0.443	—	0.000	—
G. Grouping Variable	0.000	0.000	—	—

Finally, as the loglinear coefficients in Table 12 clearly indicate, the complete homogeneity hypothesis involves restricting to zero all parameters that include the grouping variable (G). As a consequence, the two rightmost columns of Table 12 are all zero, as are all of the entries in the final row of Table 12. Only those loglinear coefficients that exclude the grouping variable are significantly different from zero.

Before this section is completed, it is important to note that all of the restrictions that are possible with the single-sample model are also possible in the multisample instance (McCutcheon and Hagenaars, 1997). Equality and deterministic restrictions may be used in both instances, as can linear restrictions on the loglinear parameters (McCutcheon, 1996). As with the single-sample instance, however, one must be cautious about imposing the restrictions, as these may influence the identification of the model. Moreover, the usual variant of the EM algorithm for the estimation of LCMs does not always work properly when models are restricted (Mooijaart and Van der Heijden, 1992).[10]

6. CONCLUSION

In this brief overview of the LCM, it has been possible to illustrate only a few of the reasons for its growing popularity as a research tool. A number of popular software programs are making LCMs easily accessible to social and behavioral researchers worldwide. Clearly, the two parameterizations make this approach a highly flexible and attractive method for the analysis of categorical data. The ability to impose a variety of restrictions on these parameterizations and the range of hypotheses that can be explicitly tested have also played an important part. Also, the extension of the LCM to the multisample instance has played an important role in its application in comparative research. As the remaining chapters of this volume clearly illustrate, the extensions of this powerful model continue to push it to the forefront in new areas of the analysis of categorical data.

REFERENCES

Agresti, A. (1990). *Categorical Data Analysis*. New York: Wiley.

Akaike, H. (1974). "A new look at statistical model identification," *IEEE Transactions on Automatic Control*, **AC-19**, 716–23.

Clogg, C. C. (1981). "New developments in latent structure analysis." In D. M. Jackson & E. F. Borgatta (eds.), *Factor Analysis and Measurement in Sociological Research*. Beverly Hills: Sage, pp. 214–80.

Clogg, C. C. (1988). "Latent class models for measuring." In R. Langeheine & J. Rost (eds.), *Latent Trait and Latent Class Models*. New York: Plenum, pp. 173–205.

Clogg, C. C. (1995). "Latent class models." In G. Arminger, C.C. Clogg, & M. Sobel (eds.), *Handbook of Statistical Modeling for the Social and Behavioral Sciences*. New York: Plenum, pp. 311–59.

Clogg, C. C., & Sawyer, D. O. (1981). "A comparison of alternative models for analyzing the scalability of response patterns." In S. Leinhart (ed.), *Sociological Methodology, 1981*. San Francisco: Jossey–Bass, pp. 240–80.

Clogg, C. C., & Goodman, L. A. (1984). "Latent structure analysis of a set of multidimensional contingency tables," *Journal of the American Statistical Society*, **79**, 762–71.

Clogg, C. C., & Goodman, L. A. (1985). "Simultaneous latent structure analysis in several groups." In N. B. Tuma (ed.), *Sociological Methodology*. San Francisco: Jossey–Bass, pp. xxx–xxx.

Clogg, C. C., & Goodman, L. A. (1986). "On scaling models applied to data from several groups," *Psychometrika*, **51**, 123–35.

Clogg, C. C., & Shihadeh, E. S. (1994). *Statistical Models for Ordinal Variables*. Newbury Park, CA: Sage.

Davis, J. A., & Smith, T. W. (1982). *General Social Surveys, 1972–1982*. [Machine-Readable Data File]. Chicago: National Opinion Research Center.

Everitt, B. S. (1987). *Introduction to Optimization Methods and Their Application in Statistics*. London: Chapman & Hall.

Everitt, B. S., & Hand, D. J. (1981). *Finite Mixture Models*. New York: Chapman & Hall.

Goodman, L. A. (1968). "The analysis of cross-classified data: independence, quasi-independence and interactions in contingency tables with or without missing entries," *Journal of the American Statistical Association*, **63**, 1091–131.

Goodman, L. A. (1970). "The multivariate analysis of qualitative data: interactions among multiple classifications," *Journal of the American Statistical Association*, **65**, 225–56.

Goodman, L. A. (1974a). "The analysis of systems of qualitative variables when some of the variables are unobservable. Part I – A modified latent structure approach," *American Journal of Sociology*, **79**, 1197–59.

Goodman, L. A. (1974b). "Exploratory latent structure analysis using both identifiable and unidentifiable models," *Biometrika*, **61**, 215–31.

Goodman, L. A. (1978). *Analyzing Qualitative/Categorical Data: Log-Linear Models and Latent-Structure Analysis*. Jay Magidson (ed.). Cambridge, MA: Abt Books.

Goodman, L. A. (1979). "Simple models for the analysis of association in cross-classifications having ordered categories," *Journal of the American Statistical Association*, **74**, 537–52.

Haberman, S. J. (1974). "Log-linear models for frequency tables with ordered classification." *Biometrics*, **30**, 589–600.

Haberman, S. J. (1979). *Analysis of Qualitative Data, Vol. 2, New Developments*. New York: Academic Press.

Hagenaars, J. A. (1990). *Categorical Longitudinal Data: Log-Linear Panel, Trend, and Cohort Analysis*. Newbury Park, CA: Sage.

Heinen, T. (1996). *Latent Class and Discrete Latent Trait Models: Similarities and Differences*. Thousand Oaks, CA: Sage.

Lindsay, B. G. (1995). *Mixture Models: Theory, Geometry and Applications*. Hayward, CA: Institute of Mathematical Science.

McCutcheon, A. L. (1986). "Sexual morality, pro-life values and attitudes toward abortion," *Sociological Methods and Research*, **16**, 256–75.

McCutcheon, A. L. (1987). *Latent Class Analysis*. Newbury Park, CA: Sage.

McCutcheon, A. L. (1996). "Multiple group association models with latent variables: an analysis of secular trends in abortion attitudes, 1972–1988." In A. Raftery (ed.), *Sociological Methodology, 1996*. Cambridge, MA: Blackwell, pp. xxx–xx.

McCutcheon, A. L., & Hagenaars, J. A. (1997). "Simultaneous latent class models for comparative social research." In R. Langeheine & J. Rost (eds.), *Applications of Latent Trait and Latent Class Models*. New York: Waxman, pp. 266–77.

McCutcheon, A. L., & Nawojczk, M. (1995). "Making the break: popular sentiment toward legalized abortion among Polish and American Catholic laities," *International Journal for Public Opinion Research*, **7**, 232–52.

McLachlan, G. J., & Basford, K. E. (1988). *Mixture Models: Inference and Applications to Clustering*. New York: Marcel Dekker.

McLachlan, G. J., & Krishnan, T. (1997). *The EM Algorithm and Extensions*. New York: Wiley.

McLachlan, G., & Peel, D. (2000). *Finite Mixture Models*. New York: Wiley.

Mooijaart, A., & Van der Heijden, P. G. M. (1992). "The EM algorithm for latent class models with constraints," *Psychometrica*, **57**, 261–71.

Parsons, T. (1949). *Essays in Sociological Theory, Pure and Applied*. Glencoe, IL: Free Press.

Raftery, A. E. (1995). "Baysian model selection in social research." In P. V. Marsden (ed.), *Sociological Methodology, 1995*. Cambridge, MA: Blackwell, pp. xxx–xx.

Stouffer, S. A., & Toby, J. (1951). "Role conflict and personality," *American Journal of Sociology*, **89**, 395–406.

Titterington, D. M., Smith, A. F. M., & Makov, U. E. (1985). *Statistical Analysis of Finite Mixture Distributions*. New York: Wiley.
Vermunt, J. K. (1997). *Log-Liner Models for Event Histories*. Thousand Oaks, CA: Sage.
Vermunt, J. K. (1997). *LEM: A General Program for the Analysis of Categorical Data* (users' manual). Tilburg, The Netherlands: Tilburg University.
Wedel, M., & Kamakura, W. A. (1998). *Market Segmentation: Conceptual and Methodological Foundations*. Boston: Kluwer Academic.

NOTES

1. In some instances, it is plausible that the observed associations reflect cause–effect relationships. In such instances, latent class analysis is inappropriate. As Goodman (1974a) and Hagenaars (1990, this volume) show, however, the LCM can be integrated into models that have such causal associations.
2. The LCM is part of the larger family of mixture models (see, e.g., McLaughlin and Peel, 2000; Lindsay, 1995; McLachlan and Basford, 1988; Titterington et al., 1985; Everitt and Hand, 1981). Within the area of mixture models, the latent class probability is often referred to as the *mixing proportion*.
3. We must be cautious in drawing the analogy between latent class probabilities and factor loadings because, unlike factor loadings, the interpretation of conditional probabilities depends on their size relative to those in other latent classes. For example, if Class 1 has a conditional probability of .70 of getting a right answer for item *A*, .70 is high if for Class 2 the conditional probability of getting it right is .25, but it is low if the conditional probability in Class 2 is .99.
4. It is important to note that an absolute 0.000 or 1.000 estimate of a conditional probability for an indicator is referred to as a *boundary estimate* because probabilities are bounded by 0.00 and 1.00. Boundary estimates are problematic within latent class analysis. During the iterative process for finding the parameter estimates, if one or more of the estimates goes to a boundary (i.e., either 0 or 1), all of the other parameters are maximum-likelihood solutions only if the parameters estimated to be at the boundary of the parameter space are truly zero or one in the population. Fortunately, most of the available software programs permit the researcher to specify multiple start values for the iterative process; it is highly recommended that this option be used in any instance in which a boundary value is estimated for an LCM.
5. An alternative approach is for the iterations to end after some prespecified number have been completed. This criterion is risky, however, as substantial changes in the parameter estimates may still occur after a very large number of iterations.
6. An additional concern regarding boundary values has to do with the determination of the degrees of freedom. When the boundary value is established as an a priori hypothesis (i.e., as a deterministic restriction), no parameter is

estimated. When the boundary value is the result of the estimation procedure, however, there is some concern about the number of degrees of freedom to associate with the model.

7. An alternative view of parsimony is to define the simplest model in terms of the logic of the model instead of the number of model parameters. Within the context of LCMs, for example, this approach is used to examine the scalability of a set of indicator variables (see, e.g., Clogg and Sawyer, 1981; Clogg and Goodman, 1986).
8. Once again, we must note that this approach to post-hoc hypothesis testing is for illustrative purposes only.
9. In general, if $j = 1, \ldots, J$, then each of the linear coefficients has the values that are computed as $j - (J + 1)/2$.
10. The LEM program (Vermunt, 1997) was used to estimate all of the models presented in this chapter. LEM corrects for the errors common to problems discussed by Mooijaart and van der Heijden (1992).

CLASSIFICATION AND MEASUREMENT

THREE

Latent Class Cluster Analysis

Jeroen K. Vermunt and Jay Magidson

1. INTRODUCTION

Kaufman and Rousseeuw (1990) define *cluster analysis* as the classification of similar objects into groups, in which the number of groups as well as their forms are unknown. The *form of a group* refers to the parameters of cluster; that is, to its cluster-specific means, variances, and covariances that also have a geometrical interpretation. A similar definition is given by Everitt (1993), who speaks about deriving a useful division into a number of classes, in which both the number of classes and the properties of the classes are to be determined. These could also be definitions of exploratory latent class (LC) analysis, in which objects are assumed to belong to one of a set of K latent classes, with the number of classes and their sizes not known a priori. In addition, objects belonging to the same class are similar with respect to the observed variables in the sense that their observed scores are assumed to come from the same probability distributions, whose parameters are, however, unknown quantities to be estimated. Because of the similarity between cluster and exploratory LC analysis, it is not surprising that the latter method is becoming a more popular clustering tool.

In this paper, we describe the state-of-the-art in the field of LC cluster analysis. Most of the work in this field involves continuous indicators assuming (restricted) multivariate normal distributions within classes. Although authors seldom refer to the work of Gibson (1959) and Lazarsfeld and Henry (1968), actually they are using what these authors called *latent profile analysis:* that is, latent structure models with a single categorical latent variable and a set of continuous indicators. Wolfe (1970) was the first one who made an explicit connection between LC and cluster analysis.

Throughout the 1990s there was a renewed interest in the application of LC analysis as a cluster analysis method. Labels that are used to describe such a use of LC analysis are as follows: mixture-likelihood approach to clustering (McLachlan and Basford, 1988; Everitt, 1993), model-based clustering (Banfield and Raftery, 1993; Bensmail et al., 1997; Fraley and Raftery, 1998a, 1998b), mixture-model clustering (Jorgensen and Hunt, 1996; McLachlan et al., 1999), probabilistic clustering (Bacher, 2000), Bayesian classification (Cheeseman and Stutz, 1995), unsupervised learning (McLachlan and Peel, 1996), and latent class cluster analysis (Vermunt and Magidson, 2000). Probably the most important reason of the increased popularity of LC analysis as a statistical tool for cluster analysis is the fact that currently high-speed computers make these computationally intensive methods practically applicable. Several software packages are available for the estimation of LC cluster models.

An important difference between standard cluster analysis techniques and LC clustering is that the latter is a model-based clustering approach. This means that a statistical model is postulated for the population from which the sample under study is taken. More precisely, it is assumed that the data are generated by a mixture of underlying probability distributions. When using the maximum-likelihood method for parameter estimation, the clustering problem involves maximizing a log-likelihood function. This is similar to standard nonhierarchical cluster techniques in which the allocation of objects to clusters should be optimal according to some criterion. These criteria typically involve minimizing the within-cluster variation and/or maximizing the between-cluster variation. An advantage of using a statistical model is, however, that the choice of the cluster criterion is less arbitrary. Nevertheless, the log-likelihood functions corresponding to LC cluster models may be similar to the criteria used by certain nonhierarchical cluster techniques like *k* means.

LC clustering is very flexible in the sense that both simple and complicated distributional forms can be used for the observed variables within clusters. As in any statistical model, restrictions can be imposed on the parameters to obtain more parsimony and formal tests can be used to check their validity. Another advantage of the model-based clustering approach is that no decisions need be made about the scaling of the observed variables; for instance, when working with normal distributions with unknown variances, the results will be the same irrespective of whether the variables are normalized. This is very different from standard nonhierarchical cluster methods, in which scaling is always an issue. Other advantages are that it is relatively easy to deal with variables of mixed measurement levels (different scale types) and that there are more formal criteria to make decisions about the number of clusters and other model features.

LC analysis yields a probabilistic clustering approach. This means that although each object is assumed to belong to one class or cluster, it is taken into account that there is uncertainty about an object's class membership. This makes LC clustering conceptually similar to fuzzy clustering techniques. An important difference between these two approaches is, however, that in fuzzy clustering an object's grades of membership are the "parameters" to be estimated (Kaufman and Rousseeuw, 1990), whereas in LC clustering an individual's posterior class-membership probabilities are computed from the estimated model parameters and his or her observed scores. This makes it possible to classify other objects belonging to the population from which the sample is taken, which is not possible with standard fuzzy cluster techniques.

The remainder of this chapter is organized as follows. The next section discusses the LC cluster model for continuous variables. Subsequently, attention is paid to models for sets of indicators of different measurement levels, also known as *mixed-mode data*. Then an explanation is given of how to include covariates in an LC cluster model. After a discussion of estimation and testing, two empirical examples are presented. The paper ends with a short discussion. An appendix describes computer programs that implement the various kinds of LC clustering methods presented in this chapter.

2. CONTINUOUS INDICATOR VARIABLES

The basic LC cluster model has the form

$$f(\mathbf{y}_i \mid \theta) = \sum_{k=1}^{K} \pi_k f_k(\mathbf{y}_i \mid \theta_k) .$$

Here, $\mathbf{y}_i$ denotes an object's scores on a set of observed variables, K is the number of clusters, and π_k denotes the prior probability of belonging to latent class or cluster k or, equivalently, the size of cluster k. Alternative labels for the y's are indicators, dependent variables, outcome variables, outputs, endogenous variables, or items. As can be seen, the distribution of $\mathbf{y}_i$ given the model parameters of θ, $f(\mathbf{y}_i \mid \theta)$, is assumed to be a mixture of class-specific densities, $f_k(\mathbf{y}_i \mid \theta_k)$.

Most of the work on LC cluster analysis has been done for continuous variables. Generally, these continuous variables are assumed to be normally distributed within latent classes, possibly after applying an appropriate nonlinear transformation (Lazarsfeld and Henry, 1968; Banfield and Raftery, 1993; McLachlan, 1988; McLachlan et al., 1999; Cheeseman and Stutz, 1995). Alternatives for the normal distribution are student, Gompertz, or gamma distributions (see, e.g., McLachlan et al., 1999).

The most general Gaussian distribution of which all restricted versions discussed later are special cases is the multivariate normal model with parameters μ_k and Σ_k. If no further restrictions are imposed, the LC clustering problem involves estimating a separate set of means, variances, and covariances for each latent class. In most applications, the main objective is finding classes that differ with respect to their means or locations. The fact that the model allows classes to have different variances implies that classes may also differ with respect to the homogeneity of the responses to the observed variables. In standard LC models with categorical variables, it is generally assumed that the observed variables are mutually independent within clusters. This is, however, not necessary here. The fact that each class has its own set of covariances means that the y variables may be correlated with clusters, as well as that these correlations may be cluster specific. So, the clusters not only differ with respect to their means and variances, but also with respect to the correlations between the observed variables.

It will be clear that as the number of indicators and/or the number of latent classes increases, the number of parameters to be estimated increases rapidly, especially the number of free parameters in the variance–covariance matrices, Σ_k. Therefore, it is not surprising that restrictions that are imposed to obtain more parsimony and stability typically involve constraining the class-specific variance–covariance matrices.

An important constraint model is the local independence model obtained by assuming that all within-cluster covariances are equal to zero or, equivalently, by assuming that the variance–covariance matrices, Σ_k, are diagonal matrices. Models that are less restrictive than the local independence model can be obtained by fixing some but not all covariances to zero or, equivalently, by assuming certain pairs of y's to be mutually dependent within latent classes.

Another interesting type of constraint is the equality or homogeneity of variance–covariance matrices across latent classes, that is, $\Sigma_k = \Sigma$. Such a homogeneous or class-independent error structure yields clusters having the same forms but different locations. Note that these kinds of equality constraints can be applied in combination with any structure for Σ.

Banfield and Raftery (1993) proposed reparameterizing the class-specific variance–covariance matrices by an eigenvalue decomposition:

$$\Sigma_k = \lambda_k D_k A_k D_k{}^T .$$

The parameter λ_k is a scalar, D_k is a matrix with eigenvectors, and A_k is a diagonal matrix whose elements are proportional to the eigenvalues of Σ_k. More precisely, $\lambda_k = |\Sigma_k|^{1/d}$, where d is the number of observed variables and A_k is scaled such that $|A_k| = 1$.

A nice feature of this decomposition is that each of the three sets of parameters has a geometrical interpretation: λ_k indicates what can be called the *volume* of cluster k, D_k is the *orientation* of cluster k, and A_k is the *shape* of cluster k. If we think of a cluster as a clutter of points in a multidimensional space, the volume is the size of the clutter, whereas the orientation and shape parameters indicate whether the clutter is spherical or ellipsoidal. Thus, restrictions imposed on these matrices can directly be interpreted in terms of the geometrical form of the clusters. Typically, matrices are assumed to be class-independent, and/or simpler structures (diagonal or identity) are used for certain matrices. See Bensmail et al. (1997) and Fraley and Raftery (1998b) for overviews of the many possible specifications.

Rather than by a restricted eigenvalue decomposition, the structure of the Σ_k matrices can also be simplified by means of a covariance-structure model. Several authors have proposed using LC models for dealing with unobserved heterogeneity in covariance-structure analysis (Arminger and Stein, 1997; Dolan and Van der Maas, 1997; Jedidi, Jagpal, and DeSarbo, 1997). The same methodology can be used to restrict the error structure in LC cluster analysis with continuous indicators. An interesting structure for Σ_k, which is related to the eigenvalue decomposition described earlier, is a factor analytic model (Yung, 1997; McLachlan and Peel, 1999); that is,

$$\Sigma_k = \Lambda_k \Phi_k \Lambda_k + U_k . \tag{1}$$

Here, Λ_k is a matrix with factor loadings, Φ_k is the variance–covariance matrix of the factors, and U_k is a diagonal matrix with unique variances. Restricted versions can be obtained by limiting the number of factors (e.g., to one) and/or fixing some factor loading to zero. Such specifications make it possible to describe the correlations between the y variables within clusters or, equivalently, the structure of local dependencies, by means of a small number of parameters.

3. MIXED INDICATOR VARIABLES

In the previous section, we concentrated on LC cluster models for continuous indicators by assuming a (restricted) multivariate normal distribution for $\mathbf{y}_i$ within each of the classes. Often however, we are, confronted with other types of indicators, such as nominal or ordinal variables or counts. LC cluster models for nominal and ordinal variables assuming (restricted) multinomial distributions for the items are equivalent to standard exploratory LC models (Goodman, 1974; Clogg, 1981, 1995). Böckenholt (1993) and Wedel et al. (1993) proposed LC models for Poisson counts.

With the use of the general structure of the LC model, it is straightforward to specify cluster models for sets of indicators of different scale types or, as Everitt (1988, 1993) called it, for *mixed-mode data* (see also Lawrence and Krzanowski, 1996; Jorgensen and Hunt, 1996; Bacher, 2000; and Vermunt and Magidson, 2000; pp. 147–52). With an assumption of local independence, the LC cluster model for mixed y's is of the form

$$f(\mathbf{y}_i \mid \theta) = \sum_{k=1}^{K} \pi_k \prod_{j=1}^{J} f_k(y_{ij} \mid \theta_{jk}), \tag{2}$$

where J denotes the total number of indicators and j is a particular indicator.

Rather than specifying the joint distribution of $\mathbf{y}_i$ given class membership by using a single multivariate distribution, we now have to specify the appropriate univariate distribution function for each element y_{ij} of $\mathbf{y}_i$. Possible choices for continuous y_{ij} are univariate normal, student, gamma, and log-normal distributions. A natural choice for discrete nominal or ordinal variables is the (restricted) multinomial distribution. Suitable distributions for counts are, for instance, Poisson, binomial, or negative binomial.

In the previously mentioned specification, we assumed that the ys are conditionally independent within latent classes. This assumption can easily be relaxed by using the appropriate multivariate rather than univariate distributions for sets of locally dependent y variables. It is not necessary to present a separate formula for this situation; we merely think of the index j in Equation (2) as denoting a set of indicators rather than a single indicator. For sets of continuous variables, we can again work with a multivariate normal distribution. A set of nominal/ordinal variables can combined into a (restricted) joint multinomial distribution. Correlated counts could be modeled with a multivariate Poisson model. More difficult is the specification of the mixed multivariate distributions. Krzanowski (1983) described two possible ways of modeling the relationship between a nominal/ordinal and a continuous y: by means of a conditional Gaussian or by means of a conditional multinomial distribution, which means either using the categorical variable as a covariate in the normal model or the continuous one as a covariate in the multinomial model.

Lawrence and Krzanowski (1996) and Hunt and Jorgensen (1999) used the conditional Gaussian distribution in LC clustering with combinations of categorical and continuous variables. Local dependencies with a Poisson variable could be dealt with in the same way, that is, by allowing its mean to depend on the relevant continuous or categorical variable(s). The possibility of including local dependencies between indicators is very important when using LC analysis as a clustering tool. First, it prevents that one ends with a solution that contains too many clusters. Often, a simpler solution with less clusters is obtained by including a few direct effects between y variables. It should be stressed that there is also a risk of allowing for within-cluster associations: direct effects may hide relevant clusters.

A second reason for relaxing the local independence assumption is that it may yield a better classification of objects into clusters. Saying that two variables are locally dependent is conceptually the same as saying that they contain some overlapping information that should not be used when determining to which class an object belongs. Consequently, if we omit a significant bivariate dependency from an LC cluster model, the corresponding locally dependent indicators get a too-high weight in the classification formula [see Equation (3)] compared with the other indicators.

4. COVARIATES

The LC cluster modeling approach described previously is quite general: It deals with mixed-mode data and it allows for many different specifications of the (correlated) error structure. An important extension of this model is the inclusion of covariates to predict class membership. Conceptually, it makes quite a bit of sense to distinguish (endogenous) variables that serve as indicators of the latent variable from (exogenous) variables that are used to predict to which cluster an object belongs. This idea is, in fact, the same as in Clogg's (1981) latent class model (LCM) with external variables.

Note that in certain situations we may want to use the LC variable as a predictor of an observed response variable rather than as a dependent variable. For such situations, we do not need special arrangements such as those needed with covariates. A model in which the cluster variable serves as predictor can be obtained by using the response variable as one of the y variables.

With the use of the same basic structure as in Equation (2), this yields the following LC cluster model:

$$f(\mathbf{y}_i \mid \mathbf{z}_i, \theta) = \sum_{k=1}^{K} \pi_{k|\mathbf{z}_i} \prod_{j=1}^{J} f_k(y_{ij} \mid \theta_{jk}) .$$

Here, $\mathbf{z}_i$ denotes object i's covariate values. Alternative terms for the zs are concomitant variables, grouping variables, external variables, exogenous variables, and inputs. For the number of parameters to be reduced, the probability of belonging to class k given covariate values $\mathbf{z}_i$, $\pi_{k|\mathbf{z}_i}$, will generally be restricted by a multinomial logit model, that is, a logit model with "linear effects" and no higher-order interactions.

An even more general specification is obtained by allowing covariates to have direct effects on the indicators, which yields

$$f(\mathbf{y}_i \mid \mathbf{z}_i, \theta) = \sum_{k=1}^{K} \pi_{k|\mathbf{z}_i} \prod_{j=1}^{J} f_k(y_{ij} \mid \mathbf{z}_i, \theta_{jk}) .$$

The conditional mean of the y variables can now be related directly to the covariates. This makes it possible to relax the implicit assumption in the previous specification that the influence of the zs on the ys goes completely by the latent variable. For an example, see Vermunt and Magidson (2000, p. 155).

The possibility to have direct effects of zs on ys can also be used to specify direct effects between indicators of different scale types by means of a simple trick: one of the two variables involved should be used both as covariate (not influencing class membership) and as indicator. We use this trick next in our second example.

5. ESTIMATION

The two main methods to estimate the parameters of the various types of LC cluster models are the maximum-likelihood (ML) method and the maximum-posterior (MAP) method. Wallace and Dowe (forthcoming) proposed a minimum message length (MML) estimator, which in most situations is similar to the MAP method. The log-likelihood function required in the ML and MAP approaches can be derived from the probability density function defining the model. Bayesian MAP estimation involves maximizing the log-posterior distribution, which is the sum of the log-likelihood function and the logs of the priors for the parameters.

Although generally there is not much difference between ML and MAP estimates, an important advantage of the latter method is that it prevents the occurrence of boundary or terminal solutions; probabilities and variances cannot become zero. With a very small amount of prior information, the parameter estimates are forced to stay within the interior of the parameter space. Typical priors are Dirichlet priors for multinomial probabilities and inverted-Wishart priors for the variance–covariance matrices in multivariate normal models. For more details on these priors, see Vermunt and Magidson (2000, pp. 164–65).

Most software packages use the expectation–maximization (EM) algorithm or some modification of it to find the ML or MAP estimates. In our opinion, the ideal algorithm is starting with a number of EM iterations and, when close enough to the final solution, switching to Newton–Raphson. This is a way to combine the advantages of both algorithms, that is, the stability of EM even when far away from the optimum and the speed of Newton–Raphson when close to the optimum.

A well-known problem in LC analysis is the occurrence of local solutions. The best way to prevent ending with a local solution is to use multiple sets of starting values. Some computer programs for LC clustering have automated the search for good starting values by using several sets of random starting values as well as solutions obtained with other cluster methods.

In the application of LC analysis to clustering, we are not only interested in the estimation of the model parameters; another important "estimation" problem is the classification of objects into clusters. This can be based on the posterior class-membership probabilities

$$\pi_{k|\mathbf{y}_i,\mathbf{z}_i,} = \pi_{k|\mathbf{z}_i} \prod_j f_k(y_{ij} \mid \mathbf{z}_i, \theta_{jk}) \Big/ \sum_k \pi_{k|\mathbf{z}_i} \prod_j f_k(y_{ij} \mid \mathbf{z}_i, \theta_{jk}) . \quad (3)$$

The standard classification method is modal allocation, which amounts to assigning each object to the class with the highest posterior probability.

6. MODEL SELECTION

The model selection issue is one of the main research topics in LC clustering. Actually, there are two issues: the first concerns the decision about the number of clusters; the second concerns the form of the model given the number of clusters. For an overview on this topic, see Celeux, Biernacki, and Govaert (1997).

Assumptions with respect to the forms of the clusters given their number can be tested by using standard likelihood-ratio tests between nested models, for instance, between a model with an unrestricted covariance matrix and a model with a restricted covariance matrix. Wald tests and Lagrange multiplier tests can be used to assess the significance of certain included or excluded terms, respectively. It is well known that these kinds of chi-squared tests cannot be used to determine the number of clusters.

The most popular set of model selection tools in LC cluster analysis are information criteria such as Akaike, Bayesian, and consistent Akaike information criteria, or AIC, BIC, and CAIC (Fraley and Raftery, 1998b). The most recent development is the use of computationally intensive techniques such as parametric bootstrapping (McLachlan et al., 1999) and Markov chain Monte Carlo methods (Bensmail et al., 1997) to determine the number of clusters and their forms. Cheeseman and Stutz (1995) proposed a fully automated model selection method using approximate Bayes factors (different from BIC).

Another set of methods for evaluating LC cluster models is based on the uncertainty of classification or, equivalently, the separation of the clusters. Aside from the estimated total number of misclassifications, the Goodman–Kruskal lambda, the Goodman–Kruskal tau, or entropy-based measures can be used to indicate how well the indicators predict class membership. Celeux et al. (1997) described various indices that combine information on model fit and information on classification errors, two of which are the classification likelihood (C) and the approximate weight of evidence (AWE).

7. TWO EMPIRICAL EXAMPLES

Next, LC cluster modeling is illustrated by means of two empirical examples. The analyses are performed with the LC analysis (LCA) program Latent GOLD (Vermunt and Magidson, 2000), which implements both ML and MAP estimation with Dirichlet and inverted-Wishart priors for multinomial probabilities and error variance–covariance matrices, respectively. A feature of the program that was used extensively in the analyses described next is the possibility to add local dependencies by using information on bivariate residuals. Model selection was based on BIC; it should be noted that the BIC we use is computed by using the log-likelihood value and the number of parameters rather than by using the L^2 value and the number of degrees of freedom.

A. Diabetes Data

The first empirical example concerns a three-dimensional data set involving 145 observations used for diabetes diagnosis (Reaven and Miller, 1979). The three continuous variables are labeled *glucose* (y_1), *insulin* (y_2), and *sspg* (steady-state plasma glucose, y_3). The data set also contains information on the clinical classification in three groups (normal, chemical diabetes, and overt diabetes), which makes it possible to compare the clinical classification with the classification obtained from the cluster model. The substantive question of interest is whether the three indirect diagnostic measures yield a reliable diagnosis; that is, whether they yield a classification that is close to the clinical classification.

This data set comes with the MCLUST program and is also used by Fraley and Raftery (1998a, 1998b) to illustrate their model-based cluster analysis based on the eigenvalue decomposition described in Equation (1). The final model they selected on the basis of the BIC criterion was the unrestricted three-class model, which means that none of the restrictions that can be specified with their approach holds for this data set.

We used six different specifications for the variance–covariance matrices: class-dependent and class-independent unrestricted, class-dependent and class-independent diagonal, as well as class-dependent and class-independent with only the y_1–y_2 error covariance free. The specification *unrestricted* means that all covariances are free; the specification *diagonal* means that all covariances are assumed to be zero. The models with only the y_1–y_2 error covariance free were used because the bivariate residuals of both diagonal models indicated that there was only a local dependency between these two variables. Moreover, the results from the unrestricted models indicated that the y_1–y_3 and y_2–y_3 covariances did not differ significantly from zero.

Table 1 reports the BIC values for the estimated one to five class models. The three-class model that only includes the error covariance between y_1 and y_2 and with class-dependent variances and covariances has the lowest BIC value. Its BIC value is slightly lower than of the class-dependent unrestricted three-class model, which is Fraley and Raftery's final model for this data set. The BIC values in Table 1 show clearly that models with too-restrictive error structures for a particular data set overestimate the number of clusters. Here, this applies to the models

Table 1. BIC Values for Diabetes Example

Model	No. of Clusters 1	2	3	4	5
1. Class-dep. unrestricted Σ_k	5138	4819	4762	4788	4818
2. Class-ind. unrestricted Σ_k	5138	5014	4923	4869	4858
3. Class-dep. diagonal Σ_k	5530	4957	4833	4805	4815
4. Class-ind. diagonal Σ_k	5530	5170	4999	4938	4895
5. Class-dep. Σ_k with only σ_{12k} free	5156	4835	4756	4761	4784
6. Class-ind. Σ_k with only σ_{12k} free	5156	5008	4920	4862	4859

with class-independent error variances and the class-dependent diagonal model. Therefore, it is important to be able to work with different types of error structures. Note that the most restrictive model that we used – the model with a class-independent diagonal error structure – can be seen as a probabilistic variant of *k*-means cluster analysis (McLachlan and Basford, 1988).

Table 2 reports the parameters estimates for the three-class model with class-dependent variance–covariance matrices and with only a local dependence between y_1 and y_2. These parameters are the cluster sizes (π_k), the cluster-specific means (μ_{jk}), the cluster-specific variances ($\sigma_{jk}{}^2$), and the cluster-specific covariance between y_1 and y_2 (σ_{12k}). The overt diabetes group (Cluster 3) has much higher means on glucose and insulin and a much lower mean on sspg than the normal group (Cluster 1). The chemical diabetes group (Cluster 2) has somewhat lower means on glucose and insulin and a much lower mean on sspg than the normal group. The reported error variances show that the overt diabetes cluster is much

Table 2. Parameter Estimates for Diabetes Example

	Cluster					
	1 = Normal		2 = Chemical		3 = Overt	
Parameter	Estimate	SE	Estimate	SE	Estimate	SE
π_k	0.27	0.05	0.54	0.05	0.19	0.03
μ_{1k}	104.00	2.85	91.23	1.06	234.76	14.87
μ_{2k}	495.06	22.74	359.22	6.63	1121.09	58.70
μ_{3k}	309.43	28.06	163.13	6.37	76.98	9.47
$\sigma_{1k}{}^2$	230.09	62.96	76.48	12.93	5005.91	1414.43
$\sigma_{2k}{}^2$	14844.55	3708.65	2669.75	506.55	73551.09	22176.29
$\sigma_{3k}{}^2$	22966.52	5395.90	2421.45	476.65	2224.50	616.43
σ_{12k}	1279.92	420.93	96.46	60.30	17910.71	5423.37

Table 3. Clinical vs. LC Cluster Classification in Diabetes Example

	LC Cluster Class.			
Clinical Class.	**Normal**	**Chemical**	**Overt**	**Total**
Normal	26	10	0	36
Chemical	4	72	0	76
Overt	5	0	28	33
Total	35	82	28	145

more heterogeneous with respect to glucose and insulin and much more homogeneous with respect to sspg than the normal cluster. The chemical diabetes group is the most homogeneous cluster on all three measures. The error covariances are somewhat easier to interpret if we transform them into correlations. Their values are 0.69, 0.21, and 0.93 for Clusters 1, 2, and 3, respectively. This indicates that in the overt diabetes group there is a very strong association between glucose and insulin, whereas in the chemical diabetes group this association is very low, and even not significantly different from zero ($\widehat{\sigma}_{12k}/\mathrm{SE}_{\widehat{\sigma}_{12k}} = 1.60$). Note that the within-cluster correlation of 0.93 is very high, which indicates that, in fact, the two measures are equivalent in Cluster 3.

Not only is the BIC of our final model somewhat better than that of Fraley and Raftery, but also our classification is more in agreement with the clinical classification: our model "misclassifies" 13.1% of the patients whereas the unrestricted model misclassifies 14.5%. Table 3 reports the cross-tabulation of the clinical and the LC cluster classification based on the posterior class-membership probabilities. As can be seen, some normal patients are classified as cases with chemical diabetes and vice versa. The other type of error is that some overt diabetes cases are classified as normal.

B. Prostate Cancer Data

Our second example concerns the analysis of a mixed-mode data set with pretrial covariates from a prostate cancer clinical trial (Byar and Green, 1980). Jorgensen and Hunt (1996) and Hunt and Jorgensen (1999) used this data set containing information on 506 patients to illustrate the use of the LC cluster model implemented in their MULTIMIX program. The eight continuous indicators are age (y_1), weight index (y_2), systolic blood pressure (y_5), diastolic blood pressure (y_6), serum hemoglobin (y_8), size of primary tumor (y_9), index of tumor stage and histolic grade (y_{10}), and serum prostatic acid phosphatase (y_{11}). The four categorical observed variables are performance rating (y_3, four levels), cardiovascular disease history (y_4, two levels), electrocardiogram code (y_7, seven levels), and bone metastases (y_{12}, two levels). The research question of interest is whether on the basis of these pretrial covariates it is possible to identify subgroups that differ with respect to the likelihood of success of the medical treatment of prostate cancer.

The categorical variables are treated as nominal, and for the continuous variables we assumed normal distributions with class-specific variances. We estimated models from one to four latent classes. The first model for each number of classes assumes local independence. The other four specifications are obtained by subsequently adding the direct relationships between y_5 and y_6, y_2 and y_8, y_8 and y_{12}, and y_{11} and y_{12}. This exploratory improvement of the model fit was guided by Latent GOLD's bivariate residuals information, as well as the results reported by Hunt and Jorgensen (1999).

An indication about the computation time needed for these kinds of models is that all two-class models took less than 5 seconds to converge, and all four class models took less than 20 seconds on a Pentium II 350 MHz. Note that here we have a data set with almost 500 cases and 12 indicators. The estimation time increases linearly with the number of cases and, as long as we do not include too many local dependencies, also almost linearly with the number of indicators.

Table 4 presents the BIC values for the estimated models. As can be seen, the two-class model that includes all four direct relationships has the lowest BIC. A comparison of the various models given a certain number of classes shows that inclusion of the direct relationship between y_5 and y_6 (the two blood pressure measures) improves the fit in all situations. The other bivariate terms improve the fit in the one-, two-, and three-class models, but not in the four-class model. If we compare the models with

Table 4. BIC Values for Cancer Example

	No. of Clusters			
Model	**1**	**2**	**3**	**4**
1. Local independence	23,762	23,112	23,089	23,088
2. Model 1 + σ_{56k}	23,529	22,889	22,883	22,887
3. Model 2 + σ_{28k}	23,502	22,872	22,875	22,893
4. Model 3 + $\beta_{8.12}$	23,473	22,861	22,866	22,895
5. Model 4 + $\beta_{11.12}$	23,322	22,845	22,855	22,888

different number of classes for a given error structure, the four-class model performs best when assuming local independence, the three-class model when including the y_5 and y_6 covariance, and the two-class model when including additional bivariate terms. Thus, if we are willing to include the y_5–y_6 effect, a model with no more than three classes should be selected. If we are willing to include more direct effects, the two-class model is the preferred one. This shows again that the possibility to work with more local dependencies may yield a simpler final model.

Table 5 reports the parameter estimates for the two-class model containing all four direct effects. Wald tests for the difference of the means and probabilities between classes indicate that only the mean ages (μ_{1k}) are not significantly different between classes. Cluster 2 turns out to have somewhat higher means on weight (μ_{2k}), blood pressure (μ_{5k} and μ_{6k}), and serum hemoglobin (μ_{8k}), and lower means on size of tumor (μ_{9k}), index of tumor stage (μ_{10k}), and serum prostatic acid phosphatase (μ_{11k}). If we look at the nominal indicators, we see a large difference between the two classes in the distribution of bone metastases (y_{12}), somewhat smaller differences in performance rating (y_3) and cardiovascular disease history (y_4), and a very small difference in electrocardiogram code (y_7). The direct effects between the indicators are quite strong. They all have a positive sign except for the effect of y_{12} on y_{11}.

To investigate the usefulness of the applied technique, Jorgensen and Hunt (1996) and Hunt and Jorgensen (1999) investigated the strength of the relationship between the obtained classification and the outcome of the medical trial. They showed that their two-class solution, which is similar to the two-class model with local dependencies obtained here, predicted very well the success of the medical treatment.

8. CONCLUSIONS

This paper described the state-of-art in the field of cluster analysis by using LC models. Two important recent developments are the possibility of using various kinds of meaningful restrictions on the covariance structure in mixtures of multivariate normal distributions and the possibility of working with mixed-mode data.

The first example demonstrated the use of different types of specifications for the covariance structure. It showed that models that are too restrictive may yield too many latent classes. The second example illustrated LC clustering with mixed-mode data by using models with and without local dependencies.

Table 5. Parameter Estimates for Prostate Cancer Example

	Cluster 1		Cluster 2	
Parameter	**Estimate**	**SE**	**Estimate**	**SE**
π_k	0.45	0.03	0.55	0.03
μ_{1k}	71.38	0.51	71.70	0.43
μ_{2k}	97.51	0.98	100.26	0.83
$\pi_{1,3k}$	0.85	0.02	0.94	0.02
$\pi_{2,3k}$	0.09	0.02	0.05	0.01
$\pi_{3,3k}$	0.05	0.02	0.01	0.01
$\pi_{4,3k}$	0.01	0.01	0.00	0.00
$\pi_{1,4k}$	0.65	0.03	0.49	0.03
$\pi_{2,4k}$	0.35	0.03	0.51	0.03
μ_{5k}	14.18	0.16	14.54	0.16
μ_{6k}	8.00	0.09	8.29	0.10
$\pi_{1,7k}$	0.35	0.03	0.33	0.03
$\pi_{2,7k}$	0.05	0.02	0.05	0.01
$\pi_{3,7k}$	0.14	0.02	0.07	0.02
$\pi_{4,7k}$	0.04	0.01	0.06	0.02
$\pi_{5,7k}$	0.30	0.03	0.31	0.03
$\pi_{6,7k}$	0.12	0.02	0.17	0.02
$\pi_{7,7k}$	0.00	0.00	0.00	0.00
μ_{8k}	128.01	1.38	132.21	1.80
μ_{9k}	4.11	0.12	2.88	0.08
μ_{10k}	12.02	0.11	8.88	0.08
μ_{11k}	4.00	0.12	2.11	0.11
$\pi_{1,12k}$	0.65	0.03	0.99	0.01
$\pi_{2,12k}$	0.35	0.03	0.01	0.01
σ^2_{1k}	52.35	5.36	43.97	4.15
σ^2_{2k}	186.60	19.82	166.73	15.89
σ^2_{5k}	4.98	0.50	6.60	0.59
σ^2_{6k}	1.79	0.18	2.40	0.21
σ^2_{8k}	355.82	35.44	325.52	29.47
σ^2_{9k}	2.91	0.29	1.40	0.14
σ^2_{10k}	2.05	0.21	1.25	0.13
σ^2_{11k}	2.56	0.25	0.25	0.03
σ_{28k}	61.98	19.14	47.56	15.12
σ_{56k}	1.82	0.25	2.52	0.30
$\beta_{8.12}$	5.76	1.35	5.76	1.35
$\beta_{11.12}$	−0.49	0.11	−0.49	0.11

REFERENCES

Arminger, G., & Stein, P. (1997). "Finite mixture of covariance structure models with regressors: loglikehood function, distance estimation, fit indices, and a complex example," *Sociological Methods and Research*, **26**, 148–82.

Bacher, J. (2000). "A probabilistic clustering model for variables of mixed type," *Quality and Quantity*, **34**, 223–35.

Banfield, J. D., & Raftery, A. E. (1993). "Model-based Gaussian and non-Gaussian clustering," *Biometrics*, **49**, 803–21.

Bensmail, H., Celeux, G., Raftery, A. E., & Robert, C. P. (1997). "Inference in model based clustering," *Statistics and Computing*, **7**, 1–10.

Böckenholt, U. (1993). "A latent class regression approach for the analysis of recurrent choices," *British Journal of Mathematical and Statistical Psychology*, **46**, 95–118.

Byar, D. P., & Green, S. B. (1980). "The choice of treatment for cancer patients based on covariate information: application to prostate cancer," *Bulletin of Cancer*, **67**, 477–90.

Celeux, G., Biernacki, C., & Govaert, G. (1997). *Choosing Models in Model-Based Clustering and Discriminant Analysis*. Technical Report. Rhone-Alpes: INRIA.

Cheeseman, P., & Stutz, J. (1995). "Bayesian classification (Autoclass): theory and results." In U. M. Fayyad, G. Piatetsky-Shapiro, P. Smyth, & R. Uthurusamy (eds.), *Advances in Knowledge Discovery and Data Mining*. Menlo Park: The AAAI Press, pp. XXX–XX.

Clogg, C. C. (1981). "New developments in latent structure analysis." In D. J. Jackson & E. F. Borgotta (eds.), *Factor Analysis and Measurement in Sociological Research*. Beverly Hills: Sage, pp. XXX–XX.

Clogg, C. C. (1995). "Latent class models." In G. Arminger, C. C. Clogg, & M. E. Sobel (eds.), *Handbook of Statistical Modeling for the Social and Behavioral Sciences*. New York: Plenum Press, pp. 311–59.

Dolan, C. V., & Van der Maas, H. L. J. (1997). "Fitting multivariate normal finite mixtures subject to structural equation modeling," *Psychometrika*, **63**, 227–253.

Everitt, B. S. (1988). "A finite mixture model for the clustering of mixed-mode data," *Statistics and Probability Letters*, **6**, 305–309.

Everitt, B. S. (1993). *Cluster Analysis*. London: Edward Arnold.

Fraley, C., & Raftery, A. E. (1998a). *MCLUST: Software for Model-Based Cluster and Discriminant Analysis*. Technical Report No. 342, Department of Statistics, University of Washington.

Fraley, C., & Raftery, A. E. (1998b). *How Many Clusters? Which Clustering Method? Answers via Model-Based Cluster Analysis*. Technical Report No. 329, Department of Statistics, University of Washington.

Gibson, W. A. (1959). "Three multivariate models: factor analysis, latent structure analysis, and latent profile analysis," *Psychometrika*, **24**, 229–52.

Goodman, L. A. (1974). "Exploratory latent structure analysis using both identifiabe and unidentifiable models," *Biometrika*, **61**, 215–31.

Hunt, L, & Jorgensen, M. (1999). "Mixture model clustering using the MULTIMIX program," *Australian and New Zeeland Journal of Statistics*, **41**, 153–72.

Jedidi, K., Jagpal, H. S., & DeSarbo, W. S. (1997). "Finite-mixture structural equation models for response-based segmentation and unobserved heterogeneity," *Marketing Science*, **16**, 39–59.

Jorgensen, M., & Hunt, L. (1996). "Mixture model clustering of data sets with categorical and continuous variables." In *Proceedings of the Conference ISIS '96, Australia, 1996*, pp. 375–84.

Kaufman, L., & Rousseeuw, P. J. (1990). *Finding Groups in Data: An Introduction to Cluster Analysis*. New York: Wiley.

Krzanowski, W. J. (1983). "Distance between populations using mixed continuous and categorical variables," *Biometrika*, **70**, 235–43.

Lawrence C. J., & Krzanowski, W. J. (1996). "Mixture separation for mixed-mode data," *Statistics and Computing*, **6**, 85–92.

Lazarsfeld, P. F., & Henry, N. W. (1968). *Latent Structure Analysis*. Boston: Houghton Mill.

McLachlan, G. J., & Basford, K. E. (1988). *Mixture Models: Inference and Application to Clustering*. New York: Marcel Dekker.

McLachlan, G. J., & Peel, D. (1996). "An algorithm for unsupervised learning via normal mixture models." In D. L. Dowe, K. B. Korb, & J. J. Oliver (eds.), *Information, Statistics and Induction in Science*. Singapore: World Scientific, pp. XXX–XX.

McLachlan, G. J., & Peel, D. (1999). *Modelling Nonlinearity by Mixtures of Factor Analysers via Extension of the EM Algorithm*. Technical Report, Australia, Center for Statistics, University of Queensland.

McLachlan, G. J., Peel, D., Basford, K. E., & Adams, P. (1999). "The EMMIX software for the fitting of mixtures of normal and *t*-components," *Journal of Statistical Software*, **4**, No. 2.

Reaven, G. M., & Miller, R. G. (1979). "An attempt to define the nature of chemical diabetes using multidimensional analysis," *Diabetologia*, **16**, 17–24.

Vermunt, J. K., & Magidson, J. (2000). *Latent GOLD's User's Guide*. Boston: Statistical Innovations, Inc.

Wallace, C. S., & Dowe, D. L. (Forthcoming). "MML clustering of multi-state, Poisson, von Mises circular and Gaussian distributions," *Statistics and Computing*, **X**, X–X.

Wedel, M., DeSarbo, W. S., Bult, J. R., & Ramaswamy, V. (1993). "A latent class Poisson regression model for heterogeneous count data with an application to direct mail," *Journal of Applied Econometrics*, **8**, 397–411.

Wolfe, J. H. (1970). "Pattern clustering by mulltivariate cluster analysis," *Multivariate Behavioral Research*, **5**, 329–50.

Yung, Y. F. (1997). "Finite mixtures in confirmatory factor-analysis models," *Psychometrika*, **62**, 297–330.

FOUR

Some Examples of Latent Budget Analysis and Its Extensions

Peter G. M. van der Heijden, L. Andries van der Ark, and Ab Mooijaart

1. INTRODUCTION

Latent budget analysis is a tool for the analysis of two-way contingency tables. The idea was initiated by Goodman (1974). Clogg (1981) extended this idea to an asymmetrical latent class model for the analysis of social mobility tables. Clogg used the following example: Let profession of the father be variable A, with categories indexed by i $(i = 1, \dots, I)$; let profession of the son be variable B, with categories indexed by j $(j = 1, \dots, J)$; let the latent social class variable be X, with categories indexed by t $(t = 1, \dots, T)$. Let π_{ij} be the joint probability of profession i of the son and profession j of the father. Let π_t^X be the probability that a son belongs to the tth latent social class; $\pi_{it}^{\bar{A}X}$ the conditional probability that a son has a father with profession i given that he belongs to latent social class t; and $\pi_{jt}^{\bar{B}X}$ the conditional probability that a son has profession j given that he belongs to latent social class t.

The latent class model with T latent classes for a two-way table with probabilities p_{ij} is

$$\pi_{ij} = \sum_{t=1}^{T} \pi_t^X \pi_{it}^{\bar{A}X} \pi_{jt}^{\bar{B}X}, \tag{1}$$

with all parameters nonnegative and restricted by

$$\sum_{t=1}^{T} \pi_t^X = 1, \quad \sum_{I=1}^{I} \pi_{it}^{\bar{A}X} = 1, \quad \sum_{j=1}^{J} \pi_{jt}^{\bar{B}X} = 1.$$

In this example, the explanatory variable is profession of the father and the response variable is profession of the son. Clogg assumed that there was a mediating (latent) variable, which he interpreted as social class.

He assumed that this latent variable was categorical. By rescaling the parameters $\pi_{it}^{\bar{A}X}$ into parameters $\pi_{it}^{A\bar{X}}$ by

$$\pi_{it}^{A\bar{X}} = \pi_t^X \pi_{it}^{\bar{A}X} \Big/ \sum_{t=1}^{T} \pi_t^X \pi_{it}^{\bar{A}X},$$

Goodman (1974) and Clogg (1981) noticed that it is possible to rewrite Equation (1) into

$$\frac{\pi_{ij}}{\pi_{i+}} = \sum_{t=1}^{T} \pi_{it}^{A\bar{X}} \pi_{jt}^{\bar{B}X}, \tag{2}$$

with parameter restrictions

$$\sum_{t=1}^{T} \pi_{it}^{A\bar{X}} = 1, \qquad \sum_{j=1}^{J} \pi_{jt}^{\bar{B}X} = 1. \tag{3}$$

Compared with Model 1, in Model 2 the probabilities that are decomposed are conditional probabilities rather than joint probabilities. That is, the conditional probability π_{ij}/π_{i+} is the probability that the son has profession j given that the father has profession i. The parameters are interpreted as follows: The parameters $\pi_{it}^{A\bar{X}}$ are the probabilities that a father with profession i belongs to the tth latent social class, and $\pi_{jt}^{\bar{B}X}$ are the probabilities that the son has profession j given that he belongs to the tth social latent class. It may be noted that the parameters $\pi_{jt}^{\bar{B}X}$ have the same interpretation in Model 1 and Model 2.

Model 2 is illustrated graphically in Figure 1. In the social sciences, the representation in this figure is known as a MIMIC model (i.e., the Multiple Indicator Multiple Cause model; Goodman, 1974). It may be noted that the squares in Figure 1 represent the levels of the professions, whereas the T circles represent the levels of the latent variable. (This should not be confused with representations of structural equations models often used in the social sciences, where both circles and squares always represent variables, and not levels of variables.)

Independently, de Leeuw and van der Heijden (1988) reinvented Model 2 in the context of an analysis of time budgets. A time budget of an individual i is the distribution of time over J mutually exclusive activities. Hence, the J elements add up to 1 and they are nonnegative, just like the conditional probabilities (π_{ij}/π_{i+}) in Model 2. The word *budget* emphasizes that if time is spent on one activity, it cannot be spent on another activity at the same time. Therefore, they termed Model 2 the *latent budget model* (LBM). The T vectors of parameters $(\pi_{1t}^{\bar{B}X}, \ldots, \pi_{Jt}^{\bar{B}X})$ are

Figure 1. Graphic representation of a MIMIC model.

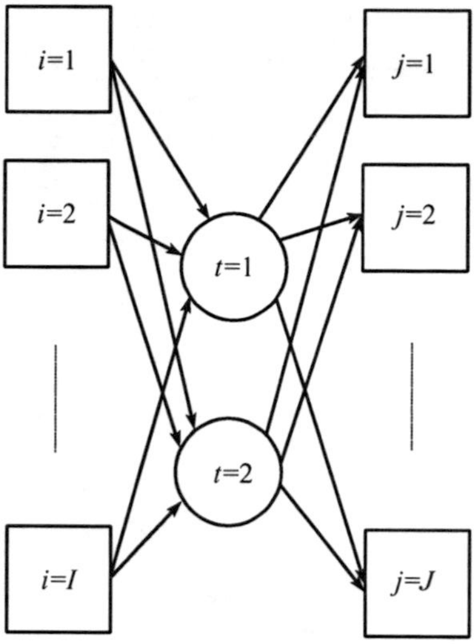

MIMIC model

called *latent budgets*. Similarly, the I vectors of conditional probabilities $(\pi_{i1}/\pi_{i+}, \ldots, \pi_{iJ}/\pi_{i+})$ are called *expected budgets*.

In 1988, the authors were unaware of the fact that the idea of the LBM had been introduced much earlier by Goodman (1974). Van der Heijden, Mooijaart, and de Leeuw (1992) pointed out the equivalence between the LBM and Goodman's and Clogg's work. However, they emphasized a mixture-model interpretation of the LBM. The expected budgets are mixtures of T latent budgets. The mixture interpretation is illustrated graphically in Figure 2. In Figure 2 only the expected budgets i and the latent budgets t are shown. The figure shows that an expected budget i is a mixture of the T latent budgets. The T *mixing parameters* for row i are provided by the parameters $\pi_{it}^{A\bar{X}}$. These mixing parameters show for which proportion the expected budgets are built up from the latent budgets. The mixing parameters are not revealed by Figure 2.

The LBM with T latent budgets has $(I - T)(J - T)$ degrees of freedom. For $T = 1$, the LBM is equivalent to the independence model because then $p_{ij}/p_{i+} = \pi_{jt}^{\bar{B}X} = p_{+j}$. For $T = \min(I, J)$, the LBM is saturated, and estimates of expected proportions are equal to observed proportions.

The LBM is usually estimated by the method of maximum likelihood under the assumption that the frequencies are generated by a product-multinomial distribution (although we have also been working on other estimation methods; see Mooijaart, van der Heijden, and van der Ark,

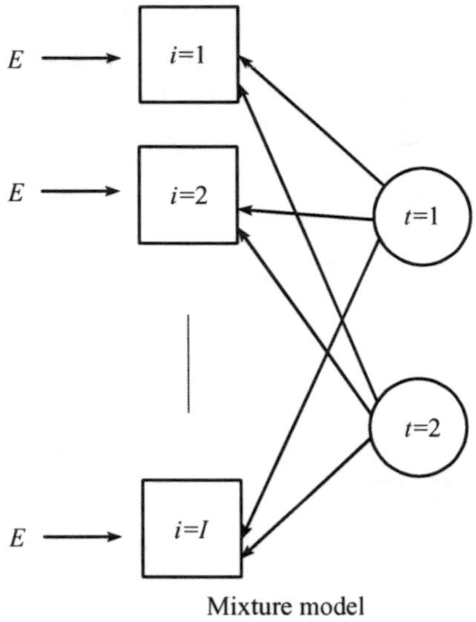

Figure 2. Graphic representation of a mixture model.

1999; van der Ark, 1999). Likelihood ratio tests are used to assess the fit of the LBM against the data and to determine the number of latent budgets (i.e., T) needed to describe the data adequately.

Clogg (1981) noted that Model 2 is not identified. De Leeuw, van der Heijden, and Verboon (1990) also discussed the identification problem of the LBM, and they worked out the situation for $T = 2$ in some detail. Writing Model 2 in matrix notation shows the identification problem: Collect the conditional proportions π_{ij}/π_{i+} in a matrix $\mathbf{\Pi}$, the mixing parameters $\pi_{it}^{A\bar{X}}$ in a matrix $\mathbf{A}$, and the latent budget parameters $\pi_{jt}^{\bar{B}X}$ in a matrix $\mathbf{B}$; then Model 2 equals

$$\mathbf{\Pi} = \mathbf{AB}'. \tag{4}$$

It is always possible to rewrite Equation 4 into $\mathbf{\Pi} = (\mathbf{AS}^{-1})(\mathbf{SB}') = \mathbf{A}^*\mathbf{A}^{*'}$, where $\mathbf{S}$ is a $K \times K$ matrix with each row adding up to 1. The parameters $\mathbf{A}$ and $\mathbf{B}$ yield the same expected budgets as $\mathbf{A}^*$ and $\mathbf{B}^*$. Because the elements of each row of $\mathbf{S}$ add up to 1, the parameters $\mathbf{A}^*$ and $\mathbf{B}^*$ are also subject to the equality restrictions in Equation (3). Furthermore, $\mathbf{S}$ can be chosen freely as long as all elements of $\mathbf{A}^*$ and $\mathbf{B}^*$ are nonnegative. De Leeuw et al. (1990) choose $\mathbf{S}$ such that as many parameters as possible from either $\mathbf{A}^*$ or $\mathbf{B}^*$ are zero, because this facilitated the interpretation. Van der Ark, van der Heijden, and Sikkel (1999) extended this work for $T > 2$. Their view of the identification problem for the LBM is similar to the identification problem in factor analysis, in which unidentified solutions are usually rotated to simplify interpretation. The common factor model is called *identified* because the varimax-rotated solutions are always unique in practical situations. Similarly, Van der Ark et al. (1999) called the LBM identified for some specific choices of **S**. They proposed an *inner extreme* solution, that is, choosing **S** such that the mixing parameters are as distinct as possible, which facilitates the interpretation in terms of the explanatory variable (e.g., the example of Section 2), and *outer extreme* solution, that is, choosing **S** such that the latent budgets are as distinct as possible, which facilitates the interpretation in terms of the response variable (e.g., the example in Section 3).

Van der Heijden et al. (1992) discuss various ways in which the parameters of the LBM can be constrained. They distinguish fixed-value constraints (e.g., some parameters are fixed to some constant), equality constraints (see, for some estimation problems, Mooijaart and van der Heijden, 1992), and situations in which the parameters $\pi_{it}^{A\bar{X}}$ and $\pi_{jt}^{\bar{B}X}$ are functions of external information. Sometimes these constraints can also be used as well to identify the LBM (e.g., the example in Section 4). A later development was to study how latent budget analyses of different groups could be compared; this was termed *simultaneous latent budget analysis* (see Siciliano and van der Heijden, 1994).

The LBM is closely related to correspondence analysis, and de Leeuw and van der Heijden (1991) describe under what circumstances the LBM is equivalent to correspondence analysis (see van der Ark and van der Heijden, 1997; van der Ark et al., 1999; van der Heijden, Gilula, and van der Ark, 1999). Latent budget analysis is also used in geology, where it is known as *end-member analysis* (see Renner, 1993; Weltje, 1997; van der Heijden, 1994). Many other results, in particular concerning least-squares estimates, standard errors, and testing procedures, can be found in van der Ark (1999).

In this chapter, by discussing some examples, we demonstrate some of the possibilities of the LBM and its extensions. Section 2 shows an example of latent budget analysis of a two-way table dealing with sentence endings of the books of Plato. Section 3 illustrates the possibilities of the LBM for comparing contingency tables in the context of trades started by different ethnic groups; here, the city of Amsterdam is compared with the city of Rotterdam. Section 4 shows the possibilities of the LBM for studying how the school success of pupils is related to explanatory variables such as IQ, sex, and the profession of the father.

2. THE WORKS OF PLATO

We start with a straightforward application of the LBM. The Greek philosopher Plato wrote forty-five books. The exact order in which these works were written is known approximately, except for the books *Critias*, *Philebus*, *Politicus*, *Sophist*, and *Timaeus*. The objective of this example is to show that the LBM can be used for seriation, that is, to find the chronological order in which all 45 books were written. For this purpose, we used data obtained by Kaluscha (1904), who collected all "sentence endings" in the 45 books. Each of the last five syllables of a sentence ending is scored as being "short" or "long," so that each sentence of each book belongs to one of $2^5 = 32$ categories.

The idea underlying the determination of the chronological order of the books from the distributions of sentence endings is that the style and rhythm of the texts changed through time, and that sentence endings are considered highly relevant with regard to rhythm (Boneva, 1970). For each book, we had the frequencies of sentence endings, yielding a matrix of 45 books by 32 sentence endings. The data are in Table 1, where the chronological order of the 40 "known" books is preserved. The 45 books are considered to be 45 budgets, each containing 32 categories. The frequencies in these 32 categories express the writing style of the particular book.

The LBM takes typical styles of writing as latent budgets, and the different books are then approximated by a mixture of these typical styles. The mixture-model interpretation (see Figure 2) is most appropriate in this context. The data, or an aggregated version, were studied earlier by, for example, Cox and Brandwood (1959), Atkinson (1970), and Greenacre (1984).

Latent budget analysis considers the frequencies of sentence endings of each book as a sample from a multinomial distribution. The LBM with $T = 1$ latent budget (independence of works and sentence endings) has a likelihood ratio chi square of $L^2 = 3{,}678$ (the degrees of freedom, df, is 1,364). This model implies that the writing styles in all books are identical. The LBM with $T = 2$ latent budgets has a fit of $L^2 = 2{,}022$ (df is 1,290). This model implies that there are two typical writing styles. The two estimated latent budgets show what these typical writing styles are. For each book i, the two mixing parameters $\pi_{it}^{A\bar{X}}$ $(t = 1, 2)$ show how the budget of book i is built up from these two typical writing styles. This model described $(3{,}678 - 2{,}022)/3{,}678 = 0.45$ of the departure from independence. The LBM with $T = 3$ latent budgets assumes that there are three typical writing styles. The fit was $L^2 = 1{,}661$ (df is 1,218), and this explained 0.55 of the departure from independence. For the LBM with $T = 4$ latent budgets, the fit was $L^2 = 1{,}440$ (df is 1,148), and this explained 0.61 of the departure from independence. The model with $T = 2$ described a considerable part of the departure from independence.

Not much more information was extracted from the data by consideration of more latent budgets.

For seriation, the LBM with $T = 2$ latent budgets is most appropriate. This model yields a unidimensional chronological order of the books because each book has two mixing parameters, $\pi_{i1}^{A\bar{X}}$ and $\pi_{i2}^{A\bar{X}}$, where $\pi_{i1}^{A\bar{X}} + \pi_{i2}^{A\bar{X}} = 1.0$. Therefore, it suffices to interpret only the 45 estimates $\hat{\pi}_{i1}$ for studying the differences between the books. We give a graphical representation of these 45 mixing parameter estimates in Figure 3 because this simplifies the interpretation. A graphical interpretation of the LBM with two latent budgets is that the 45 books are on a line segment. The latent budgets are the endpoints of the line segment: If the writing style of book i matches the typical writing style of latent budget 1 exactly, then $\pi_{i1}^{A\bar{X}} = 1.0$, and $\pi_{i2}^{A\bar{X}} = 0.0$. Book i is plotted on the endpoint of the line segment that coincides with latent budget 1. If the writing style of book i' is built up for 0.5 from the first typical writing style and for 0.5 from the second typical writing style, then book i' is plotted in the middle of the line segment, exactly in between the two latent budgets. If one writing style is typical for the earlier years of Plato's writings, and the second writing style is typical for the later years, then the line segment represents the chronological order of the books.

Not all the individual books could be printed into Figure 3. Therefore, two shaded areas are given. In the shaded area on the left side (close to the older writing style budget) are all the works with known chronological order up until *Republic 10*, with the exception of *Laches* and *Cratylus*. In the shaded area on the right side (close to the newer writing style budget) are the works with known chronological order from *Laws 2* onward. Figure 2 shows that there are clearly two distinct

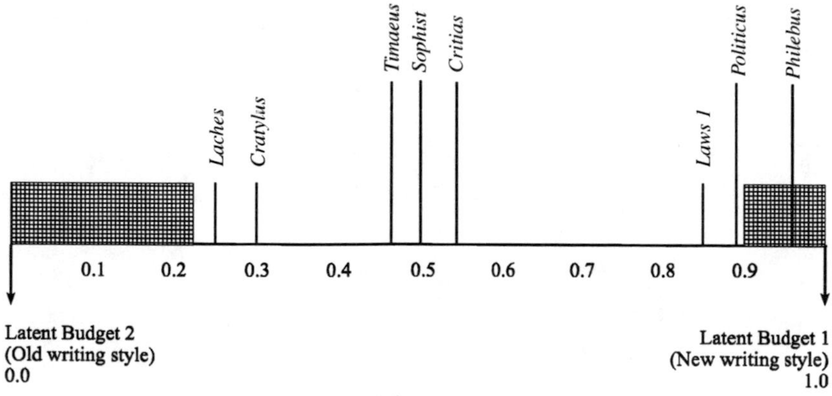

Figure 3. Graphic representation of mixing-parameter estimates for Plato data.

Table 1. Sentence Endings in Plato's Books

	Sentence Endings															
	·····	····−	···−·	···−−	··−··	··−·−	··−−·	··−−−	·−···	·−··−	·−·−·	·−·−−	·−−··	·−−·−	·−−−·	·−−−−
Books	−····	−···−	−··−·	−··−−	−·−··	−·−·−	−·−−·	−·−−−	−−···	−−··−	−−·−·	−−·−−	−−−··	−−−·−	−−−−·	−−−−−
Charmides	5	6	11	11	8	20	8	15	6	17	17	9	14	19	6	20
	4	10	10	14	9	31	16	20	4	17	8	21	9	23	8	19
Laches	4	5	8	11	1	8	6	13	4	16	6	9	14	19	6	24
	7	16	12	13	8	22	7	18	8	9	22	12	10	30	8	20
Lysis	3	14	17	17	11	15	4	6	10	15	17	10	15	19	3	10
	7	6	5	16	13	19	9	5	11	11	16	8	13	14	4	4
Euthyphro	10	3	14	9	5	10	4	11	3	10	6	10	7	9	3	9
	7	10	5	12	8	9	5	12	2	9	5	10	5	11	3	9
Gorgias	16	27	22	44	19	39	27	45	24	38	29	40	36	73	24	42
	18	42	30	52	21	61	34	58	22	59	32	43	39	46	37	46
Hippas Minor	2	6	4	15	8	9	1	6	2	6	10	7	7	8	3	11
	5	5	4	11	3	11	5	7	1	2	8	5	2	9	6	9
Euthydemus	8	9	13	24	11	37	21	27	10	21	35	25	12	26	18	28
	11	12	20	20	17	51	12	21	9	32	20	26	16	36	16	28
Cratylus	29	26	23	42	17	25	13	32	17	29	28	34	38	47	9	39
	27	22	18	36	21	35	17	25	25	22	52	31	27	34	21	38
Meno	5	5	11	11	9	14	8	27	8	23	18	23	21	22	12	24
	4	25	13	24	6	26	16	17	5	19	13	15	14	18	11	20
Menexenus	1	3	3	0	1	9	6	6	6	17	5	6	5	8	3	11
	1	6	4	6	6	4	4	6	3	8	4	6	6	14	9	7
Phaedrus	9	9	10	23	18	28	10	21	17	33	23	40	18	26	9	26
	13	22	25	42	29	29	18	19	14	39	17	37	20	39	11	19

Symposium	10	11	27	26	16	33	13	22	11	28	16	39	19	35	22	27
	11	20	25	29	10	48	26	24	12	23	26	25	18	45	13	32
Phaedo	17	15	21	30	10	32	23	30	15	38	21	34	15	32	15	30
	16	41	10	43	19	44	18	34	16	28	28	33	26	47	11	32
Theaetetus	19	20	21	35	20	39	29	40	18	25	32	35	39	37	17	37
	23	28	38	43	23	36	37	45	16	39	34	37	42	55	18	49
Parmenides	22	15	19	31	9	23	7	15	17	26	21	30	15	16	9	26
	18	12	25	28	13	24	13	18	14	24	19	26	13	21	7	25
Protagoras	9	13	14	8	10	11	15	21	12	15	14	19	20	16	13	24
	11	9	21	17	11	31	17	25	8	25	18	22	12	28	13	20
Crito	2	2	4	6	1	5	4	6	1	4	6	9	5	10	3	8
	0	3	5	6	3	9	6	5	2	6	3	7	3	13	1	10
Apology	2	10	11	13	2	13	11	12	8	14	11	13	7	13	17	11
	3	10	14	17	5	22	16	15	6	16	16	11	6	14	9	15
Republic 1	12	11	10	9	10	15	13	22	10	27	15	20	28	23	10	25
	15	17	13	26	7	30	7	14	11	11	11	18	11	18	13	29
Republic 2	3	6	9	11	7	13	11	12	8	23	9	18	12	13	5	19
	9	12	6	10	7	20	5	12	16	12	13	15	15	13	10	15
Republic 3	2	2	4	12	6	19	14	13	6	10	7	13	15	19	19	15
	2	13	9	12	9	25	19	13	7	11	10	18	21	22	4	22
Republic 4	3	13	9	10	4	16	7	11	4	24	7	12	10	18	11	11
	3	17	10	18	8	22	8	11	10	11	6	18	14	6	6	17
Republic 5	5	13	9	14	13	19	10	15	13	22	9	18	19	18	10	15
	10	19	17	23	9	27	7	16	9	19	15	18	14	22	11	18
Republic 6	5	8	9	11	10	13	19	7	5	16	12	17	7	21	5	6
	7	19	13	16	10	22	6	18	4	16	13	17	8	10	6	17

(*continued*)

Table 1 (*continued*)

	Sentence Endings															
	.....	....-	...-.	...—	..-..	..-.-	..—.	..——	.-...	.-..-	.-.-.	.-.—	.—..	.—.-	.——.	.——
Books	-....	-...-	-..-.	-..—	-.-..	-.-.-	-.—.	-.——	—...	—..-	—.-.	—.—	——..	——.-	——.	——
Republic 7	3	8	6	8	6	16	3	8	6	15	12	15	8	15	3	5
	3	24	5	14	10	29	7	16	9	18	6	12	6	21	11	6
Republic 8	2	5	9	9	5	23	5	6	5	11	9	18	5	15	7	8
	1	17	12	15	6	21	5	25	3	7	11	12	8	9	7	10
Republic 9	1	4	6	3	1	19	7	8	2	16	6	13	13	15	11	12
	5	17	12	16	8	21	5	8	3	11	16	9	6	20	4	9
Republic 10	5	4	8	6	10	14	11	5	6	9	12	15	9	23	8	15
	5	19	8	9	7	23	8	19	7	11	13	19	6	15	5	15
Laws 1	6	13	13	21	5	10	6	15	11	3	4	4	14	24	5	15
	5	25	3	3	5	8	13	13	16	3	8	12	13	27	8	10
Laws 2	6	11	10	20	7	3	3	5	4	4	3	1	10	17	1	19
	7	17	0	3	6	6	8	6	5	2	6	10	11	26	8	21
Laws 3	8	19	10	11	8	11	7	8	6	7	3	3	13	22	7	14
	16	37	2	3	4	5	14	8	6	1	5	5	15	22	7	12
Laws 4	7	11	12	17	10	9	5	3	4	2	3	5	7	22	5	8
	13	20	3	2	9	6	1	15	9	4	5	2	15	17	17	13
Laws 5	8	14	9	15	11	8	1	5	6	3	2	0	5	18	5	3
	11	19	1	1	5	9	3	10	9	1	3	6	13	15	13	11
Laws 6	9	20	12	32	6	13	9	14	4	1	1	3	15	29	5	15
	22	30	1	2	4	5	7	16	11	8	3	5	12	22	9	22

Laws 7	13	17	13	24	10	11	7	20	9	1	7	6	12	46	9	17
	13	43	3	3	12	13	6	13	11	1	5	11	21	32	12	29
Laws 8	4	10	10	10	8	8	3	11	3	1	2	3	3	21	6	17
	8	23	1	1	5	4	3	14	3	3	6	6	11	28	4	12
Laws 9	10	7	6	19	7	11	3	7	4	5	3	6	15	36	9	15
	17	31	3	1	7	9	4	14	7	2	2	8	8	28	16	13
Laws 10	4	9	9	21	14	9	4	8	8	6	3	2	11	28	3	9
	13	40	2	3	4	11	9	9	8	3	8	7	19	46	8	20
Laws 11	8	6	3	13	6	8	5	5	4	3	1	5	13	17	13	13
	7	20	1	0	2	7	9	9	9	7	2	4	7	40	5	15
Laws 12	7	8	8	12	8	12	5	9	8	5	4	6	12	29	5	11
	12	28	2	1	4	7	9	12	8	3	0	4	12	29	7	17
Critias	5	6	10	10	2	5	4	2	3	3	4	4	4	8	5	3
	3	9	2	7	3	2	3	7	5	1	3	9	4	3	5	6
Philebus	24	46	38	52	25	18	30	25	20	7	6	12	64	51	32	32
	27	62	7	7	14	27	11	32	41	4	7	23	53	86	28	47
Politicus	13	22	25	33	20	23	24	35	24	8	7	8	41	34	19	29
	19	31	3	6	18	14	26	38	25	5	21	16	53	52	22	56
Sophist	26	23	22	47	24	28	28	31	31	28	15	12	30	42	23	27
	33	21	19	28	37	19	34	32	30	21	24	38	48	43	24	31
Timaeus	18	27	26	30	14	17	23	23	46	20	17	25	26	23	17	18
	30	25	13	21	26	23	25	23	26	25	25	49	23	29	17	14

subgroups within the books with known chronological order: the earlier works (up until *Republic*), and the later works (*Laws*). Within these two subgroups, however, the chronological order was not clearly shown by the LBM. From the five undated books, *Philebus* and *Politicus* are mostly built up from the newer writing style budget. From their sentence endings, these books appear to be similar to the later works. The remaining books, *Critias*, *Sophist*, and *Timeaus*, do not belong to the later works or to the earlier works. Their writing style is a mixture of the older writing style and the newer writing style. This may suggest that these books were written in between.

In this example, we did not interpret the latent budgets because we lack the knowledge of sentence endings.

3. ETHNIC DIFFERENCES AMONG PEOPLE STARTING A TRADE: AN EXAMPLE OF SIMULTANEOUS LATENT BUDGET ANALYSIS

In The Netherlands, trades are registered with Chambers of Commerce. Kloosterman and van der Leun (1998), who investigated the way ethnic groups differ in the types of trades they start, concentrated on the so-called sheltered sector and on two large cities in The Netherlands, namely Rotterdam and Amsterdam. The data are presented in Table 2a.

Table 2a. Trades Started in Amsterdam and Rotterdam: Cross-Classification by Ethnic Group and Type of Trade

	Amsterdam						Rotterdam					
Group	**1**	**2**	**3**	**4**	**5**	**Total**	**1**	**2**	**3**	**4**	**5**	**Total**
Dutch	382	367	788	113	28	1933	323	209	459	91	153	1235
Turks	14	21	3	8	10	56	29	30	2	15	14	90
Moroccans	12	36	2	5	7	62	8	17	2	13	5	45
Antilleans	8	6	2	1	2	19	5	4	3	4	3	19
Surinamese	44	33	33	17	24	151	35	31	28	19	33	146
Cape Verd.	0	0	0	0	0	0	5	1	0	0	3	9
Ghanaians	23	4	4	2	4	37	3	1	0	0	1	5
Other	185	93	82	24	35	419	74	16	19	16	8	133
No info.	146	116	119	39	61	481	42	23	31	7	7	110
Total	814	676	1033	209	426	3158	524	332	544	165	227	1792
Proportions	0.257	0.214	0.327	0.066	0.135	1.000	0.292	0.185	0.304	0.092	0.127	1.000

Note: Types of trade are 1 = wholesale trade; 2 = retail trade; 3 = producer services; 4 = catering and restaurants; 5 = personal services.

Surinam was a former Dutch colony, and the Antilles are still closely linked administratively to The Netherlands. By their educational system (language, history), this makes it easier for members of these groups to integrate into Dutch society. The Turks and Moroccans are large ethnic groups that originally came in the 1960s and 1970s as so-called guest workers. The Cape Verdeans and the Ghanaians are relatively small ethnic minorities. The trades speak for themselves. Amsterdam and Rotterdam differ in that the port of Rotterdam generates considerable employment, specifically in the wholesale trade and catering services (compare the marginal column proportions in Table 2a), whereas Amsterdam is both a tourist and an industrial center. The two cities thus provide the ethnic groups with a different opportunity structure. One could argue that the success of the different ethnic groups with respect to the opportunities offered depends on their network in specific trades, for example, the number of clients of the same ethnic group, and on their human capital, for instance, knowledge of the Dutch language or knowing how the trade as a whole operates in The Netherlands. These different types of human capital and networks ensure that some ethnic groups are more likely to start certain specific trades rather than others.

This is where the usefulness of latent budget analysis becomes apparent: As shown in Figure 1, the LBM assumes the existence of a categorical latent variable, with T states between ethnic group i and trade j, and these latent states could very well be reflecting human capital and the networks. In terms of Figure 2, the LBM approximates the distribution of each ethnic group (observed budget) by a mixture of a number of latent distributions (latent budgets). The latent budgets may be interpreted as typical extreme distributions that deviate from the marginal distribution of trades started in Rotterdam and Amsterdam. The way in which they deviate reveals how typical sources of human capital and networks create specific opportunities to start specific trades.

It should be noted that the absolute sizes of the ethnic groups are not reflected in the parameter estimates. For completeness, absolute sizes are provided for some of the groups in Table 2b. We concentrated here on the type of trade that people from ethnic groups choose when they start a trade, that is, the information provided in Table 2a. Another study would be to look at the relative proportions of ethnic groups that start trades at all and then to compare Amsterdam and Rotterdam. The relevant data are shown in Table 2b.

For Amsterdam, the LBM with $T = 1$ latent budget (i.e., the independence model) has $L^2 = 299$ (df is 28); for $T = 2$, $L^2 = 69$ (df is 18); for

Table 2b. Trades Started in Amsterdam and Rotterdam: Absolute Sizes of the Ethnic Groups

	Amsterdam			Rotterdam		
	Trades			Trades		
Group	Sample	Prop.	No. of Inhab.	Sample	Prop.	No. of Inhab.
Dutch	1,933	0.612	419,698	1,235	0.689	358,425
Turks	56	0.018	30,992	90	0.050	35,598
Moroccans	62	0.020	47,202	45	0.025	24,550
Antilleans	19	0.006	10,501	19	0.011	11,708
Surinamese	151	0.048	69,011	146	0.081	46,679
Cape Verd.	0	0.000	not spec.	9	0.005	not spec.
Ghanaians	37	0.012	not spec.	5	0.003	not spec.
Other	419	0.133	not spec.	133	0.074	not spec.
No info.	481	0.152	not spec.	110	0.061	not spec.

$T = 3$, $L^2 = 13$ (df is 10). For Rotterdam, for $T = 1$, $L^2 = 218$ (df is 32); for $T = 2$, $L^2 = 75$ (df is 21); for $T = 3$, $L^2 = 22$ (df is 12; $0.025 < p < .05$). The fit of the LBMs with $T = 3$ therefore seems adequate. In terms of Figure 1, the latent states represent three types of human capital and networks that lead to specific patterns of trade that are started. The fit indices should be interpreted with care because many observed frequencies equal zero. We studied the parameter estimates for the solutions with $T = 3$, given in Table 3. We have identified the solution by making the latent budgets as extreme as possible, that is, by making as many latent budget parameters ($\pi_{jt}^{B\bar{X}}$) equal to zero as possible (see van der Ark et al., 1999).

The latent budgets are most easily interpreted by comparing parameter estimates with the marginal proportions p_{+j}. This shows that for Amsterdam, the first latent budget is characterized by wholesale trade (i.e., estimate 0.933 is greater than the marginal proportion 0.257). In terms of human capital and networks, the first latent state represents *knowledge of the supply side*. The second latent budget is characterized by retail trade ($0.635 > 0.214$), catering industry ($0.175 > 0.066$), and personal services ($0.190 > 0.135$); this latent state represents *knowledge of the demand side of economy*. The third latent budget is characterized by producer services ($0.805 > 0.327$) and personal services ($0.184 > 0.135$); this latent state represents a *good education and access to relevant Dutch networks*.

We interpreted the mixing-parameter estimates $\hat{\pi}_{it}^{A\bar{X}}$ from graphical displays similar to Figure 3. Because $T = 3$, we now use *ternary diagrams*

Table 3. Parameter Estimates for LBMs with $T = 3$ for Amsterdam and Rotterdam

	Amsterdam				Rotterdam			
Mixing Parameters	$T = 1$	$T = 2$	$T = 3$		$T = 1$	$T = 2$	$T = 3$	
Dutch	0.212	0.282	0.506		0.329	0.144	0.527	
Turks	0.267	0.661	0.071		0.407	0.561	0.032	
Moroccans	0.207	0.755	0.038		0.240	0.701	0.058	
Antilleans	0.449	0.420	0.131		0.341	0.446	0.213	
Surinamese	0.313	0.399	0.288		0.292	0.428	0.280	
Cape Verd.					0.582	0.418	0.000	
Ghanaians	0.661	0.179	0.160		0.661	0.339	0.000	
Other	0.474	0.292	0.235		0.707	0.095	0.198	
No information	0.326	0.367	0.308		0.488	0.110	0.402	
Latent budgets	$T = 1$	$T = 2$	$T = 3$	p_{+j}	$T = 1$	$T = 2$	$T = 3$	p_{+j}
Wholesale trade	0.933	0.000	0.000	0.257	0.795	0.000	0.000	0.292
Retail trade	0.045	0.635	0.000	0.214	0.096	0.431	0.146	0.185
Producer serv.	0.000	0.000	0.805	0.327	0.000	0.000	0.705	0.304
Catering & rest.	0.022	0.175	0.011	0.066	0.108	0.259	0.000	0.092
Personal serv.	0.000	0.190	0.184	0.135	0.000	0.310	0.149	0.127

(see van der Ark and van der Heijden, 1997). Figure 4(a) gives the plot of the parameter estimates for the LBM with $T = 3$ latent budgets for the ethnic groups in Amsterdam. The vertices of the triangle represent the latent budgets. The upper vertex represents the first latent budget, the right-hand vertex represents the second latent budget, and the left-hand vertex represents the third latent budget. The side opposite a vertex represents the area where the corresponding mixing parameters ($\pi_{it}^{A\bar{X}}$) are zero. The expected budgets can be depicted in the diagram, and their mixing parameters determine the position in the diagram; that is, the position of an expected budget in the diagram is $\pi_{i1}^{A\bar{X}}$ times the distance from the bottom side to the upper vertex, $\pi_{i2}^{A\bar{X}}$ times the distance from the left-hand side to the right-hand vertex, and $\pi_{i3}^{A\bar{X}}$ times the distance from the right-hand side to the left-hand vertex.

Figure 4(a) reveals that, more than average, the Dutch currently start in the third latent budget (latent state for good education and access to Dutch networks), whereas Ghanaians, Antilleans, Turks, and Moroccans are ordered between the first latent budget (latent state for supply side) and the second latent budget (latent state for the demand side). The Surinamese are intermediate between the Dutch and the other ethnic groups. This might be explained by the fact that the Surinamese form an ethnic group that are reasonably well integrated in Dutch society.

In Rotterdam [Figure 4(b)], the graphical representation is very similar to that in Amsterdam. The first latent budget is characterized by wholesale trade and to some extent by catering; the second latent budget by retail trade, catering, and personal services; and the third latent budget by producer services and some personal services. Again, the Dutch start trades predominantly in the third latent budget; the Ghanaians, Cape Verdeans, Turks, and Moroccans are ordered between the first and second latent budgets; and the Surinamese and now also the Antilleans are intermediate between the Dutch and the other ethnic groups.

Although there are differences between the solutions of Amsterdam and Rotterdam, the similarities are striking. Therefore, we investigated whether a more parsimonious solution, obtained by imposing equality restrictions to the parameter estimates, could describe the data. This is done in simultaneous latent budget analysis (Siciliano and van der Heijden, 1994). Because the Cape Verdeans did not start any trades in Amsterdam, we deleted them from the table of Rotterdam, and we analyzed a table of 2 (cities) $\times$ 8 (ethnic groups) $\times$ 5 (trades).

In a first analysis, we imposed the latent budget parameters $(\pi_{jt}^{\bar{B}X})$ to be equal for Rotterdam and Amsterdam. Thus, the latent budgets for Amsterdam and Rotterdam are equivalent, but the way in which ethnic groups make use of them may differ. In terms of Figure 1, this implies that the ethnic groups in Amsterdam have different sources of human capital and networks than the ethnic groups in Rotterdam, but the way in which this human capital leads to starting trades is the same in both cities. The LBM with $T = 3$ has a fit of $L^2 = 48.3$ (df is 26).

In a second analysis, we imposed equality of the mixing parameters $(\pi_{it}^{A\bar{X}})$ for both Rotterdam and Amsterdam. Thus, the latent budgets of Amsterdam and Rotterdam are different, but the way in which they are mixed by $\pi_{it}^{A\bar{X}}$ is identical. In terms of Figure 1, this means that the ethnic groups in both cities have the same human capital and networks, but this leads to different trades in Amsterdam than in Rotterdam. Because the opportunities of the two cities differ (compare their marginal proportions), the specific latent budget estimates for Amsterdam and Rotterdam are not expected to be equal when we define the estimates of the mixing probabilities as equal. The LBM with $T = 3$ latent budgets has an adequate fit of $L^2 = 41.8$ (df is 30; $p > .05$). Given the worse fit of the solution with equality restrictions on latent budget parameters,

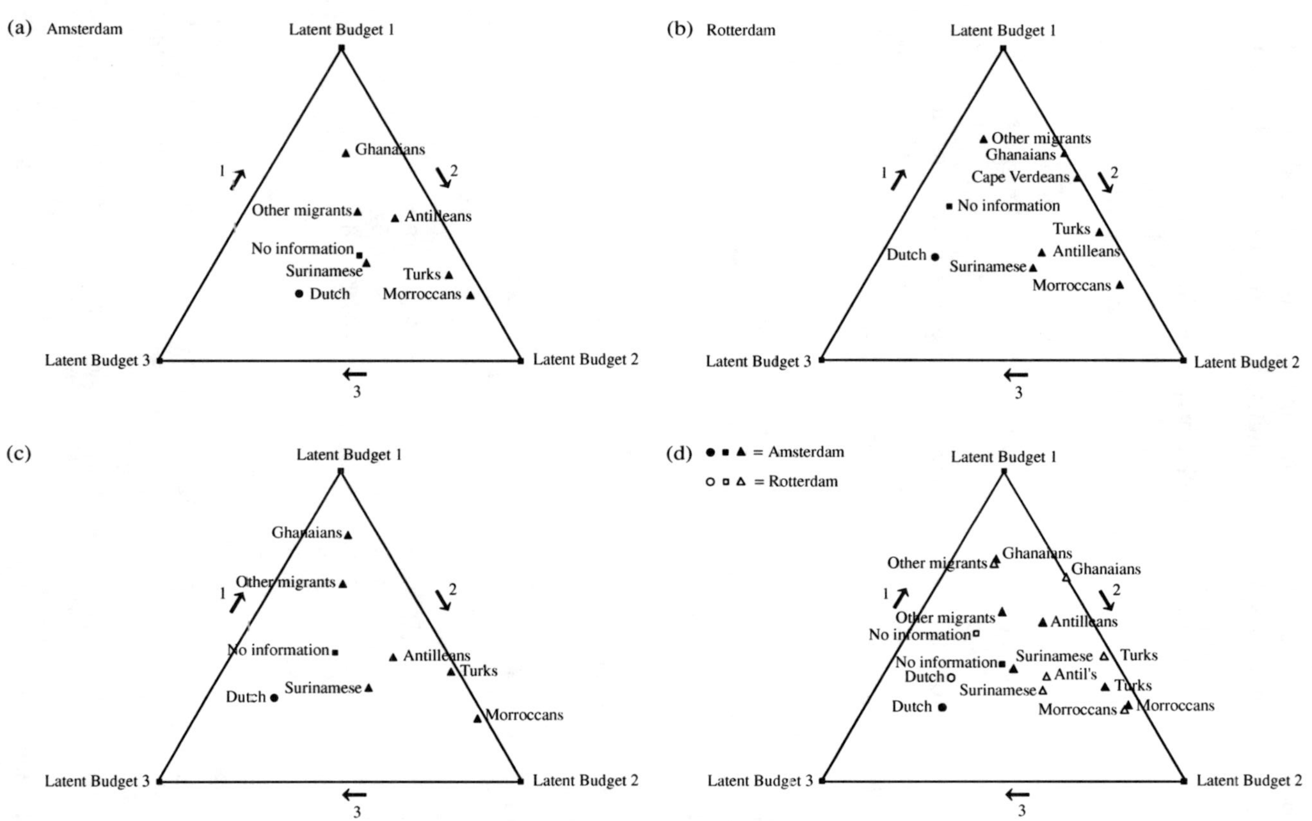

Figure 4. LBM of ethnic differences among people starting a trade in (a) Amsterdam; (b) Rotterdam; (c) Amsterdam and Rotterdam, with homogeneous mixing parameters; (d) Amsterdam and Rotterdam, with homogeneous latent budgets.

it comes as no surprise that the fit for $T = 3$ was not adequate if we imposed the restriction that both the mixing parameters and the latent budget parameters are equal in Amsterdam and Rotterdam: $L^2 = 84.1$ (df is 42).

First, we interpreted the solution with equality restrictions on the mixing parameters ($\pi_{it}^{A\bar{X}}$; see Table 4a), and next the solution with equality restrictions on the latent budget parameters ($\pi_{jt}^{\bar{B}X}$; see Table 4b).

In Table 4a, the first latent budget is characterized by wholesale trade, although to a larger extent for Rotterdam than for Amsterdam. In Amsterdam, this is compensated for by larger estimates for all other trades, except for catering. In the second latent budget, retail dominates, particularly in Amsterdam (together with wholesale trade), whereas in Rotterdam catering and personal services are larger. The third latent budget is characterized by producer services, with personal services a bit larger in Amsterdam, whereas retail is a bit larger in Rotterdam. We found it difficult to interpret these small (but significant) differences between Amsterdam and Rotterdam substantively. Figure 4(c), which shows the mixing-parameter estimates, is quite similar to Figures 4(a) and 4(b). The

Table 4a. Homogeneous Mixing Parameters in Amsterdam and Rotterdam

	Amsterdam				Rotterdam			
Mixing Parameters	$T=1$	$T=2$	$T=3$		$T=1$	$T=2$	$T=3$	
Dutch	0.264	0.185	0.551		0.264	0.185	0.551	
Turks	0.349	0.627	0.024		0.349	0.627	0.024	
Moroccans	0.196	0.775	0.029		0.196	0.775	0.029	
Antilleans	0.396	0.444	0.160		0.396	0.444	0.160	
Surinamese	0.295	0.416	0.289		0.295	0.416	0.289	
Ghanaians	0.793	0.120	0.086		0.793	0.120	0.086	
Other	0.635	0.159	0.206		0.635	0.159	0.206	
No info.	0.409	0.262	0.329		0.409	0.262	0.329	
Latent budgets	$T = 1$	$T = 2$	$T = 3$	p_{+j}	$T = 1$	$T = 2$	$T = 3$	p_{+j}
Wholesale trade	0.682	0.102	0.000	0.257	0.855	0.000	0.062	0.292
Retail trade	0.164	0.543	0.083	0.214	0.052	0.461	0.126	0.185
Producer serv.	0.072	0.000	0.701	0.327	0.000	0.020	0.674	0.304
Catering & rest.	0.044	0.167	0.031	0.066	0.093	0.259	0.000	0.092
Personal serv.	0.038	0.188	0.185	0.135	0.000	0.259	0.139	0.127

Note: Table gives parameter estimates for simultaneous latent budget analysis with $T = 3$ for Rotterdam and Amsterdam.

Table 4b. Homogeneous Latent Budgets in Amsterdam and Rotterdam

	Amsterdam				Rotterdam			
Mixing Parameters	$T=1$	$T=2$	$T=3$		$T=1$	$T=2$	$T=3$	
Dutch	0.243	0.215	0.542		0.326	0.189	0.486	
Turks	0.307	0.622	0.072		0.398	0.572	0.030	
Moroccans	0.240	0.720	0.041		0.233	0.714	0.054	
Antilleans	0.510	0.349	0.141		0.334	0.461	0.205	
Surinamese	0.358	0.346	0.296		0.290	0.437	0.274	
Ghanaians	0.716	0.112	0.172		0.657	0.343	0.000	
Other	0.542	0.201	0.257		0.702	0.122	0.186	
No info.	0.374	0.299	0.327		0.480	0.168	0.352	
Latent budgets	$T=1$	$T=2$	$T=3$	p_{+j}	$T=1$	$T=2$	$T=3$	p_{+j}
Wholesale trade	0.811	0.000	0.000	0.257	0.811	0.000	0.000	0.292
Retail trade	0.113	0.545	0.078	0.214	0.113	0.545	0.078	0.185
Producer serv.	0.000	0.000	0.756	0.327	0.000	0.000	0.756	0.304
Catering & rest.	0.076	0.205	0.000	0.066	0.076	0.205	0.000	0.092
Personal serv.	0.000	0.250	0.167	0.135	0.000	0.250	0.1670	0.127

Note: Table gives parameter estimates for simultaneous latent budget analysis with $T = 3$ for Rotterdam and Amsterdam.

Dutch predominate particularly in the third latent budget; Surinamese are situated between the Dutch and the other ethnic groups, ordered as Moroccans, Turks, Antilleans, then Ghanaians.

Table 4b shows the parameter estimates for the LBM with homogeneous latent budgets. Again, latent budget 1 is characterized by wholesale trade; latent budget 2 by retail trade, catering and restaurants, and personal services; and latent budget 3 by producer services and personal services. The mixing-parameter estimates are displayed in Figure 4(d). For each ethnic group, we found the Amsterdam label close to the Rotterdam label. For interpreting small distinctions, we concentrated on more specific characterizations by specific budgets. The Amsterdam Dutch are to a larger extent characterized by latent budget 3, the Amsterdam Turks to a larger extent by latent budget 2, and the Amsterdam Antilleans by latent budget 1, whereas the Rotterdam Antilleans are characterized more by latent budget 2, the Rotterdam Surinamese by latent budgets 2 and 3, the Amsterdam Ghanaians by latent budget 1, Rotterdam other migrants more by latent budget 1, and Rotterdam "no information" more by latent budgets 1 and 3. For more information, we refer to Kloosterman and van der Leun (1998).

4. SOCIAL MILIEU AND SECONDARY EDUCATION: AN EXAMPLE OF CONSTRAINED LATENT BUDGET ANALYSIS

At the age of 11–12 years, children in The Netherlands go from primary school to secondary school. Distinct types of secondary education can be chosen, with two main types: vocational types of education and general types of education. Choice depends on such aspects as capacities of children, interests, advice of the primary school teacher, and advice of parents. In educational research, much interest is directed to the way in which the social milieu of a child influences this choice. In this example, we investigated this question by using the LBM. The best interpretation of the LBM in this context is in terms of the MIMIC model (see Figure 2). Three explanatory variables, that is, sex, social milieu, and IQ, yield (as shown by the mixing parameters, $\pi_{it}^{A\bar{X}}$) an individual's human capital (the latent variable, having T classes), and this human capital provides opportunities to go to a specific level of education (as shown by the latent budget parameters, $\pi_{jt}^{\bar{B}X}$).

In 1977 and 1981, data were collected from more than 37,000 children about their social milieu and aspects regarding their secondary education. Distinct variables were collected; for a description, see Statistics Netherlands (1982) and Meester and de Leeuw (1983). The variables we used in our analysis are the scores on an intelligence test, social milieu (profession of father), sex, and the level of education attained in 1981, that is, after 4 years of secondary education. The intelligence test used was the (Dutch) Test for Intellectual Capacity (TIC), a figure exclusion test that consists of 33 items. The TIC scores were recoded by Meester and de Leeuw (1983) as 1 for 1–14 correct items, 2 for 15–17 correct, 3 for 18–20 correct, 4 for 21–23 correct, 5 for 24–26 correct, 6 for 27–29 correct, and 7 for 30–33 correct items. The social milieu of the family is measured by the profession of the father, in six categories: category 1 is skilled and unskilled laborers, 2 is farmers and farm laborers, 3 is shopkeepers, 4 is lower employees, 5 is middle employees, and 6 is higher employees and scientific and free professions. The last explanatory variable is the dichotomous variable of sex. The response variable is the level of education attained after 4 years, and these levels are 1, dropped out; 2, junior vocational education (LBO); 3, general education, medium level (MAVO); 4, general education, high level (HAVO); 5, general education, preparatory to university (VWO); and 6, senior vocational training ((M)BO). Meester and de Leeuw (1983) eliminated all children having no TIC score (16,433 children). According to them, this elimination is not crucial because having no TIC score seemed to have been a random process. Furthermore, children with a value missing on the level of education attained (38) or on a type of education called *extraordinary lower education* (646) were eliminated from the sample. Children having a father who is unemployed or medically unfit for work were also eliminated (6,190). This last elimination is more crucial, and it should be kept in mind that our analysis does not discuss children having these characteristic. Following these selections there remained a sample of 16,236 children. The data are given in Table 5.

We analyzed the data with the LBM by coding the levels of the explanatory variables sex, social milieu, and TIC as $2 \times 6 \times 7 = 84$ rows and the levels of the response variable level of education attained as six columns. Let the variable sex be A, indexed by i; let social milieu be C, indexed by k; let TIC be F, indexed by p; and let the response variable level of education attained be B, indexed by j. Thus, the LBM can be rewritten by replacing the index i in Model 2 by ikp, so that the LBM becomes

$$\frac{\pi_{ikpj}}{\pi_{ikp+}} = \sum_{t=1}^{T} \pi_{ikpt}^{ACF\bar{X}} \pi_{jt}^{\bar{B}X}.$$

The LBM with $T = 1$ (independence) is equivalent to the model in which the variables sex, social milieu, and TIC are dependent, and independent from level of education attained. This model may be considered as our baseline model. It has a fit of $L^2 = 4{,}612$, with df = 415. The LBM with two or more latent budgets can be interpreted as a MIMIC model (Figure 1). The MIMIC model emphasizes that each child has T probabilities $\pi_{ikpt}^{ACF\bar{X}}$ of falling into the latent classes, which can be interpreted as the individual's human capital. These T probabilities are determined by the levels of explanatory variables A, C, and F. Once a child is in one of the T latent classes, there are J probabilities $\pi_{jt}^{\bar{B}X}$ of attaining each of the levels of education.

A sensible approach to the analysis is first to determine the number of latent classes T that is needed to give an adequate description of the data. For $T = 2$, $L^2 = 1{,}113$ (df is 328); for $T = 3$, $L^2 = 441$ (df is 243); for $T = 4$, $L^2 = 226$ (df is 160); and for $T = 5$, $L^2 = 116$ (df is 79). All the models have to be rejected at $p = .05$. To check whether this could be due to the specific form of our models, we studied the residuals of the least restricted LBM, that is, the LBM with $T = 5$. We found no intelligible patterns in the residuals or specific outlier cells, so we assumed that the misfit of the models is due to a large sample size.

Table 5. The SMVO Data

School Type		Boys						Girls					
SES	**TIC**	**1**	**2**	**3**	**4**	**5**	**6**	**1**	**2**	**3**	**4**	**5**	**6**
1	1	43	126	23	5	2	17	28	87	24	13	3	35
	2	41	172	58	20	9	28	29	131	57	15	0	74
	3	50	271	83	58	24	87	67	209	128	59	6	141
	4	64	268	131	93	44	111	64	200	157	95	34	194
	5	43	202	121	113	47	109	35	163	177	105	39	201
	6	11	78	60	62	43	78	20	54	106	92	48	103
	7	4	15	20	23	27	19	2	10	22	40	38	28
2	1	3	13	1	1	1	8	2	8	5	1	0	5
	2	3	18	9	0	0	10	2	14	10	4	0	12
	3	2	18	12	15	3	23	5	18	16	19	3	26
	4	8	25	15	14	9	47	0	18	23	21	8	46
	5	5	25	16	12	16	35	0	13	28	21	15	39
	6	2	4	7	20	11	22	5	6	19	37	15	30
	7	0	3	2	5	7	9	0	4	4	12	17	10
3	1	11	17	6	1	1	10	7	12	11	2	0	8
	2	9	37	11	6	2	10	6	29	11	5	1	11
	3	23	59	26	12	6	29	16	43	30	19	4	38
	4	12	72	34	23	14	38	18	39	39	36	13	49
	5	11	40	26	37	25	36	16	32	54	54	25	39
	6	7	20	26	25	30	25	11	12	28	41	20	24
	7	3	1	7	9	12	9	2	3	3	16	7	3
4	1	9	29	13	4	1	4	3	15	6	3	0	10
	2	9	38	21	5	4	13	10	24	26	7	2	29
	3	12	56	47	37	15	27	12	54	40	37	15	35
	4	11	62	52	54	26	43	15	39	64	56	27	61
	5	12	48	62	55	37	30	9	31	54	87	44	52
	6	6	15	33	40	45	24	7	11	35	49	39	39
	7	3	4	7	17	23	7	2	3	5	23	26	9
5	1	5	25	14	9	3	9	6	20	8	3	1	12
	2	8	26	30	23	7	11	9	22	24	19	4	30
	3	13	60	65	39	35	50	10	42	50	44	33	59
	4	20	79	91	94	71	70	17	58	97	82	55	79
	5	11	58	70	95	95	63	11	44	89	103	101	70
	6	9	39	44	71	107	40	5	17	46	117	104	47
	7	4	7	9	28	57	12	2	3	28	49	70	21
6	1	4	6	10	6	4	3	5	2	6	1	1	5
	2	7	14	15	11	5	12	4	3	6	18	2	11
	3	5	31	34	39	21	23	5	16	24	33	16	21
	4	10	16	45	54	52	36	9	16	44	83	46	29
	5	7	16	44	71	105	28	7	7	40	80	83	27
	6	3	12	24	40	85	19	8	7	32	66	100	15
	7	3	4	9	16	52	9	1	3	10	29	51	1

Notes: School types are 1, drop out; 2, LBO; 3, MAVO; 4, HAVO; 5, VWO; 6, (M)BO. Social milieu is 1, skilled and unskilled laborers; 2, farmers and farm laborers; 3, shopkeepers; 4, lower employees; 5, middle employees; 6, higher employees. TIC scores are number of items correct: 1, 1–14; 2, 15–17; 3, 18–20; 4, 21–23; 5, 24–26; 6, 27–29; 7, 30–33.

Given the large sample size, we were satisfied with the description that the LBM offers with three latent budgets. Although significant, the discrepancy between the $L^2 = 441$ and df $= 243$ is not enormous; the model describes 0.904 of the departure from the independence model ($T = 1$) (i.e., $0.904 = (4{,}612 - 441)/4{,}612$). Because the gain in percentage moving from the LBM with three to the LBM with four latent budgets is relatively small, we choose the LBM with three latent budgets to examine more carefully.

The latent budget parameter estimates ($\pi_{jt}^{\hat{B}X}$) are shown in Table 6. In the first latent budget children go predominantly into *lower vocational training (LBO) or drop out*, and to a lesser extent they go into medium general education (MAVO) and (M)BO. In the second latent budget, children go predominantly into *higher general education (HAVO and VWO)*, and in the third latent budget they go predominantly into *medium and higher general education (MAVO and HAVO) and higher vocational training (MBO)*, but not to general, university preparatory education (VWO).

For the study of the mixing-parameter estimates, we give plots of the estimates separately for each TIC score p and each sex i. This gives $7 \times 2 = 14$ plots, shown in Figure 5. In each plot, we have set out horizontally the six levels of social milieu k and vertically the probability of going to one of the latent budgets t. Each plot has 18 points; namely, children in each of the six levels of social milieu can go to each of the three latent budgets; points belonging to the same latent budgets are connected, so that each plot has three lines. In Figure 5, the first latent budget is indicated by the line with the circles, the second latent budget is indicated by the

Table 6. Latent Budgets Estimates for $T = 1$ (Independence) and $T = 3$ for Educational Level after 4 Years of Secondary School

		Panel 1			Panel 2		
Group	$T = 1$	$t = 1$	$t = 2$	$t = 3$	$t = 1$	$t = 2$	$t = 3$
1. Drop out	0.063	0.160	0.014	0.011	0.177	0.025	0.005
2. LBO	0.226	0.658	0.000	0.000	0.701	0.038	0.006
3. MAVO	0.192	0.121	0.090	0.325	0.092	0.090	0.331
4. HAVO	0.188	0.000	0.367	0.232	0.000	0.337	0.228
5 VWO	0.142	0.000	0.530	0.000	0.015	0.500	0.000
6. (M)BO	0.189	0.061	0.000	0.432	0.015	0.011	0.430
π_t^X	1.000	0.343	0.267	0.389	0.304	0.274	0.422

Note: Panel 1, unconstrained estimates $T = 3$; panel 2, estimates $T = 3$ constrained by the multinomial logit model.

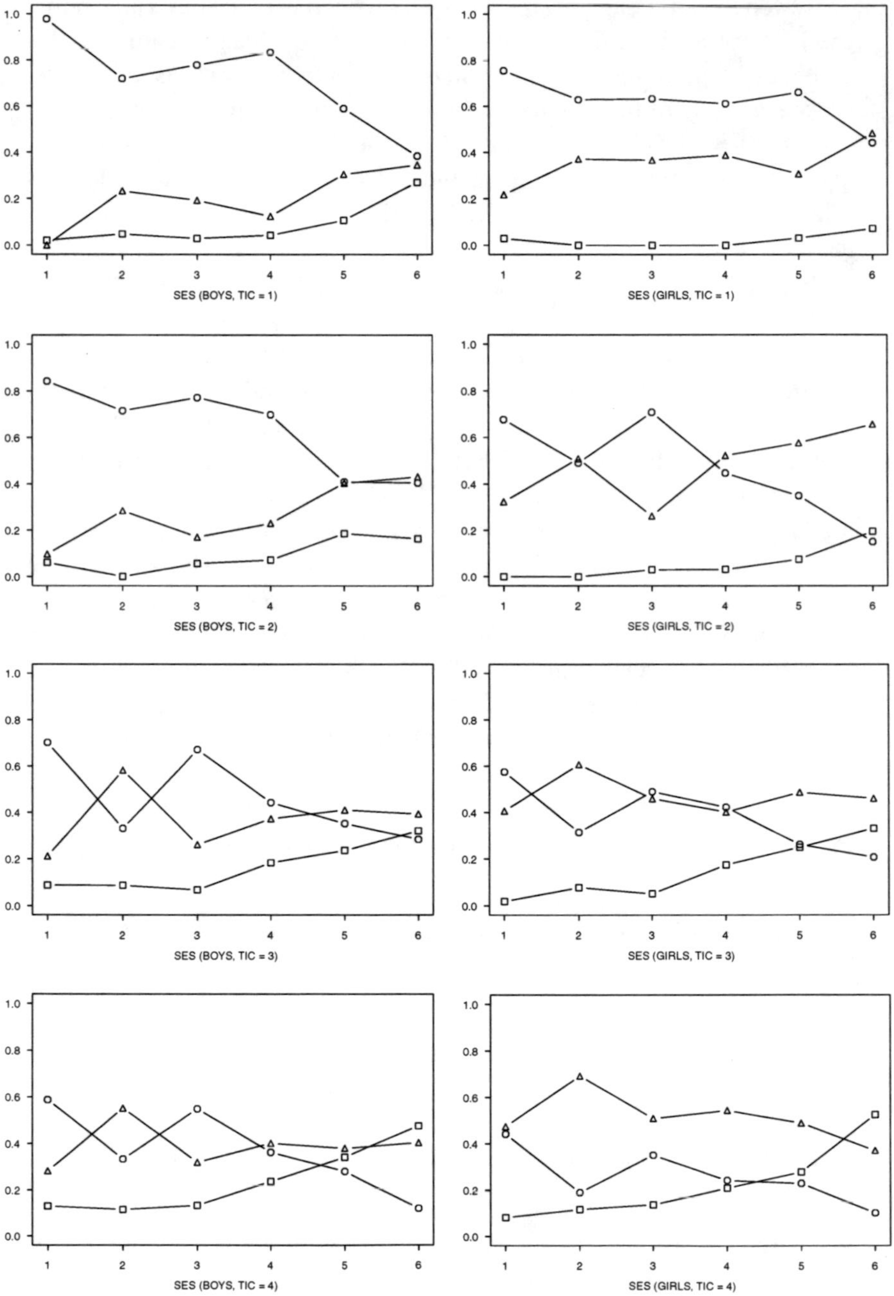

Figure 5. Unconstrained mixing-parameter estimates for each TIC score group and each sex.

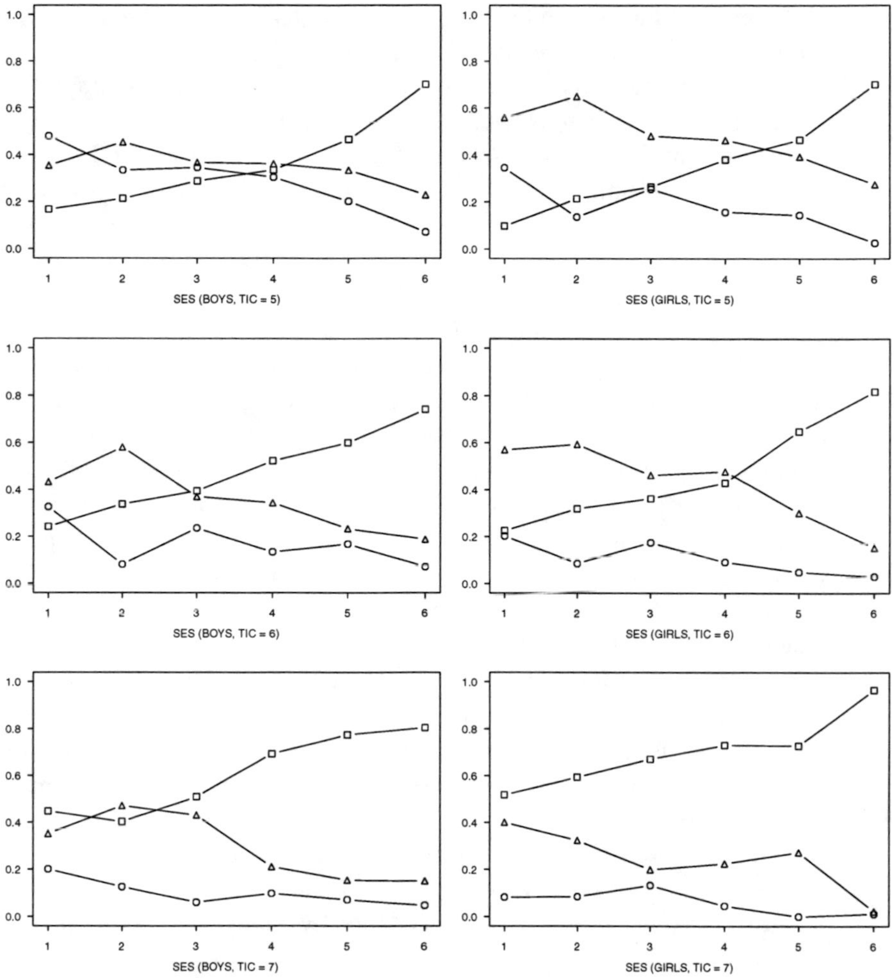

Figure 5 (*continued*)

line with the squares, and the third latent budget is indicated by the line with the triangles.

We chose to display the parameters in this way for the following reasons: First, if sex had no influence on the probability of going to latent budgets, then plots on the left (boys) would be identical to plots on the right (girls). This way of displaying the parameters clearly shows the influence of sex if we look at how each pair of plots differs. Second, if the social milieu had no influence on the probability of going to the latent budgets, then all lines would be horizontal, and departures from this would be easily displayed.

It is clear that the probability of going to a latent budget will be strongly influenced by the TIC score because the levels attained not only reveal differences between different types of education (general vs. vocational), but also between higher and lower types of education. Therefore, in going from the plots at the top (TIC score equals 1) to the bottom (TIC score equals 7), the line with circles drops generally; this is not surprising because this line shows the probability of going to latent budget 1, which is the budget in which 0.658 of the children go to LBO and 0.160 drop out: children more often drop out or go to LBO when their TIC score is lower.

There are many interesting aspects in these plots. For instance, in all levels of TIC, children with fathers who are medium or higher employees (5 and 6) have a much higher probability than average of going to latent budget 2, which is the budget for higher general education (HAVO, 0.367) and university preparatory education (VWO, 0.530). Their probability of going to budget 1 (drop out and LBO) is much lower. The reverse holds for children whose father is a skilled or unskilled laborer: Given their TIC score, their probability of going to budget 1 is in general the highest. On average, children whose fathers are farmers (2) are more likely than average, given their TIC score, to go to latent budget 3, where they have a high probability of following medium vocational training ((M)BO). It may be noted that the latent budget parameters, being probabilities, can be interpreted easily; they not only show tendencies in the data (e.g., girls go on average less to budget 1 than boys), but also show how strong the effects are.

Van der Heijden et al. (1992) showed how the factorial structure in the explanatory variables could be used to investigate the effects of each of the factors and their interactions. This is done by means of a multinomial logit model for the mixing parameters $\pi_{ikpt}^{ACF\bar{X}}$, that is,

$$\pi_{ikpt}^{AXF\bar{X}} = \exp\left(\sum_{m=1}^{M} x_{ikpm}\gamma_{mt}\right) \Big/ \sum_{t=1}^{T} \exp\left(\sum_{m=1}^{M} x_{ikpm}\gamma_{mt}\right). \qquad (5)$$

The design matrix $\mathbf{X}$ has $I \times K \times P$ rows and M columns, and these M columns represent dummy variables for the main effects for factors A, C, and F; for their two-way interaction effects $A \times C$, $A \times F$, and $C \times F$; and their three-way interaction $A \times C \times F$. The elements γ_{mt} are parameters for column m and latent budget t. To identify the model, $\gamma_{m1} = 0$. For more details, see van der Heijden et al. (1992), who also explain the relationship between these models and loglinear models with latent variables.

We have systematically imposed all possible constraints on the mixing parameters ($\pi_{ikpt}^{ACF\bar{X}}$). The most restrictive LBM has only main effects

A, C, and F. This LBM turns out to fit reasonably well, with $L^2 = 627$ (df is 379). Forsaking the model with unconstrained mixing parameters ($\pi_{ikpt}^{ACF\bar{X}}$) for the model with only main effects thus gains us $379 - 243 = 136$ df, at the expense of a loss of fit of $627 - 441 = 186$.

The latent budget parameter estimates are similar to those for the unconstrained model with three latent budgets (see Table 6). We studied the estimates by deriving averages of mixing parameters $\hat{\pi}_{it}^{A\bar{X}} \equiv \hat{\pi}_{i+++t}/\hat{\pi}_{i++++}$, $\hat{\pi}_{kt}^{C\bar{X}} \equiv \hat{\pi}_{+k++t}/\hat{\pi}_{+k+++}$, and $\hat{\pi}_{pt}^{F\bar{X}} \equiv \hat{\pi}_{++p+t}/\hat{\pi}_{++p++}$. Thus, we obtained parameters for sex only, for social milieu only, and for TIC only. Plots of these parameter estimates are given in Figure 6. The plot for TIC score shows that the probability of going to latent budget 1 (mainly LBO, drop out) decreased as TIC increased; the probability of going to latent budget 2 (mainly VWO, HAVO) increased as TIC increased; and the probability of going to latent budget 3 (mainly MAVO, HAVO, (M)BO) increased from TIC 1 to 4, and then decreased smoothly. In the plot for social milieu, the probability of going to budget 1 is low for children of farmers (2) and medium and higher employees (5, 6), the probability to go to budget 2 increased rapidly for children of lower to higher employees, and the probability of going to budget 3 was above average for children of farmers and below average for children of higher employees. In the plot for sex, we found that there is no difference in the probability of boys and girls going to latent budget 1. However, there was a difference in their probability of going to latent budgets 2 and 3: for boys, these probabilities were approximately equal, whereas girls went more often to latent budget 3 and less often to latent budget 2.

Latent budget analysis offered considerable insight into these data. The MIMIC-model interpretation that we used showed with which probabilities children, given a specific background, go to specific latent budgets. These latent budgets specified the probabilities of reaching specific final levels of education. In this example, the parameters were also very easy to interpret, so that it was easy to indicate the processes that operate in the relationship between explanatory variables such as TIC, sex, and social milieu on the one hand, and secondary education on the other. The constraints allowed a simplified interpretation.

5. CONCLUSIONS

The LBM closely answered specific research questions that are interesting from a substantive point of view. For Plato's data, we assumed that there were a few typical (latent) writing styles, and each book is a

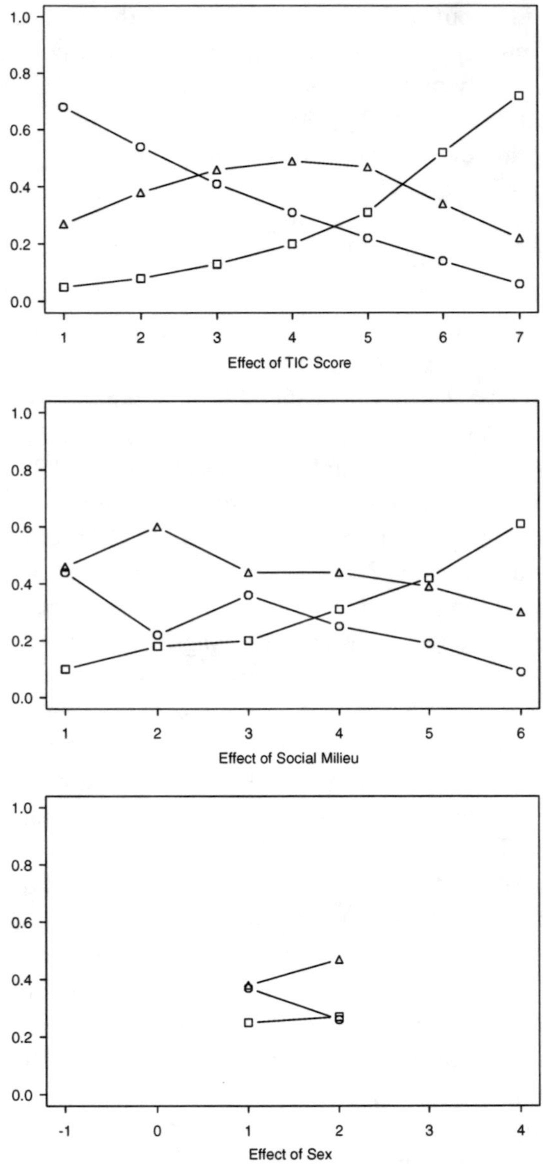

Figure 6. Constrained mixing-parameter estimates for the TIC score groups, for the social milieu groups, and the sexes.

mixture of these typical styles. This assumption is related directly to the parameterization of LBM. In this example, the LBM was interpreted as a mixture model. In the ethnic entrepreneur example and the secondary education example, we assumed the existence of a latent variable mediating between the explanatory variable and the response variable.

In both examples, the interpretation of this latent variable was human capital. The LBM was interpreted as a MIMIC model.

An important asset of the LBM is the simple interpretation of the parameters, also for nonstatisticians. The merits of the LBM are most evident when the row variable can be interpreted as the explanatory variable and the column variable can be interpreted as the response variable. Otherwise, a (symmetrical) latent class model is more suitable.

REFERENCES

Atkinson, A. C. (1970). "A method for discriminating between models," *Journal of the Royal Statistical Society, Series B*, **32**, 323–53.

Boneva, L. (1970). "A new approach to a problem of chronological seriation associated with the works of Plato." In F. R. Hodson, D. G. Kendall, & P. Tautu, *Mathematics in the Archeological and Historical Sciences*. Edinburgh: Edinburgh University Press, pp. 173–85.

Clogg, C. C. (1981). "Latent structure models of mobility," *American Journal of Sociology*, **86**, 836–68.

Cox, D. R., & Brandwood, L. (1959). "On a discriminatory problem connected with the works of Plato," *Journal of the Royal Statistical Society, Series B*, **21**, 195–200.

de Leeuw, J., & van der Heijden, P. G. M. (1988). "The analysis of time-budgets with a latent time-budget model." In E. Diday et al. (eds.), *Data Analysis and Informatics 5*. Amsterdam: North-Holland, pp. 159–66.

de Leeuw, J., & van der Heijden, P. G. M. (1991). "Reduced rank models for contingency tables," *Biometrika*, **78**, 229–32.

de Leeuw, J., van der Heijden, P. G. M., & Verboon, P. (1990). "A latent time budget model," *Statistica Neerlandica*, **44**, 1–22.

Goodman, L. A. (1974). "The analysis of systems of qualitative variables when some of the variables are unobservable. I. A modified latent structure approach," *American Journal of Sociology*, **79**, 1179–1259.

Greenacre, M. J. (1984). *Theory and Applications of Correspondence Analysis*. New York: Academic Press.

Kaluscha, W. (1904). "Zur Chronologie der Platonischen Dialoge." *Wiener Studien*, pp. 25–7.

Kloosterman, R., & van der Leun, J. (1998). "The same musical chairs? Urban opportunity structures and business starters by immigrant entrepreneurs in Amsterdam and Rotterdam." In J. Rath & R. Kloosterman (eds.), *Immigrant Entrepreneurs in the Netherlands* (in Dutch). Amsterdam: het Spinhuis, pp. XXX–XX.

Meester, A., & de Leeuw, J. (1983). *Intelligence, Social Milieu and the School Career* (in Dutch). Leiden: Department of Data Theory.

Mooijaart, A., & van der Heijden, P. G. M. (1992). "The EM-algorithm for latent class analysis with constraints," *Psychometrika*, **57**, 261–69.

Mooijaart, A., van der Heijden, P. G. M., & van der Ark, L. A. (1999). "A least-squares algorithm for a mixture model for compositional data," *Computational Statistics and Data Analysis*, **30**, 359–79.

Renner, R. M. (1993). "The resolution of a compositional data set into mixtures of fixed source compositions," *Applied Statistics*, **42**, 615–31.

Siciliano, R., & van der Heijden, P. G. M. (1994). "Simultaneous latent budget analysis of a set of two way tables with constant row sum data," *Metron*, **53**, 155–79.

Statistics Netherlands (1982). "School career and origin of pupils in secondary education. Part 2: cohort 1977, choice of school type" (in Dutch). The Hague: Staatsuitgeverij.

van der Ark, L. A. (1999). *Contributions to Latent Budget Analysis. A Tool for the Analysis of Compositional Data*. Leiden: DSWO Press.

van der Ark, L. A., & van der Heijden, P. G. M. (1997). "Graphical display of latent budget analysis and latent class analysis, with special reference to correspondence analysis." In M. Greenacre & J. Blasius (eds.), *Visualization of Categorical Data*. San Diego: Academic Press, pp. 489–508.

van der Ark, L. A., van der Heijden, P. G. M., & Sikkel, D. (1999). "On the identifiability in the latent budget model," *Journal of Classification*, **16**, 117–37.

van der Heijden, P. G. M. (1994). "End-member analysis and latent budget analysis," *Applied Statistics*, **43**, 527–28.

van der Heijden, P. G. M., Gilula, Z., & van der Ark, L. A. (1999). "An extended study into the relationships between correspondence analysis and latent class analysis." In M. Sobel & M. Becker (eds.), *Sociological Methodology 1999*, Vol. 29. Cambridge: Basil Blackwell, pp. 147–86.

van der Heijden, P. G. M., Mooijaart, A., & de Leeuw, J. (1992). "Constrained latent budget analysis." In C. C. Clogg (ed.), *Sociological Methodology 1992*, Vol. 22. Cambridge: Basil Blackwell, pp. 279–320.

Weltje, G. J. (1997). "End member modeling of compositional data: numerical-statistical algorithms for solving the explicit mixing problem," *Mathematical Geology*, **29**, 503–49.

FIVE

Ordering the Classes

Marcel Croon

1. INTRODUCTION

A. The Problem of Ordering the Classes in Latent Class Analysis

Latent class analysis was conceived by Lazarsfeld (1950a, 1950b, 1954) as a method for analyzing the association among manifest qualitative variables in terms of an unobserved nominal variable whose values represent different subpopulations. The latent class model assumes that the heterogeneous population from which the respondents were sampled can be partitioned in a restricted number of homogeneous subpopulations: the latent classes. Individual respondents belong to one of these classes, and respondents belonging to the same class share with each other the same set of response probabilities for the manifest variables. Furthermore, within each latent class the responses to different manifest variables are supposed to be stochastically independent. This assumption of *local indepedence*, as it is called, implies that the classes in the latent class model can be considered to be the values of a nominal latent variable in such a way that if one conditions on the value of the latent variable, the association between the manifest variables disappears completely. The association among the manifest variables is then completely explained by their relationships with the latent variable.

In this basic formulation of the latent class model, nothing impels the classes to be ordered along some continuum because the set of latent classes only represents a partition of the original population into disjoint subpopulations. The latent variable, whose levels correspond to the classes, is clearly a nominal variable. In social research, however, the response categories of the manifest variables can often be ordered on some common attributional continuum. In many applications of the latent class model, the data are obtained by having the respondents indicate how strong they agree or disagree with a particular statement, and most often response scales with two, three, five, or seven ordered categories are used. Analyzing the association among variables that are obviously measured on an ordinal level in terms of a latent variable on a lower measurement level is probably somewhat deficient in the sense that the full information in the data is not efficiently used by the data analysis method. When the manifest variables have ordered response categories, one should consider the development of latent class models in which the latent classes themselves are ordered, that is, models in which the latent variable, just like the manifest variables, can be conceived as an ordinal variable.

The development of models in which the latent classes can be thought of as being ordered along an underlying continuum becomes especially important if, following Clogg (1988), one considers the latent class model as a means to solve certain measurement problems in the social sciences. If latent class analysis is used to assess the status of individual respondents on an underlying latent continuum, and if the manifest variables themselves are measured on an ordinal scale, a minimal requirement one may formulate for such a measurement procedure is that it provides a scale on which the respondents can be ordered with respect to the latent property that is being measured. The construction of such an ordinal latent scale requires that the latent classes themselves can be ordered.

B. Some Previous Solutions

The problem of how to order the classes in the latent class model was already considered by Lazarsfeld and Henry (1968), who devoted an entire chapter of their book to this problem. Starting from the observation that in the general latent class model the classes "only define a typology, a segmentation of the population, but nothing in the assumptions of the model permits us to say that one class ranks higher than another" (p. 122), they propose two different solutions for the problem of ordering the classes. Both their solutions are formulated only for dichotomous items and are not easily generalized to the case of polytomous items.

Their first solution, the *latent distance model*, is a probabilistic generalization of Guttman's perfect scale analysis. Individual items are assumed to correspond to points on an underlying continuum and to divide it in an item-specific "positive" and "negative" part. In this way, a set of n different dichotomous items defines a partition of the continuum in $n + 1$

segments, each of which can be considered to be a latent class containing a particular proportion of the respondents. Classes that are below the cut-off point of a particular item have a constant item-specific probability of a positive response, which is expected to be lower than the response probability for the same item in classes that are above its cut-off point. Note, however, that the latter expectation is not explicitly imposed as an inequality constraint on the unknown response probabilities. It is clear that the classes obtained in this way are ordered along the latent continuum, and that the ordering of the classes derives from the ordering of the items that can be determined on the basis of their observed marginal response probabilities. However, the number of latent classes cannot be varied independently of the number of items: n dichotomous items invariably define no more than $n+1$ segments on the continuum, and, hence, $n+1$ latent classes are needed. In contrast to the general latent class model in which each class has a specific response probability for each item, the latent distance model only requires two parameters for each item: one parameter representing the item response probability in classes below the item cut-off point, and the other parameter representing the item response probability in classes above the cut-off point. Although the latent distance model can be generalized to the case of polytomous items, its application to this type of items becomes quite cumbersome as a consequence of the fact that the number of latent classes increases as a linear function of the number of response categories. Because each item with m response categories partitions the continuum in $m-1$ segments, $n(m-1)$ different latent classes have to be defined whose ordering, moreover, has to be inferred from the marginal response probabilities for each item category.

The second solution to the problem of ordered classes considered by Lazarsfeld and Henry (1968) is called the *located classes model.* In this approach, the classes are represented by scores on a continuous latent dimensions, and the probability of a positive response to an item is assumed to be a polynomial function of the latent class scores. The coefficients of these polynomials are item specific and are, in addition to the latent scores, unknown parameters that must be estimated on the basis of the data. Lazarsfeld and Henry (1968) derive algebraic solutions to the estimation problem only for the relatively simple cases of linear and quadratic tracelines. More recently, a similar approach has been proposed by Rost (1988, 1991), who derived a latent class model by letting the classes correspond to scores on a latent continuum. Instead of working with polynomial tracelines, he starts from a continuous latent trait model for rating data that is actually an extension of the well-known Rasch model for polytomous items. Rost's approach is not limited to dichotomous items but is, in contrast to the Lazarsfeld and Henry proposal for located classes analysis, easily applicable to general polytomous items. Both approaches solve the problem of how to order the classes by letting each class correspond to a latent score, and the ordering of the classes mimics the numerical ordering of the latent sores. Both approaches can be considered to be discretized versions of continuous models.

Still a third approach to the problem of ordered classes was presented by Clogg (1979). Restricting himself to the case of items with three response categories (Low, Medium, and High) analyzed by means of a latent class model with three latent classes (Low, Medium, High), Clogg (1979) imposed the constraints that first, in the Low class, the probability of a High response is zero, and that second, in the High class, the probability of a Low response is zero. In the Medium class, all three types of responses are allowed. As a general rule, in Clogg's approach one requires that for a member of the tth latent class, responses only to the same or immediately adjacent response categories ($t-1, t$, and $t+1$) are permissible. This rule also applies when there are more than three categories per item. A serious drawback of Clogg's approach is, however, that the number of latent classes is equal to the number of response categories of the manifest variables, and such a direct link between these two quantities is probably difficult to defend on logical grounds. Finally, Clogg's approach is not clear about what should be done if items with different numbers of response categories are involved in the same latent class analysis.

C. What Is There to Come?

In this chapter, the approach advocated by the present author in a series of papers (Croon, 1990, 1991, 1993) is discussed and applied to some real data examples. As will be seen shortly, in this approach an order relation is defined on the set of latent classes by imposing inequality constraints on the item response probabilities in such a way that one may say that the probability of a "positive" response increases when one runs through the set of latent classes from the "lowest" to the "highest" one. Sometimes the term *ordinal* is also used when linear constraints are imposed on certain loglinear parameters, such as, for instance, in the linear-by-linear association models for variables with ordered categories (Agresti, 1990). Croon's approach differs from this by imposing inequality rather than equality constraints on response probabilities, and is in a sense more in agreement with the true nature of ordinal variables, whose categories can only be ordered from "low" to "high."

In Section 2, the principles of what will be called *ordinal latent class analysis* (OLCA) will be explained. An illustration of OLCA using data from a large cross-national social survey will be given in Section 3. In Section 4, the basic OLCA model will be extended to check the "double-monotony" condition for dichotomous items. This will result in a latent class model by means of which the Mokken scalability of a set of dichotomous items can be investigated. An illustration of such a Mokken analysis will be given in Section 5. Section 6 will discuss some other extensions of the latent class model with ordered latent classes.

In this chapter, principles and models will be stressed. For information on the technical details of the estimation procedures, the reader will be referred to the appropriate literature.

2. ORDERING THE CLASSES BY RESTRAINING RESPONSE PROBABILITIES

A. Notation and Terminology

Suppose that the association among n items, each having m response categories, is analyzed by means of a latent class model with T latent classes. The manifest variables are represented by $A, B, C, \ldots$; the latent variable is denoted by X. The probability that a respondent from latent class t gives response i to item A is represented by $\pi_{it}^{\bar{A}X}$; the proportion of respondents belonging to latent class t is given by π_t^X. For four items A, B, C, and D, the probability of response patterns (i, j, k, l) in latent class t is equal to

$$\pi_{it}^{\bar{A}X}\pi_{jt}^{\bar{B}X}\pi_{kt}^{\bar{C}X}\pi_{lt}^{\bar{D}X}.$$

The probability of observing this response pattern in the population is then given by

$$\sum_t \pi_t^X \pi_{it}^{\bar{A}X}\pi_{jt}^{\bar{B}X}\pi_{kt}^{\bar{C}X}\pi_{lt}^{\bar{D}X}.$$

B. Dichotomous Items

Suppose now that the set $\{1, 2, \ldots, t, \ldots, T\}$ of latent classes is ordered by a weak order relation $\preceq$. Without loss of generality, one may suppose that the classes are numbered in such a way that

$$1 \preceq 2 \preceq 3 \preceq \cdots \preceq t \preceq t+1 \preceq \cdots \preceq T.$$

So, the first latent class is the "lowest" in this ordering of classes, the second class is the second "lowest," and so on. Finally, latent class T is the "highest." Being a weak order relation, $\preceq$ has some important properties. First, the order relation $\preceq$ is complete; this means that any two latent classes can be compared under the order relation. Second, it is transitive, which means that if class t_1 is lower than class t_2, which itself is lower than class t_3, then class t_1 is also lower than class t_3.

In order to avoid the definition of $\preceq$ having no empirical meaning, one should require that the ordering of the classes has some implication for their response probabilities. Suppose for a moment that all items are dichotomous variables ($m = 2$), with score 2 representing the positive and score 1 representing the negative response to each item. The corresponding response probabilities for item A in class t are then $\pi_{2t}^{\bar{A}X}$ and $\pi_{1t}^{\bar{A}X}$. For dichotomous items, one has $\pi_{2t}^{\bar{A}X} + \pi_{1t}^{\bar{A}X} = 1$. If the set of latent classes is ordered by $\preceq$ along a latent continuum that underlies the ordered response to the manifest items, then it seems reasonable to assume that for each item, the probability of a positive response should increase, or at least should not decrease, as one runs through the set of classes from "low" to "high." More formally, one should expect that for each item A, the response probabilities $\pi_{2t}^{\bar{A}X}$ satisfy the following system of inequalities:

$$\pi_{21}^{\bar{A}X} \leq \cdots \leq \pi_{2t}^{\bar{A}X} \leq \pi_{2,t+1}^{\bar{A}X} \leq \cdots \leq \pi_{2T}^{\bar{A}X}.$$

C. Polytomous Items

For polytomous items with $m > 2$ response categories, the previously described approach can be generalized by considering all possible ways in which the set of response categories $\{1, 2, \ldots, m\}$ can be dichotomized into two cumulative, nonoverlapping subsets: $\{1, 2, \ldots, k-1\}$ and $\{k, k+1, \ldots, m\}$. The first subset represents the negative or low response under this dichotomization; the second subset represents the positive or high response. Because for each value $k : 2 \leq k \leq m$ a different dichotomization of the same item is obtained, each polytomous item with m response categories corresponds with $m-1$ derived dichotomous items. The implications of the weak order relation $\preceq$ defined on the set of latent classes for a polytomous item can now be stated in the following way: For each cumulative dichotomization of a polytomous item, the probability of a positive response should increase when going through the set of classes in the direction of $\preceq$. For $1 \leq t \leq T-1$ and for $2 \leq k \leq m$, this implies that the item response probabilities for item A satisfy the following system of inequalities:

$$\sum_{g=k}^{m} \pi_{g,t}^{\bar{A}X} \leq \sum_{g=k}^{m} \pi_{g,t+1}^{\bar{A}X}.$$

For $m = 3$, one obtains in this way for $1 \leq t \leq T-1$ two sets of constraints of the following form:

$$\pi_{3t}^{\bar{A}X} \leq \pi_{3,t+1}^{\bar{A}X},$$

$$\pi_{2t}^{\bar{A}X} + \pi_{3t}^{\bar{A}X} \leq \pi_{2,t+1}^{\bar{A}X} + \pi_{3,t+1}^{\bar{A}X}.$$

For $m = 4$, three sets of constraints have to be considered:

$$\pi_{4t}^{\bar{A}X} \leq \pi_{4,t+1}^{\bar{A}X},$$

$$\pi_{3t}^{\bar{A}X} + \pi_{4t}^{\bar{A}X} \leq \pi_{3,t+1}^{\bar{A}X} + \pi_{4,t+1}^{\bar{A}X},$$

$$\pi_{2t}^{\bar{A}X} + \pi_{3t}^{\bar{A}X} + \pi_{4t}^{\bar{A}X} \leq \pi_{2,t+1}^{\bar{A}X} + \pi_{3,t+1}^{\bar{A}X} + \pi_{4,t+1}^{\bar{A}X}.$$

The preceding examples clearly show that this approach to the problem of ordered classes implies certain monotonicity constraints on the cumulative response probabilities for each item. Note that these constraints may be formulated separately for each item: at no point are the response probabilities pertaining to different items compared with each other. Note also that this approach only make sense if the response categories of each item can be meaningfully ordered: for purely nominal items, the definition of the successive cumulative response probabilities would be arbitrary and would not convey the meaning of a progressively more and more positive response.

D. Estimating and Testing the Model

Croon (1990) showed how the maximum-likelihood estimates of the item response probabilities under the monotonicity constraints can be obtained by an adaptation of the basic expectation–maximization (EM) algorithm that is used in unconstrained latent class analysis. Only the M step of this algorithm must be modified: by using techniques from isotonic regression analysis, the maximum-likelihood estimates of the item response probabilities can be obtained under the inequality constraints implied by the order relation $\preceq$ on the set of classes.

An aspect of the solutions obtained by a latent class analysis with ordered classes is the following: In general, some of the inequalities will be made active in the form of an equality that is imposed on the parameters. By making an inequality active in this sense, one free parameter is lost. The optimization procedure used guarantees that a minimal set of inequalities will be made active in the final solution. Contrary to intuition perhaps, a strategy in which, first, an unconstrained solution is obtained to decide, second, which inequalities are violated and, hence, should be made active in a equality-constrained solution is not necessarily optimal. It quite often happens that a first subset of inequalities, which are violated in the unconstrained analysis, have not to be made active in the constrained analysis because a second subset of constraints that do become active in the constrained analysis induce the first subset of constraints to be strictly satisfied.

As yet, the problem of how to test statistically whether a model with a specified number of ordered classes provides an acceptable fit to the data is not adequately solved. In a traditional latent class analysis the hypothesis that T latent classes are sufficient to explain the data can be tested by means of a log-likelihood ratio test that compares the fit of the model with T classes to the fit of the general multinomial alternative. Under quite general regularity conditions, log-likelihood ratio statistics of this kind are asymptotically chi-squared distributed with specified numbers of degrees of freedom.

In a latent class analysis with inequality constraints imposed on the parameters, such a straightforward statistical interpretation of the log-likelihood ratio test is not warranted. In problems of statistical inference under order restrictions, likelihood ratio tests follow weighted chi-square distributions in which the weights depend on the null hypothesis considered and the region of the parameter space defined by the order restrictions (Robertson, Wright, and Dykstra, 1988). Only in the most simple cases can one obtain analytical expressions for those weights. At present, no numerical algorithm has been developed to determine the weights in the case of an ordinal latent class analysis. However, a conservative strategy, often used by the present author, to test the fit of an ordinal latent class analysis is the following. First, perform a series of unconstrained latent class analyses to obtain the optimal number T of latent classes by means of the standard log-likelihood ratio test. Second, conduct an OLCA with the same number of classes and count the number of free parameters lost by making some inequality constraints active. Compute the log-likelihood ratio statistic for the constrained and locate it under a chi-square distribution whose degrees of freedom are obtained by adding the number of lost parameters to the degrees of freedom of the unconstrained solution. The p value obtained in this way is an upper bound to the p value that one would obtain if working under the correct weighted chi-square distribution. One obtains a lower bound to this p value by locating the log-likelihood ratio statistics under a chi-square distribution with the same number of degrees of freedom as the unconstrained solution. That one so obtains a lower and an upper bound to the true p value is a consequence of the fact that the true asymptotic distribution is a probability mixture of chi-square distributions whose degrees of freedom vary between the two values used to derive the bounds.

A comparison of the log-likelihood ratio statistics for the constrained and unconstrained solution in relation to the number of free parameters lost may also give an indication of how much worse the model fit becomes when inequality constraints are imposed on the solution. Obviously, the validity of procedures of this kind also depends on how good the asymptotic chi-square distribution approximates the true sample distribution of the log-likelihood ratio statistic in the unconstrained solution. It is well known that if the data contain a large number of observed zero frequencies, this approximation may be quite bad. Then, all statistical inferences based on the likelihood ratio statistic become doubtful. In such situations, selecting the final solution should probably be better based on the ease with which it allows substantial interpretation of the results than on shaky statistical procedures whose validity can be questioned.

Before OLCA is illustrated by means of an example, it might be of some interest to relate OLCA to some other topics in statistics and psychometry.

E. OLCA and Regression Dependence

The monotonicity constraints derived from the assumption that the latent classes can be ordered are related to a particular concept of dependence among random variables as described by Lehmann (1966). Let X and Y be two random variables. Then, Lehmann (1966) defined regression dependence in the following way: Y is positively regression dependent on X if $\Pr(Y \leq y \mid X = x)$ is nonincreasing in x:

$$x_1 \leq x_2 \Rightarrow \Pr(Y \leq y \mid X = x_1) \geq \Pr(Y \leq y \mid X = x_2)$$

Of course, this implies

$$x_1 \leq x_2 \Rightarrow \Pr(Y > y \mid X = x_1) \leq \Pr(Y > y \mid X = x_2),$$

or, stated in words, the conditional probability of a large Y value is a nondecreasing function of x.

By identifying the latent variable with X and the manifest item with Y, it is clear that Croon's approach to ordering the classes implies that the manifest items are positively regression dependent on the latent variable. Note also that the concept of regression dependence is an asymmetrical one; if Y is positively regression dependent on X, the converse – X being positively regression dependent on Y – need not hold.

F. OLCA and Latent Trait Models

In recent psychometric research, latent trait models occupy a central place. In latent trait models, the item response probabilities are assumed to be continuous functions. For an item with three response categories, the "partial credit model" developed by Masters (1982) postulates the following category functions:

$$\Pr(X=1 \mid \theta) = \frac{1}{1+e^{\theta-\alpha_1}+e^{2\theta-\alpha_2}},$$

$$\Pr(X=2 \mid \theta) = \frac{e^{\theta-\alpha_1}}{1+e^{\theta-\alpha_1}+e^{2\theta-\alpha_2}},$$

$$\Pr(X=3 \mid \theta) = \frac{e^{2\theta-\alpha_2}}{1+e^{\theta-\alpha_1}+e^{2\theta-\alpha_2}}.$$

In these expressions, α_1 and α_2 are two item parameters; θ represents the respondent score and is assumed to be continuous. It is easy to show that under this latent trait model, the probabilities of a "positive" response after dichotomization, that is, the functions $\Pr(X{=}3 \mid \theta)$ and $\Pr(X{=}2 \mid \theta) + \Pr(X{=}3 \mid \theta)$, are increasing functions of θ. Hence, the item responses are positively regression dependent on the latent variable Θ.

The model underlying OLCA can now be seen as a "nonparametric" latent trait model in which the continuous latent trait is discretized by defining a restricted number of ordered classes along the continuum. Moreover, the explicit functional relationship between certain response probabilities and scores on the latent continuum is replaced by much weaker inequality constraints on the response probabilities. These constraints guarantee, however, that the response probabilities change in a way that is consistent with the order of the classes. For a more thorough discussion of the relation between latent trait and latent class models, see Heinen (1993).

3. ENVIRONMENTAL CONCERNS: AN APPLICATION OF ORDERED LATENT CLASS ANALYSIS

A. The Data

The data analyzed in this section were obtained from the cross-national comparative survey "European Values 90." This survey was designed and conducted by the European Value Systems Study Group (EVSSG), which was established in 1979 as a foundation to design and conduct an empirical study of cross-national differences and similarities in basic social values in Europe. For more information on this project and some of the more recent results, see Ester, Halman, and de Moor (1993). In this chapter, only data from The Netherlands will be used – so, no comparative analyses using data from different countries will be reported.

In this section, responses to six items that purportedly measure environmental concerns will be analyzed. The six items used in the questionnaire were as follows:

A. I would give part of my income if I were certain that the money would be used to prevent environmental pollution.
B. I would agree to an increase in taxes if the extra money is used to prevent environmental pollution.
C. The government must reduce environmental pollution but it should not cost me any money.
D. All the talk about pollution makes the people too anxious.
E. If we want to combat unemployment, we shall just have to accept environmental problems.
F. Protecting the environment and fighting is less urgent than often suggested.

The original response format had four response categories for each item: Strongly Agree, Agree, Disagree, and Strongly Disagree. Inspection of the marginal distribution of each item revealed that the most extreme anti-environment response (either Strongly Agree or Strongly Disagree, depending on the direction in which the item was worded) occurred only very rarely: only very few people bluntly reject environmental concerns. Hence, the respondents' answers were recoded into three categories, which, after appropriate mirroring of some of the items, can be labeled as Disagree (Category 1), Agree (Category 2), and Strongly Agree (Category 3) with environmental concerns.

Actually, the six items defined two different subscales. The items D, E, and F were supposed to measure how strongly a respondent is aware of environmental problems; this subscale will be called the *Awareness* subscale. In contrast, the first three items A, B, and C measured the respondents' willingness to make personal financial sacrifices in order to solve environmental pollution; this subscale will be called the *Willingness* subscale.

The following analyses were based on a sample of $N = 914$ respondents who provided answers to all six items. All analyses reported in this section were done by using the ℓEM program developed by Vermunt (1996). For more information on the technical details of the estimation procedure, see Vermunt (1999).

B. Separate Analyses for the Subscales

In a first set of analyses, classical and ordinal latent class analyses were conducted separately for the two subscales.

For the Willingness subscale, the unconstrained latent class analysis (LCA) solution with three latent classes had to be accepted: it resulted in an L^2 value of 6.5006 with which corresponds, for 6 degrees of freedom, a p value of .3695; the Pearson X^2 was equal to 5.6768. In this unconstrained solution, one of the monotonicity constraints of the ordinal model was violated. The OLCA solution with three latent classes resulted in an L^2 value of 8.3048 and a X^2 value of 7.5037. The difference in L^2 value with the unconstrained solution is equal to 1.8042; this difference seems to compare favorably with the one free parameter that is lost in the ordinal solution by making one inequality constraint active.

The three class solutions were also optimal for the Awareness subscale. The unconstrained solution with three latent class resulted in $L^2 = 9.124$ with which corresponds, for 6 degrees of freedom, a p value equal to .1667; the Pearson X^2 was equal to 8.6380. Two monotonicity constraints were violated in this solution, and it was precisely these two constraints that were made active in the ordinal solution, which resulted in an L^2 value of 9.6378. The difference in L^2 value with the unconstrained solution is 0.5318, which is quite small considering that two free parameters are lost by making two inequality constraints active.

A comparison of the OLCA solution and the unconstrained LCA solution indicates that an ordered one-dimensional continuum underlies responses to the three items in each subscale. A discussion of how the response probabilities vary along this ordinal continuum will be postponed until the joint analysis of the two subscales is discussed in the next paragraph. The solutions of the separate analyses proved to be very similar to the solution obtained in that joint analysis.

C. A Joint Analysis for the Two Subscales

The joint analysis for the two scales was based on the assumption that each of the subscales measures a distinct latent variable, but that the two latent variables involved in this way may be associated in the population. Letting X denote the latent variable corresponding with the Awareness scale and Y the latent variable corresponding with the Willingness scale, we find that the probability of observing the response pattern a, b, c, d, e, f is given by

$$\pi_{abcdef}^{ABCDEF} = \sum_x \sum_y \pi_{xy}^{XY} \pi_{ay}^{\bar{A}Y} \pi_{by}^{\bar{B}Y} \pi_{cy}^{\bar{C}Y} \pi_{dx}^{\bar{D}X} \pi_{ex}^{\bar{E}X} \pi_{fx}^{\bar{F}X}.$$

Whereas latent variable X exerts an effect only on items D, E, and F, latent variable Y determines the responses to items A, B, and C. The joint distribution of X and Y is described by the set of probabilities π_{xy}^{XY}. Because the separate analyses showed that the data for each subscale could be fitted by a latent class model with three classes, the discussion here will be restricted to the case in which the latent distribution of (X, Y) is assumed to consist of a 3×3 grid.

The unconstrained LCA resulted in an L^2 value of 566.3208 and a X^2 value of 722.1917. The ordinal LCA, in which the inequality constraints for a particular item were formulated in terms of the latent variable with which it is linked, resulted in an L^2 value of 571.4349 and a X^2 value of 780.4836. In total, three inequality constraints had to be made active, which resulted in a loss of three free parameters to estimate. The difference of 5.1141 between the two L^2 values compares favorably with this loss of three parameters. A cautionary remark is in order here: Because the 3^6 observed contingency table contains many zero cells, the asymptotic theory for the distribution of L^2 probably does not apply, and one should refrain (as is done here) from associating the observed L^2 values with a p value under a chi-square distribution.

Tables 1 and 2 contain the probability of response category 3 and of response categories 2 and 3, respectively, combined for each item in relation to the latent variable with which it is linked. Note that the response

Table 1. Three-Class Solution for Awareness Scale

	D		*E*		*F*	
Variable	**3**	**2 ⊕ 3**	**3**	**2 ⊕ 3**	**3**	**2 ⊕ 3**
$X = 1$	0.044	0.225	0.066	0.484	0.093	0.426
$X = 2$	0.072	0.789	0.066	0.863	0.093	0.971
$X = 3$	0.598	0.981	0.633	0.931	0.846	1.000

probability for category 1 can be obtained by taking the complement of the probability of the combined category $2 \oplus 3$. The probability of category 2 can be obtained by substracting the probability of category 3 from that of the combined categories $2 \oplus 3$.

For the relation between the first latent variable X and its indicator variables D, E, and F, the solution was as shown in Table 1. For the relation between Y and its indicator variables A, B, and C, the solution can be summarized as shown in Table 2.

It is clear from a visual inspection of these two tables that the "positive" response probabilities for a particular items are monotonically related to the values of the latent variable it is assumed to measure. For both latent variables, the first latent class represents the anti-environmental attitude; the two other classes represents pro-environmental attitudes, but in different strengths.

The joint distribution of X and Y is given in Table 3. Although no constraints whatsoever were imposed on the probabilities in this joint distribution, computation of the conditional probabilities $p(X=x \mid Y=y)$ and $p(Y=y \mid X=x)$ shows that Y is positively regression dependent on X, and that X is positively regression dependent on Y. So, in a sense, a high score on X is indicative of an high score on Y, and the converse is also true. It is interesting to note here that if necessary, one could estimate the joint distribution of the latent variables X and Y under constraints implied by the assumption that Y is positively regression dependent on X, or vice versa.

Table 2. Three-Class Solution for Willingness Scale

	A		*B*		*C*	
Variable	**3**	**2 ⊕ 3**	**3**	**2 ⊕ 3**	**3**	**2 ⊕ 3**
$Y = 1$	0.048	0.341	0.016	0.141	0.032	0.345
$Y = 2$	0.105	0.942	0.016	0.793	0.148	0.941
$Y = 3$	0.900	1.000	0.728	0.950	0.499	0.963

Table 3. Joint Distribution of Two Latent Variables

	Y			
Variable	**1**	**2**	**3**	
X				
1	0.146	0.112	0.017	0.275
2	0.052	0.324	0.090	0.466
3	0.009	0.107	0.143	0.259
	0.207	0.543	0.250	

D. A One-Dimensional Analysis for the Total Scale

That the latent variables X and Y are strongly related is also shown by the value of the association coefficient γ_{XY}, which is equal to 0.729. The strong association between the latent variables in the two-dimensional analysis might suggest that a model with one ordered latent variable might represent the data in a more parsimonious way. For this conjecture to be tested, latent class analyses were conducted under the assumption that the same latent variable determines the responses to all six items. The number of latent classes was varied from one to six. Table 4 contains some global results for unconstrained latent class analyses with four, five, and six latent classes as applied to the data for all six items. For each of these analyses, the values of the log-likelihood ratio statistics L^2 and the Pearson chi-square statistic X^2 together with the probability levels associated with them under the appropriate asymptotic chi-square distribution are given in Table 4.

Although both statistics are asymptotically equivalent, their values may markedly differ for sparse data containing a large number of

Table 4. Goodness-of-Fit Statistics for Unconstrained Analysis on the Total Scale

T	L^2	X^2
4	642.71	850.00
	(0.824)	(0.000)
5	565.25	773.57
	(0.998)	(0.002)
6	494.13	680.15
	(1.000)	(0.208)

Table 5. Goodness-of-Fit Statistics for the Order-Constrained Solution on the Total Scale

T	L^2	X^2	Active (%)
4	697.80	1100.05	10 (28)
5	647.08	1140.44	26 (54)
6	642.12	1146.67	33 (55)

observed zeros. From the point of view of the L^2 test, the solution with four latent classes is acceptable, but this solution still yields a large value for X^2, reflecting the fact that the four-class solution leads to large standardized residuals for many observed frequencies. Although the solution with five classes is already more adequate in this respect, the solution with six classes is acceptable from both point of views. As a way to see whether these four, five, or six classes can be considered as ordered, the latent class analyses were repeated with order constraints on the response probabilities.

Table 5 contains the L^2 and X^2 values for the ordinal analysis with four, five, and six classes. The last column of this table contains the absolute number and percentage of inequality constraints, which were made active in the final solution. The ordinal analysis with a particular number of classes is clearly much worse than the unconstrained analysis with the same number of classes. Moreover, for the solutions with five and six classes, more than half of the inequality constraints had to be made active. This indicates that, although the unconstrained solutions with five or six classes provide an acceptable fit to the data, these classes cannot be considered to be ordered along a single attitudinal continuum. So, if one adheres to an interpretation of the data in terms of simple, possibly correlated, attitudinal continua, the joint analysis of the subscales in terms of two distinct ordinal latent variables should be preferred.

4. MOKKEN SCALE ANALYSIS

A. Scalogram Analysis: Deterministic and Probabilistic Formulations

Guttman's scalogram analysis is based on a simple deterministic model in which dichotomous items and respondents are both represented by points on a latent continuum. The ordering of the item points along this continuum reflects the ordering of the items with respect to difficulty (for ability items) or popularity (for attitude items). Items at the left end of the continuum are easier or more popular than items at the right end of the continuum. Given the position of the respondent point relative to this item ordering, the respondent's answers can be predicted in a completely deterministic way. Respondents answer positively to all the items that are to the left of their respondent point on the continuum, and answer negatively to all the items that are to the right of their respondent point. Hence, if a respondent gives a positive answer to a particular item, he or she should also respond positively to the easier items. If, however, the respondent fails a particular item, he or she should also fail the more difficult items. In this way, the deterministic scalogram model implies that only a restricted number of response patterns may be observed in the data. Items whose response patterns agree with this expectation are said to form a perfect scale.

The large-scale application of scalogram analysis to real data is definitely hampered by the deterministic character of the model. All too often, scalogram analysis leads to the rejection of a large number of items because their response patterns do not conform to the requirements of a perfect scale. In an attempt to develop a more flexible scale analysis, Mokken (1971) developed a probabilistic variant of the scalogram model. He assumed that the probability of a positive response to a dichotomous item j as a function of the respondent's latent score θ on the underlying continuum is given by an item response function $\pi_j(\theta)$. The functional form of this item response function was left unspecified, but the functions $\pi_j(\theta)$ are assumed to satisfy *double monotony*. This qualitative requirement consists of two independent conditions on the item response functions.

1. The item response functions $\pi_j(\theta)$ are nondecreasing functions of their arguments:

   $$\theta_1 \leq \theta_2 \Rightarrow \pi_j(\theta_1) \leq \pi_j(\theta_2).$$

   This first condition of latent monotonicity requires that the probability of a positive response increases (or, at least, does not decrease) with increasing values of θ. It also implies that the expected value of a respondent's total score on a set of items is an increasing function of θ.
2. The graphical representations of two item response functions $\pi_j(\theta)$ and $\pi_k(\theta)$ should not intersect. In formal terms, if $\pi_j(\theta_0) \leq \pi_k(\theta_0)$ for some θ_0, then $\pi_j(\theta) \leq \pi_k(\theta)$ for all θ. This second condition implies

that the ordering of two items with respect to difficulty or popularity is independent of the value of θ, and, hence, is identical for each respondent.

Note that the preceding conditions apply only to the case of dichotomous items. Their extension to the general case of polytomous items is not trivial and will not be discussed here.

B. A Latent Class Formulation of the Double-Monotony Condition

Croon (1991) showed how for dichotomous items the condition of double monotony can be checked by means of a latent class analysis with ordered classes. A prerequisite of this latent class analysis is that the ordering of the items according to popularity is known along the underlying continuum. Although this unknown ordering could be considered as an extra parameter to be estimated by the analysis, Croon suggests taking the observed order of item difficulties as the starting point of the analysis. Without loss of generality, one may assume that the order of the items with respect to difficulty or popularity is given by their alphabetic order: A, B, C, D, ... It is assumed that A is the most difficult or least popular item, that B is the second most difficult or second least popular item, and so on. Positive responses will be denoted by score 2, so that, for instance, $\pi_{2t}^{\bar{A}X}$ represents the probability of a positive response to item A for respondents belonging to latent class t.

Assume further that T latent classes are needed in the analysis, and that the set of classes is ordered by the weak order relation $\preceq$, as in Section 2. The first condition of double monotony now requires that for each item, the probability of a positive response is nondecreasing as one runs through the set of latent classes in the direction of $\preceq$. For item A, one then obtains the following system of linear inequalities:

$$\pi_{21}^{\bar{A}X} \leq \cdots \leq \pi_{2t}^{\bar{A}X} \leq \pi_{2,t+1}^{\bar{A}X} \leq \cdots \leq \pi_{2T}^{\bar{A}X}.$$

Similar systems of inequalities must be formulated for the other items. In terms of the latent class model with ordered classes, the second condition of double monotony implies that for each latent class t the positive response probabilities should satisfy

$$\pi_{2t}^{\bar{A}X} \leq \pi_{2t}^{\bar{B}X} \leq \pi_{2t}^{\bar{C}X} \leq \cdots.$$

Inequalities of this type, which have to be formulated for each of the latent classes, constrain the ordering of the items to be the same in all the classes. In this way the requirement that the item ordering is identical for all values on the underlying latent continuum is expressed in terms of the latent class model.

C. Estimating and Testing the Model

Croon (1991) showed how the maximum-likelihood estimates of the item response parameters and of the latent class proportions can be obtained by means of an EM algorithm. During the M step of each iteration, the item response probabilities are estimated under the inequality constraints derived earlier. Croon (1991) showed how this optimization problem under inequality constraints can be solved by means of a procedure developed in the context of isotonic regression analysis. More information about the technical details of this optimization algorithm can be found in that paper.

5. MOKKEN SCALE ANALYSIS OF RELIGIOUS BELIEFS

A. The Data

In the same cross-national survey "European Values 90," respondents were asked whether they believed in (A) God, (B) life after death, (C) a soul, (D) the devil, (E) Hell, (F) Heaven, (G) sin, (H) resurrection of the dead, and (I) reincarnation. The original response format consisted of three response categories: Yes, No, and Don't Know. For the present analysis, the No and Don't Know categories were combined to one negative response category, so that a positive answer to each question indicated a clear and univocal belief in the corresponding religious topic. In the sequel, only the data from The Netherlands will be used; complete response patterns were available for $N = 1{,}012$ respondents.

B. The Analyses

In a first series of analyses, classical and ordinal latent class analyses were conducted for the complete set of nine items. For the unconstrained analysis, the solution with four latent classes was acceptable, leading to an L^2 value of 399.1334, with a p value equal to .9935; the Pearson X^2 was 1,008.8232. The solution with four ordered latent classes was less optimal. Its L^2 value was equal to 444.6796; its X^2 value was 1326.2873. Although in the unconstrained four-class solution, six monotonicity constraints were violated, in the final constrained solution only four of them had to be made active. However, the loss of fit caused by this loss of four parameters was large; the difference between the L^2 values of the two solutions was equal to 45.5462, which is probably not acceptable. A comparison of the two solutions indicated that especially response probabilities of the last item (I, Reincarnation) did not conform to the expectations based on a one-dimensional ordinal scale. Two of the four violations of the monotonicity constraints pertained to this particular item. An explanation of the aberrant behavior of this item might be that it refers to a non-Christian religious belief, whereas the eight other items all refer to beliefs within Christian theology.

Note also that in these analyses the X^2 values are consistently much larger than the L^2 values. This is probably caused by the sparseness of the data. In a second series of analyses, only responses to the eight Christian belief items were considered. The unconstrained solution with four classes leads to $L^2 = 243.9359$ with $p = .1285$; the Pearson X^2 was 433.7252. For the constrained solution, it was found that $L^2 = 245.943$ and $X^2 = 521.8433$, whereas only two monotonicity constraints had to be made active. The difference between the L^2 values of the two solutions was equal to 2.0073, which is small compared with the two free parameters lost. Hence, on the basis of the second series of analyses, one may conclude that an ordinal latent continuum underlies the responses to the eight Christian belief items.

In a next step, a Mokken scale analysis was conducted to see whether the eight items formed a latent scale in which the items have a uniform ranking over the different classes. With four latent classes, the analysis in which the item response probabilities were constrained by the double-monotony conditions resulted in a value of L^2 equal to 275.9775 and a X^2 value of 718.3649. In this analysis, eleven inequality constraints had to be made active. The conclusion seems to be warranted that, although the items can be represented along an ordinal latent continuum, they do not form a latent scale. Apparently, their ranking according to popularity is not the same in all classes.

Table 6 contains the item response probabilities in the four classes of the ordinal solution. All four latent classes contain a nonnegligible proportion of people. Class I is definitely the class of nonbelievers; Class IV, in contrast, contains people who consistently accept all belief items as true. The differences between Classes III and IV are interesting. On the basis of their item response probabilities in these two classes, the items fall in three distinct groups. First, one has the items A and C, which both have

Table 6. Ordinal Four-Class Solution for Christian Items

Item	I	II	III	IV
A	0.156	0.723	0.992	0.992
B	0.034	0.312	0.874	0.906
C	0.228	0.766	0.943	0.966
D	0.000	0.048	0.149	0.970
E	0.000	0.023	0.023	0.945
F	0.000	0.218	0.763	1.000
G	0.115	0.400	0.695	0.963
H	0.000	0.077	0.771	0.889
π_t^X	0.336	0.369	0.165	0.130

relatively high response probabilities in Classes II and III. In contrast, items D and E have both low response probabilities in these classes. The remaining four items (B, F, G, and H) have a small response probability in class II, but a high one in class III.

6. SOME RELATED WORK

In this final section, two other developments in latent class models with ordered classes are briefly discussed. First, an alternative way of defining the order relation $\preceq$ on the set of latent classes will be described. Second, some attention will be given to a latent class analysis in which the item response probabilities are not necessarily increasing with latent class number, but show a single-peaked pattern.

A. Monotonic Local and Adjacent Odds Ratios

In Lehmann (1966), several concepts of dependence among random variable are discussed. In the context of LCA with ordered classes, one of these concepts, *positive regression dependence*, is equivalent to the requirement that the cumulative odds increase with increasing latent class number t; that is, they satisfy

$$\sum_{g=k}^{m} \pi_{g,t}^{\bar{A}X} \Big/ \sum_{g=1}^{k-1} \pi_{g,t}^{\bar{A}X} \leq \sum_{g=k}^{m} \pi_{g,t+1}^{\bar{A}X} \Big/ \sum_{g=1}^{k-1} \pi_{g,t+1}^{\bar{A}X}.$$

In this way, ordinality is defined by requiring that the manifest variable A is positively regression dependent on the latent variable X.

Another concept of dependence considered by Lehmann (1966) is that of *likelihood ratio dependence.* Let $f_{XY}(x, y)$ be the joint density function of two random variables X and Y. Then, X and Y are said to show positive likelihood ratio dependence if

$$x_1 \leq x_2 \text{ and } y_1 \leq y_2 \Rightarrow f_{XY}(x_1, y_2) f_{XY}(x_2, y_1) \leq f_{XY}(x_1, y_1) f_{XY}(x_2, y_2).$$

Because the latter inequality is equivalent to

$$f_{XY}(x_1, y_2)/f_{XY}(x_1, y_2) \leq f_{XY}(x_2, y_2)/f_{XY}(x_2, y_1),$$

positive likelihood ratio dependence requires that the odds of observing large values of Y increase with increasing values of X. In the context of a latent class analysis, this requirement is equivalent to the constraints that that for observed scores $a \leq a'$ on A and for latent scores $t \leq t'$ on X, the odds satisfy

$$\pi_{a't}^{\bar{A}X} / \pi_{at}^{\bar{A}X} \leq \pi_{a't'}^{\bar{A}X} / \pi_{at'}^{\bar{A}X}.$$

Hence, under this approach, the local odds are required to be monotonic with latent class number t.

Still another approach is to require that the adjacent odds are monotonic. In the context of a latent class analysis, this is equivalent to the constraint

$$\pi_{k,t}^{\bar{A}X} \Big/ \sum_{g=1}^{k-1} \pi_{g,t}^{\bar{A}X} \leq \pi_{k,t+1}^{\bar{A}X} \Big/ \sum_{g=1}^{k-1} \pi_{g,t+1}^{\bar{A}X}.$$

for $k = 2, \ldots, m$ and $t = 1, \ldots, T - 1$.

The several ways of imposing an order relation on the set of latent classes are not equivalent to each other. One may prove that monotonic local odds imply monotonic adjacent odds, which in turn imply monotonic cumulative odds. So, the requirement of monotonic local odds is the stronger; the requirement of monotonic cumulative odds is the weaker approach to defining ordinal latent classes. See Vermunt (1999) for a systematic comparison of the different definitions of ordered classes and for information on a common estimation method.

B. Ordered Classes with Nonmonotonic Items

All models with ordered classes considered thus far in this chapter assume that the probability of a positive response to an item increases along an underlying latent continuum. Items for which this model is appropriate may be called *monotonic* items. However, this kind of model may not be adequate for all attitudinal items. For instance, if someone is asked whether he or she agrees with the government's general social policy, a negative response may indicate that this person considers this policy either too conservative or too liberal to agree with it. A positive response is given only if the respondent's point is not too far away from the item point. For dichotomous items of this kind, the probability of a positive response increases up to a specific point on the latent continuum, and decreases afterward. Hence, items for which this type of traceline is adequate may be called *nonmonotomic* items. Hoijtink (1990) and Andrich (1996) developed item response models for items with single-peaked tracelines.

Croon (1993) showed how single-peaked tracelines can be handled within the context of ordinal latent class models. Let $\preceq$ once again represent the order relation on the set of latent classes. If the probabilities of a positive response to item i follow a single-peaked pattern if one runs through the set of latent classes in the direction of $\preceq$, there should an item-specific class s_i such that

$$\pi_{21}^{\bar{A}X} \leq \pi_{22}^{\bar{A}X} \leq \cdots \leq \pi_{2,s_i-1}^{\bar{A}X} \leq \pi_{2,s_i}^{\bar{A}X},$$

but

$$\pi_{2,s_i}^{\bar{A}X} \geq \pi_{2,s_i+1}^{\bar{A}X} \geq \cdots \geq \pi_{2,T-1}^{\bar{A}X} \leq \pi_{2T}^{\bar{A}X}.$$

Croon (1993) showed how the maximum-likelihood estimates of the response probabilities can be obtained under this kind of "umbrella" ordering.

7. DISCUSSION

In this chapter, it was shown how an order relation can be defined on the set of classes in the latent class model by imposing inequality constraints on the item response probabilities. For monotonic items, these constraints represent the requirement that the probability of a positive response increases as a function of increasing latent class number. Monotonic items represent the case in which a higher position on the latent continuum is reflected in a higher manifest response. For nonmonotonic items, in contrast, the requirement is that the probability of a positive response increases up to a point on the latent class continuum, and then starts to decrease again. Nonmonotonic items then represent the situation in which respondents tend to agree with those items that are close to their own position on the latent continuum, but reject the items that are farther way from it in whatever direction.

A comment that is sometimes made with respect to ordinal latent class analysis is that it is often not necessary to explicitly impose inequality constraints on the item response probablities in order to obtain neatly ordered classes. Often, so this argument goes, a simple traditional latent class analysis will yield a solution in which the item response probabilities increase (mostly after a relabeling of the classes) as a function of the latent class number. Hence, it is not necessary to consider models in which such inequality constraints are explicitly imposed. In response to this criticism of ordinal latent class analysis, one should say that most often a traditional latent class analysis will yield only "unsolicited" ordinal solutions for items that have been selected beforehand on the basis of some criterion that guarantees unidimensionality of the scale. In the present author's experience, the application of unconstrained latent class analysis to sets of unselected items often results in violations of the order constraints. Thus, ordinal latent class analysis can be used as a item selection procedure that aims at the construction of a unidimensional scale without making strong functional or statistical assumptions about how response probabilities to items vary as functions of the underlying continuum. If items that show gross violations of the order conditions are removed, one obtains a scale of items whose responses can be described in terms of a unidimensional ordinal continuum.

Analyzing data by means of ordinal latent class analysis implies both estimating unknown parameters and testing the goodness-of-fit of the model. Up until now, only the estimation part of ordinal latent class analysis has been solved adequately. The procedures described in this paper should yield the maximum-likelihood estimates of the parameters under the appropriate inequality constraints. However, obtaining these estimates requires the maximization of a complex function over a possibly irregular region of the parameter space, and the iterative optimalization procedure used in this respect may converge to a local rather than to the global maximum. There are no definite solutions to the problem of local maxima; starting the analysis from different initial points in the parameter is often recommended as a practical but not completely satisfactory strategy.

Testing the goodness-of-fit of the model still remains a problem from a purely statistical point of view. Statistical models with inequality constraints on the parameter cannot be tested by means of traditional log-likelihood ratio tests because the number of degrees of freedom involved in these tests is not a constant known in advance to the analysis, but is also a random variable. For some problems of statistical inference under order restrictions, the appropriate statistical tests have been developed (Robertson et al., 1988), but for latent class analysis with inequality constraints on the item response probabilities, such statistical tests are not yet available. In the applications discussed in this chapter, a heuristic model testing strategy was used in which both the number of inequality violations and their effect on the log-likelihood ratio statistic were taken into account. A fully Bayesian approach, using the Gibbs sampler (Hoijtink and Molenaar, 1997) or another Markov Chain Monte Carlo method, might solve some of the estimation and testing problems involved in ordinal latent class analysis.

REFERENCES

Agresti, A. (1990). *Categorical Data Analysis.* New York: Wiley.

Andrich, D. (1996). "A hyperbolic cosine latent trait model for unfolding polytomous responses: reconciling Thurstone and Likert methodologies," *British Journal of Mathematical and Statistical Psychology*, **49**, 347–65.

Clogg, C. C. (1979). "Some latent structure models for the analysis of Likert–type data," *Social Science Research*, **8**, 287–301.

Clogg, C. C. (1988). "Latent class models for measuring." In R. Langeheine & J. Rost (eds.), *Latent Trait and Latent Class Models*. New York: Plenum, pp. 173–205.

Croon, M. A. (1990). "Latent class analysis with ordered latent classes," *British Journal of Mathematical and Statistical Psychology*, **43**, 171–92.

Croon, M. A. (1991). "Investigating Mokken scalability of dichotomous items by means of ordinal latent class analysis," *British Journal of Mathematical and Statistical Psychology*, **44**, 315–31.

Croon, M. A. (1993). "Ordinal latent class analysis for single-peaked items," *Kwantitatieve Methoden*, **14**, 128–42.

Goodman, L. (1974). "The analysis of systems of qualitative variables when some of the variables are unobservable. Part I: a modified latent structure approach," *American Journal of Sociology*, **79**, 1179–1259.

Ester, P., Halman, L., & de Moor, R. (1993) *The Individualizing Society: Value Change in Europe and North America.* Tilbung: Tilburg University Press.

Heinen, A. (1993). *Latent Class and Discrete Latent Trait Models: Similarities and Differences.* Thousand Oaks, CA: Sage Publications.

Hoijtink, H. (1990). "A latent trait model for dichotomous choice data," *Psychometrika*, **55**, 641–56.

Hoijtink, H., & Molenaar, I. W. (1997). "A multidimensional item response model: constrained latent class analysis using the Gibbs sampler and posterior predictive checks," *Psychometrika*, **62**, 171–89.

Lazarsfeld, P. F. (1950a). "The logical and mathematical foundation of latent structure analysis," In S. A. Stouffer et al. (eds.), *Measurement and Prediction*. Princeton, NJ: Princeton University Press, pp. 362–412.

Lazarsfeld, P. F. (1950b). "The interpretation and mathematical foundation of latent structure analysis." In S. A. Stouffer et al. (eds.), *Measurement and Prediction*. Princeton, NJ: Princeton University Press, pp. 413–472.

Lazarsfeld, P. F. (1954). "A conceptual introduction to latent structure analysis." In P. F. Lazarsfeld (ed.), *Mathematical Thinking in the Social Sciences*. New York: The Free Press, pp. 349–97.

Lazarsfeld, P. F., & Henry, N. W. (1968). *Latent Structure Analysis*. Boston: Houghton Mifflin.

Lehmann, E. L. (1966). "Some concepts of dependence," *Annals of Mathematical Statistics*, **37**, 1137–53.

Masters, G. N. (1982). "A Rasch model for partial credit scoring," *Psychometrika*, **47**, 149–74.

Mokken, R. J. (1971) A theory and procedure of scale analysis. 's Granen hage: Moteon.

Robertson, T., Wright, F. T., & Dykstra, R. L. (1988). *Order Restricted Statistical Inference*. New York: Wiley.

Rost, J. (1988). "Rating scale analysis with latent class models," *Psychometrika*, **53**, 327–48.

Rost, J. (1991). "A logistic mixture distribution model for polychotomous item responses," *British Journal of Mathematical and Statistical Psychology,* **44**, 75–92.

Vermunt, J. K. (1996). *Log-Linear Models for Event Histories.* Thousand Oaks, CA: Sage Publications.

Vermunt, J. K. (1999). "A general class of non-parametric models for ordinal categorical data," *Sociological Methodology*, **29**, 187–223.

SIX

Comparison and Choice

Analyzing Discrete Preference Data by Latent Class Scaling Models

Ulf Böckenholt

1. INTRODUCTION

One of the major difficulties in the analysis of preference data is accounting for individual taste differences in the comparison and evaluation of different choice options. Presenting an extension of classical latent class analysis, this chapter[1] demonstrates that latent class scaling models offer a general solution to this problem. Latent class scaling models simultaneously identify subgroups with homogeneous preferences and provide a graphical representation of the similarity structure underlying the choice options. Several applications involving "pick any/J," as well as (in)complete and partial ranking data, are discussed that illustrate the versatility of latent scaling class models in the analysis of different types of choice data.

The main reason for the versatility of Lazarsfeld's (1950) latent class (LC) model is its underlying assumption that the latent variables are categorical and nominal. No restrictions are imposed on relationships among the latent classes and the nature of the manifest variables within each latent class with the exception of conditional independence. It is therefore not surprising that extensions of the LC model focused almost exclusively on the introduction of constraints to facilitate the applicability of an LC model to particular research problems. This chapter is concerned with applications in which many individuals choose among a set of items. Responses may be expressed in the form of approvals, ratings, or (incomplete and partial) rankings. Unconstrained LC models are not well suited for the analysis of such data because they do not utilize the structure of the response format and they do not incorporate hypotheses about the underlying response process that gives rise to the data. These limitations are overcome by constrained LC models. By taking into account the response process, constrained LC models have a more parsimonious structure and yield graphical representations that depict similarities and differences among individuals and items, if possible, in a joint space (Böckenholt and Böckenholt, 1990, 1991).

Constrained LC models deviate in three major ways from Lazarsfeld's (1950) LC model. First, depending on the constraints used, the concept of nominal classes is replaced by the more restrictive notion that classes are of an at least ordinal nature. Second, the LC model was originally developed for binary and polytomous items. Constrained LC models, however, can facilitate the analysis of other discrete data types. For example, this chapter discusses applications with (in)complete and partial ranking data. Third, the only assumption that is made by unconstrained LC models about the nature of the response process is that it is probabilistic. Constrained LC models are more specific by postulating a functional relationship between the probability of a response and characteristics of both items and persons. Despite these differences it is important to take into account the "nested" relationship between constrained and unconstrained LC models. Comparisons between both types of LC models can be very powerful for determining whether the imposed constraints are consistent with the data.

Although the notion of constrained LC models for the analysis of choice data is recent, a large number of models have already been proposed in the literature, and it seems likely that this trend will continue for some time (Ben-Akiva et al., 1997; Dillon et al., 1994). As a result, this chapter cannot give an exhaustive treatment of this subject. Instead, it provides an application-oriented introduction to this topic by focusing predominantly on two classes of well-known choice models originally proposed by Luce (1959) and Coombs (1964).

2. LATENT CLASS MODELS FOR PICK ANY/*J* AND RANKING DATA

Let $\pi_Y(y)$ denote the joint distribution of the J observed variables, $Y = (Y_1, \ldots, Y_J)$. Similarly, let $\pi_X(t)$ denote the probability that X takes on the integer value t. By the axiom of local independence, we obtain

$$\pi_{Y|X(t)}(y) = \prod_{j=1}^{J} \pi_{Y_j|X(t)}(y_j),$$

and Lazarsfeld's (1950) latent class model with T latent classes is written as

$$\pi_Y(y) = \sum_{t=1}^{T} \pi_X(t)\pi_{Y|X(t)}(y). \quad (1)$$

If Y_j is binary, its conditional probabilities may be written as $\pi_{Y_j|X(t)}(0)$ and $\pi_{Y_j|X(t)}(1)$ when Y_j is equal to 0 and 1, respectively.

The conditional probabilities are allowed to vary freely among classes. As a result, even when all items are binary, the number of parameters to be estimated can become large very quickly. This may complicate the interpretation of the results and adversely affect the stability of the parameter estimates. In fact, a well-known phenomenon in LC analysis is that increments in the number of classes lead to identifiability problems as well as increases in both the number of both local optima and conditional probability estimates at their boundary values of 0 and 1.

Fortunately, in most applications, sufficient information is available that can be incorporated in an LC model with the result that the interpretation about the differencces among the latent classes is much simplified and the number of parameters to be estimated is reduced (Clogg, 1988, 1995). A simple example is the following probabilistic version of Coombs' (1964) parallelogram model for pick k/J data. These data are obtained by asking persons to select the k best out of a set of J items.

Suppose there are $J = 5$ items whose order reflects a continuum from a positive to a negative attitude toward an issue. A person agrees with an item (coded as 1) if the position of the item coincides with or is close to the person's position on this issue and otherwise disagrees (coded as 0; more detail on this idea is presented in the next section). The parallelogram model expects the following response patterns to occur when $k = 2$:

$$S_1 = (1,1,0,0,0); \quad S_2 = (0,1,1,0,0);$$
$$S_3 = (0,0,1,1,0); \quad S_4 = (0,0,0,1,1).$$

To investigate whether departures from this parallelogram pattern are caused by response errors, we can postulate four classes, in which each class corresponds to one of the response pattern types.

$$\begin{aligned}\pi_{Y_1|X(1)}(1) &= \pi_{Y_2|X(1)}(1) = \pi_{Y_3|X(1)}(0) = \pi_{Y_4|X(1)}(0) = \pi_{Y_5|X(1)}(0)\\ &= \pi_{Y_1|X(2)}(0) = \pi_{Y_2|X(2)}(1) = \pi_{Y_3|X(2)}(1) = \pi_{Y_4|X(2)}(0)\\ &= \pi_{Y_5|X(2)}(0) = \pi_{Y_1|X(3)}(0) = \pi_{Y_2|X(3)}(0) = \pi_{Y_3|X(3)}(1)\\ &= \pi_{Y_4|X(3)}(1) = \pi_{Y_5|X(3)}(0) = \pi_{Y_1|X(4)}(0) = \pi_{Y_2|X(4)}(0)\\ &= \pi_{Y_3|X(4)}(0) = \pi_{Y_4|X(4)}(1) = \pi_{Y_5|X(4)}(1)\end{aligned}$$

Under the parallelogram model, these class-specific probabilities are all equal to 1. However, by allowing for an error (or response inconsistency) rate α ($0 \leq \alpha \leq 1$), we find that the probabilities have the common value $1 - \alpha$ (Proctor, 1970). As a result, this simple probabilistic version of the parallelogram model requires the estimation of only $(J - 1)$ parameters. The model can be extended in various ways to allow for error rates that are item or class specific. However, there are several shortcomings that would prevent this probabilistic version of the parallelogram model to be of much value in applied work. First, this approach requires a priori knowledge about the ordering of the items. Second, it is difficult to generalize the model to nondichotomous items. Finally, the instruction to pick a fixed and not an arbitrary number of items (pick any/J) is rather restrictive. To overcome these limitations, the next sections focus on more general approaches that impose functional constraints on the class-specific probabilities.

3. PICK ANY/*J* DATA

In a pick any/J task, individuals are asked to select those items (from a set of items) that they like or agree with most. For example, Formann (1988) investigated attitudes toward nuclear energy by asking $N = 600$ persons whether they agreed or disagreed with the following five items selected from a questionnaire:

A. In the near future, alternative sources of energy will not be able to substitute for nuclear energy.
B. It is difficult to decide between the different types of power stations if one carefully considers all their pros and cons.
C. Nuclear power stations should not be put in operation before the problem of radioactive waste has been solved.
D. Nuclear power stations should not be put in operation before it is proven that the radiation caused by them is harmless.
E. The foreign power stations now in operation should be closed.

The data are reproduced in Table 1. For example, 118 persons agreed with items C, D, and E. We may analyze this data set with an LC model to determine the number of distinct classes that differ in their probabilities of endorsing any of these five items. The goodness-of-fit of an LC model can be assessed by a likelihood ratio test, G^2:

$$G^2 = 2 \sum f_g \ln(f_g / \hat{f}_g),$$

Table 1. Response Patterns and Observed and Predicted Frequencies under Four-Class UC and IP LC Models

Items								Items							
A	B	C	D	E	Obs.	UC	IP	A	B	C	D	E	Obs.	UC	IP
0	0	0	0	0	3	4.2	4.9	0	0	0	0	1	11	11.0	9.2
0	0	0	1	0	3	3.3	4.2	0	0	0	1	1	39	40.3	34.0
0	0	1	0	0	14	17.5	15.2	0	0	1	0	1	1	1.2	3.1
0	0	1	1	0	41	41.4	45.2	0	0	1	1	1	118	115.9	115.9
0	1	0	0	0	6	4.1	3.8	0	1	0	0	1	2	2.0	2.4
0	1	0	1	0	4	3.4	2.9	0	1	0	1	1	16	16.2	17.6
0	1	1	0	0	22	19.1	22.4	0	1	1	0	1	1	2.1	2.5
0	1	1	1	0	65	63.8	57.5	0	1	1	1	1	61	61.7	65.4
1	0	0	0	0	6	4.9	5.3	1	0	0	0	1	0	0.0	0.4
1	0	0	1	0	3	2.6	1.6	1	0	0	1	1	5	3.6	5.3
1	0	1	0	0	22	19.1	15.9	1	0	1	0	1	2	1.0	1.4
1	0	1	1	0	37	34.9	37.8	1	0	1	1	1	15	19.2	21.7
1	1	0	0	0	3	4.7	4.5	1	1	0	0	1	0	0.1	0.1
1	1	0	1	0	2	3.0	1.7	1	1	0	1	1	2	1.8	2.9
1	1	1	0	0	18	20.3	23.6	1	1	1	0	1	2	1.7	2.0
1	1	1	1	0	52	54.8	55.2	1	1	1	1	1	24	21.4	14.6

Note: Obs., observed; UC, unconstrained; IP, constrained.

where f_g and $\hat{f}_g = N\hat{\pi}_Y(y)$ denote the observed and expected frequencies for the gth response pattern. Asymptotically, G^2 follows a X^2 distribution with $[2^J - T(J+1)]$ degrees of freedom (df). For this data set, the G^2 statistics obtained for the one-, two-, three-, and four-class solutions are 249.5 (df = 26), 64.0 (df = 20), 29.6 (df = 14), and 8.4 (df = 8), respectively. These results suggest that the four-class model may be best suited to describe the variability in the data. However, this solution is not well determined because 24 parameters are required to describe the differences among the 32 response patterns. In addition, when inspecting the estimated conditional "agree" probabilities, we find that approximately 30% of them are equal to their boundary values of 0 or 1. Thus, it seems desirable to investigate constrained versions of the LC model that, although more parsimonious, still capture the relevant structures in this data set.

A. Coombs' Unfolding Model

A major objective in the analysis of preferential or attitudinal data is to investigate the similarity structure that underlies the items while taking into account individual differences. Coombs' (1964) unfolding theory provides a conceptually simple yet powerful approach for accomplishing this goal. Coombs assumed that in a choice situation, persons compare each item to their ideal or most preferred item. When asked to pick, for example, the most preferred option, individuals select the one that is closest or least dissimilar to their ideal option. However, although persons may differ in terms of their most preferred item, they agree on the similarity relationships among the items. Thus, in the unidimensional case, the items' positions along some common continuum are perceived in the same way by all persons; however, the positions of the individual ideal points may differ from person to person.

When Coombs (1964) introduced the pick any/J procedure, he posited that persons select those items that are closest to the position of their ideal points. More formally, let the positions of item j and of the ath person's ideal point be δ_j and β_a, respectively. A response of person a to item j is denoted by the binary variable Y_{aj}. Item j is selected when its distance to the ideal point is smaller than some threshold τ,

$$Y_{aj} = 1 \text{ when } |\delta_j - \beta_a| \leq \tau, \tag{2}$$

and is not selected otherwise,

$$Y_{aj} = 0 \text{ when } |\delta_j - \beta_a| > \tau.$$

Because choice behavior is frequently inconsistent, it is necessary to formulate probabilistic versions of Equation (2). Hoijtink (1990) proposed the following response function under the premise that a probabilistic unfolding model should reduce to its deterministic counterpart as a boundary case:

$$\Pr(Y_{aj} = 1) = \frac{1}{1 + |\delta_j - \beta_a|^{\gamma_j}}, \tag{3}$$

where the (positive) power parameters γ_j moderates the strength of the proximity relation on the choice probability whenever the distance between the ideal point and the item position differs from 0 or 1. When γ_j is large, the relationship between the response probability and the absolute difference between the ideal point and the item position approaches a step function. An item is chosen with certainty when its position coincides with the one of the ideal point.

Instead of estimating a different ideal point position for each person, we assign individuals to classes such that members of an LC are indistinguishable in their predicted response behavior. This approach is less restrictive than it may initially seem. The number of responses, which typically is much smaller than the sample size, limits possible distinctions that can be made among the individuals (see Lindsay et al., 1991, for a related situation). As a result, the class-specific probabilities are written as

$$\pi_{Y_j|X(t)}(1) = \frac{1}{1 + |\delta_j - \beta_t|^{\gamma_j}}. \tag{4}$$

The unfolding constraint imposed on the conditional probabilities reduces the number of effective parameters from JT in the unconstrained LC model to $(J + T - 1)$ in the unfolding LC model, because T ideal point positions and $J - 1$ item parameters must be estimated.

B. Example Continued

The LC unfolding model shown in Equation (4) with an equal power parameter for all items yields a poor fit with $G^2 = 64.6$ (df = 19). A residual analysis indicated that the power parameters of items B and E are rather distinct from the remaining items. When a separate power parameter was estimated for these two items, the fit of the model improved considerably with $G^2 = 23.2$ (df = 17). Thus, in comparison with the unconstrained four-class model, the number of effective parameters is reduced by nine while at the same time a satisfactory fit of the data is provided. The predicted frequencies (together with the frequencies of the unconstrained LC model) are given in Table 1, and the parameter estimates are listed in Table 2.

A graphical summary of the results is provided in Figure 1. This figure depicts both the tracelines for the five items and the four ideal points, which are represented by triangles. The areas of the triangles are roughly

Table 2. Parameter Estimates of the Attitude toward Nuclear Energy Data Set

	Items						
Class	A	B	C	D	E	$\hat{\pi}_{X(t)}$	$\hat{\beta}_t$
t							
1	0.56	0.45	0.10	0.06	0.01	0.03	−2.92
2	0.51	0.60	0.98	0.70	0.08	0.46	−1.02
3	0.14	0.35	0.79	0.99	0.91	0.47	−0.12
4	0.04	0.20	0.13	0.26	0.94	0.03	1.16
$\hat{\delta}_j$	−2.00	−1.74	−0.76	−0.26	0.54		
$\hat{\gamma}_j$	2.84	1.28	2.84	2.84	5.66		

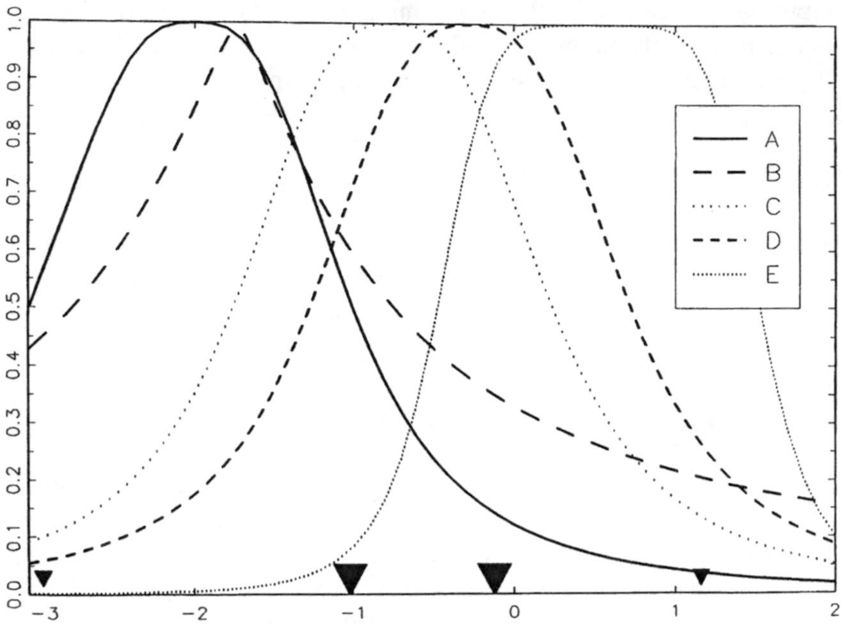

Figure 1. Item tracelines and ideal point positions for the four-class solution.

proportional to the class sizes. We note that item B does not discriminate as well as the other items among the latent ideal point classes. In contrast, item E covers a wider region of endorsement than the other items and separates well between the first two and the last two ideal point classes. The great majority of the respondents (94%) endorse items C and D or items C, D, and E.

Overall, these results show that both the items and respondents can be ordered along an unidimensional continuum ranging from passive acceptance of nuclear energy to active opposition against nuclear power stations. In addition, it was also found that items B and E differ in the way they discriminate among the groups. Obviously, item B does not elicit much disagreement regardless of one's position toward nuclear energy. However, the endorsement of item E seems possible only if one opposes the use of nuclear energy.

4. INCOMPLETE AND PARTIAL RANKING DATA

This section discusses how constrained LC models can be used for the analysis of partial and incomplete ranking data. In many applications it is not of interest or feasible to obtain a complete rank ordering of all options.

Instead, respondents may be asked to order only their k ($k \leq J$) favorite options. The method of first choices is an extreme example of the partial ranking task in which respondents are asked to identify the most preferred option only. Incomplete ranking data are obtained when respondents are instructed to rank a subset of the options. For example, in a paired comparison task with J items, a person is presented with each of the $\binom{J}{2}$ item pairs and asked to pick the more preferred one. Clearly, partial and/or incomplete rankings provide important alternatives to complete rankings because they simplify the judgmental task and provide additional information about the consistency of the respondents in their comparisons of the options.

The next sections demonstrate that constrained LC models are well suited to take into account both individual preference differences and the various response formats of a ranking task. We first focus on models for the analysis of paired comparison data (Formann, 1992) and then on (incomplete) ranking models (Croon, 1989).

A. Paired Comparisons

In a paired comparison task, persons are asked to assess the differences between items. An unconstrained LC model is of little use in the analysis of paired comparison data because it estimates the response probabilities for the $[J(J-1)/2]$ item pairs without taking into account that each response is based on a comparison between two of the J items. As a result, this model cannot provide information about the separate effects of each item. However, this information can be extracted from the data by constraining the class-specific probabilities to be a function of a choice model that captures the probabilistic nature of the decision-making process. A prominent model in this context is the one by Luce (1959), who assumed that the probability of choosing an item is proportional to its (positive) utility, which is invariant over different sets of choice alternatives. Consequently, the probability that person a in a comparison of items i and j prefers item i is given by

$$\Pr(Y_{aij} = 1) = \omega_{ia}/(\omega_{ia} + \omega_{ja}),$$

where ω_{ia} represents the utility of item i for person a, $\sum_i \omega_{ia} = 1$, and $0 < \omega_{ia} < 1$. Instead of estimating the item utilities for each person separately, we assign individuals to different classes such that members of an LC are indistinguishable in their perceived utilities of the items (Formann, 1992). As a result, the class-specific probabilities are expressed as

$$\pi_{Y_{ij}|X(t)}(1) = \omega_{it}/(\omega_{it} + \omega_{jt}). \tag{5}$$

Latent classes are a result of interindividual variability in the item utilities. We can further investigate sources of this variability by testing whether individuals are homogeneous in their perception of the items. This hypothesis can be tested by constraining an item's utility to be a function of the distance between its location and a person's ideal point position. Adopting Hoijtink's (1990) model, Böckenholt and Böckenholt (1990) proposed this unfolding extension by setting

$$\omega_{it} = |\beta_t - \delta_i|^{-\gamma}.$$

The smaller the distance between the ideal point position of class t and the location of item i, the higher is item i's utility for members of class t. As a result, the probability that item i is preferred to item j by members of class t is written as

$$\pi_{Y_{ij}|X(t)}(1) = \frac{|\beta_t - \delta_i|^{-\gamma}}{|\beta_t - \delta_i|^{-\gamma} + |\beta_t - \delta_j|^{-\gamma}}. \tag{6}$$

B. Incomplete Rankings

The same approach can be applied to incomplete ranking data by following Luce's (1959) suggestion to decompose a ranking into a set of independent first choices. Let the ordering outcome i, j, k denote that option i is ranked first, option j is ranked second, and option k is ranked last. The probability of observing this outcome in class t is given by

$$\pi_{Y_{ijk}|X(t)} = \frac{\omega_{it}}{\omega_{it} + \omega_{jt} + \omega_{kt}} \frac{\omega_{jt}}{\omega_{jt} + \omega_{kt}}. \tag{7}$$

An interesting extension of this model is obtained when the scale values are allowed to depend on the stage of the ranking process. For example, when $J = 3$, the scale values that are ranked at the second positions are not constrained to be equal to the scale values of the options ranked first, and

$$\pi_{Y_{ijk}|X(t)} = \frac{\omega_{it}^{(1)}}{\omega_{it}^{(1)} + \omega_{jt}^{(1)} + \omega_{kt}^{(1)}} \frac{\omega_{jt}^{(2)}}{\omega_{jt}^{(2)} + \omega_{kt}^{(2)}}. \tag{8}$$

This model requires the estimation of $(J - 1)^2$ parameters for each latent class. As in the context of paired comparison data, we may constrain the item utilities ω_{jt} to be a function of the distance between the ideal point position of class t and item j,

$$\pi_{Y_{ijk}|X(t)} = \frac{|\beta_t - \delta_i|^{-\gamma}}{|\beta_t - \delta_i|^{-\gamma} + |\beta_t - \delta_j|^{-\gamma} + |\beta_t - \delta_k|^{-\gamma}} \times \frac{|\beta_t - \delta_j|^{-\gamma}}{|\beta_t - \delta_j|^{-\gamma} + |\beta_t - \delta_k|^{-\gamma}}. \quad (9)$$

This model requires the estimation of $J - 2$ item parameters, J ideal points, and $J - 1$ class sizes.

The decomposition of an item's (dis)utility into an ideal point and an item parameter is nontrivial in the sense that the unfolding representation assumes that individual differences in the evaluation of the items can be explained solely by differences in the ideal point positions. Thus, although Luce's (1959) model allows for $J!$ possible rank orders of the items in the population, the *joint* scale estimated from the unidimensional unfolding model is only consistent with $[\binom{J}{2} + 1]$ of those. This shows that the unfolding model may be considerably more restrictive than Luce's model in an analysis of incomplete and/or partial ranking data because it requires individuals to be homogenous in their perception of the items. It is therefore useful to fit both Luce's model and the unfolding model because the former model provides an important benchmark for the latter.

5. PARTY PREFERENCES IN GERMANY

Ideal point models are frequently applied in the analysis of party preferences to determine the dominant underlying dimensions that voters use in choosing among parties. As pointed out by Norporth (1979), experts may select these dimensions on the basis of ideological or policy-defined issues. However, it is less clear whether average voters apply related criteria in their assessment of party similarities.

To investigate this question, party preference orders of the German electorate are analyzed with models (7) and (9). The data are taken from two preelection surveys conducted in 1972 and 1980. In both surveys, random samples of German voters were asked to state their preferences for the following five parties: CDU (Christian democrats), SPD (social democrats), FDP (liberals), NPD (neosocialists), and DKP (communists). Although complete rankings of the parties are available, the following investigation is restricted to the analysis of the first two choices to reduce the size of the data set. Table 3 contains these partial rankings, which are taken from Pappi (1983). The data from 1972 were obtained by asking the respondents to rank the parties according to their preferences. In contrast, the 1980 data were derived from pairwise preference comparisons of the five parties. Pappi (1983) reported that the paired comparison data were highly consistent; only 2.5% of the voters gave intransitive responses. Previous analyses of these data sets seem to indicate that the German parties cannot be represented on an unidimensional unfolding scale. Instead, several researchers concluded that voters differ both with respect to the perception of the parties and the criteria used in their preference judgments. For example, according to Klingemann (1972) and Pappi (1983), voters disagree on the position of FDP. Some voters represent this party to the right of CDU; others place it between SPD and CDU, or to the left of SPD (see also Borg & Staufenbiel, 1993; Norpoth, 1979). However, these conclusions are based on results obtained from *deterministic* unfolding approaches, which do not account for inconsistencies in individual judgments and have only limited provisions for assessing the goodness-of-fit of a model. It is therefore of interest to determine whether these conclusions

Table 3. Party Preference Data and Model Predictions

Position		1972			1980		
1st	2nd	Obs.	Model (7)	Model (9)	Obs.	Model (7)	Model (9)
C	S	458	457.0	473.1	172	172.8	165.0
C	F	237	234.2	214.2	303	304.4	309.6
C	N	21	21.9	22.0	15	13.6	13.4
C	D	1	0.9	0.8	0	0.1	0.4
S	C	303	303.0	296.4	89	88.6	85.5
S	F	617	613.4	620.4	497	495.5	494.3
S	N	4	5.5	3.6	0	0.3	0.3
S	D	14	16.8	17.0	13	13.2	13.0
F	C	42	45.7	52.4	79	76.9	80.0
F	S	73	76.7	73.7	139	140.9	144.6
F	N	2	0.3	0.5	0	0.6	0.6
F	D	0	0.2	0.2	2	0.9	1.0
N	C	2	3.9	4.9	2	2.9	2.9
N	S	3	1.1	1.2	1	0.3	0.3
N	F	1	0.3	0.5	0	0.5	0.5
N	D	0	0.0	0.0	0	0.0	0.0
D	C	0	0.2	0.2	0	0.1	0.1
D	S	5	1.8	1.7	2	2.9	3.0
D	F	0	0.2	0.2	1	0.7	0.8
D	N	0	0.0	0.0	0	0.0	0.0

are still valid when both preference heterogeneity and judgmental variability are taken into account. In addition, by using the likelihood ratio (LR) test, we have a more rigorous criterion for assessing the goodness-of-fit of a scaling model than is available from a deterministic analysis.

A. Results from Luce's Ranking Model

The number of distinct preference orders in the data sets was determined by fitting the latent-class versions of Luce's ranking model to the data. At least two classes were expected that favor one of the two dominant parties, CDU and SPD. It is therefore not surprising that according to Table 4, the one-class model does not fit the 1972 data ($G^2 = 602.5$, df = 15) and the 1980 data ($G^2 = 544.6$, df = 15). However, substantial improvements are obtained from the two-class model. We also fitted model (8), which allows for different scale values for parties listed at the first and second positions. However, this model did not yield a fit that is significantly different from model (7) with equal scale values at both rank positions.

The scale values (multiplied by 1,000) of the two classes are given in Table 5. Both classes are of approximately equal size and differ predominantly in their evaluation of the three major parties CDU, SPD, and FDP. In the first class of the 1972 data, SPD is most preferred, followed by FDP and CDU, and in the second class CDU is most preferred, followed by SPD and FDP. Interestingly, this latter relationship changes in the 1980 data because preferences increase strongly for the FDP. Note also that for both data sets the CDU class prefers the NPD over the DKP, whereas the SPD class shows the opposite pattern. Thus, although very little information about the preferences for these two parties is available in these data sets, the LC solutions indicate a clear ordering.

Overall, the interpretation of the results obtained by the LC ranking model is straightforward. In 1972, the SPD was the most popular party and in coalition with the FDP. In contrast, the ordering of the parties in the CDU class may reflect the fact that SPD and CDU used to be coalition

Table 4. Goodness-of-Fit Statistics of the LC Ranking Model

	1972 Data		1980 Data	
Class	G^2	df	G^2	df
1	602.5	15	544.6	15
2	14.5	10	6.1	10
3	8.9	5	5.2	5

Table 5. Parameter Estimates of the LC Ranking Model

	1972 Data		1980 Data	
Party	**Class 1**	**Class 2**	**Class 1**	**Class 2**
CDU	28	824	21	789
SPD	892	111	779	71
FDP	77	59	195	134
NPD	1	6	0	6
DKP	2	0	5	0
Size	0.53	0.47	0.54	0.46

partners until 1969. In 1980, the FDP and SPD continued to be aligned. However, the increase in the appeal of the FDP to CDU voters reflects the efforts of this party to position itself for a future coalition with the CDU, which was the most popular party in 1980.

B. Results from Unfolding Analysis

Because two classes were obtained for both data sets, it is of interest to determine whether the left–right dimension (DKP, SPD, FDP, CDU, NPD) is a dominant part of the voters' perceptual space when they ordered the parties according to their strength of preference. To obtain a comparable scale for both parties, the power parameter γ was set equal to 2. The unfolding model (9) yielded an LR tests of $G^2 = 32.9$ (df = 13) and $G^2 = 43.3$ (df = 13) for the 1980 and 1972 data sets, respectively. The item and ideal point positions for both data sets are depicted in Figure 2 by squares and triangles, respectively. The parties NPD and DKP take on the most extreme positions because very few respondents listed them as their first choice. Overall, the order of the parties seems to follow the postulated left–right dimension; however, the positions of SPD and FDP are interchanged in the 1972 data. Thus, voters may have considered more than just the ideological dimension when evaluating the parties. Norporth (1979) mentioned that in 1972 the FDP and CDU differed strongly on religious issues, which may account for the FDP–SPD–CDU order.

The fit of the unfolding models can be improved by estimating the power parameter γ from the data. For the 1980 data, we obtain $G^2 = 7.5$ (df = 12) when $\hat{\gamma} = 3.82$. For the 1972 data set, $G^2 = 18.6$ (df = 12) when the γ parameter of the DKP is allowed to vary freely ($\hat{\gamma}_{DKP} = 4.05$). This indicates that preferences for the DKP are less strongly influenced by random characteristics of the voting situation than preferences for

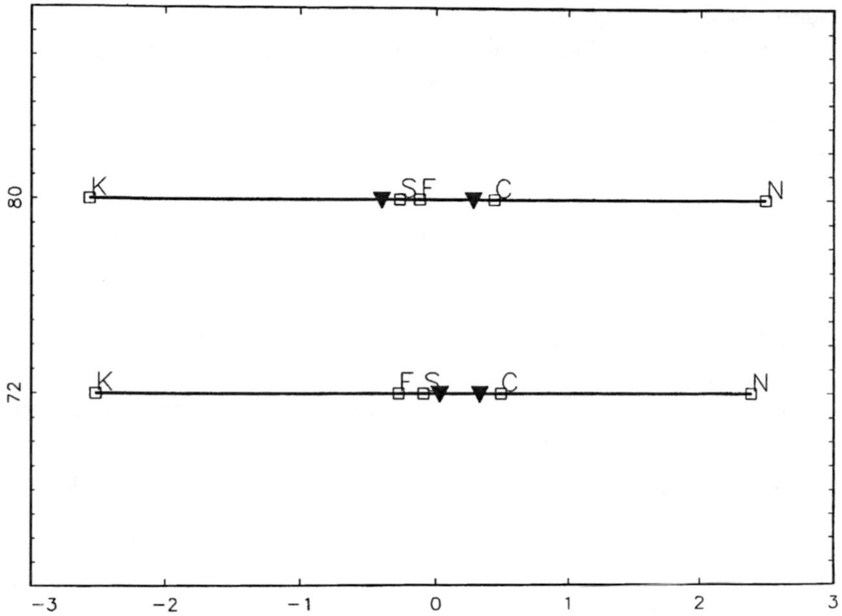

Figure 2. Joint representation of item and ideal point positions for the 1972 and 1980 data sets.

the other parties. The expected frequencies of both models are given in Table 3. We note that the unfolding model for the 1972 data seems to systematically underpredict the occurrence of the (CDU, FDP) ranking. The order of the estimated scale values for the five parties under both models is the same as depicted in Figure 2. Unfortunately, distances between the party and ideal point positions are no longer directly comparable when the power parameter γ varies among the parties. Thus, in this case it is not meaningful to display simply the item and ideal point positions as done in Figure 2.

In summary, these analyses showed that two voter groups (of approximately equal size) can be identified who differ in their evaluation of the parties. This result may not be surprising considering that the two extreme parties NPD and DKP have very little support in this data set. However, it is of interest to note that the rather restrictive choice model by Luce (1959) is sufficient to describe the within-group variability in the ordering of the parties. Thus, there are no noticeable similarity relationships among the parties after the two voter groups are accounted for. It was also found that the parties and voter groups can be ordered along a unidimensional continuum. This continuum was at least partially compatible with the hypothesized ideological dimension, indicating that voters do consider the left–right framework in their evaluation of the parties.

6. EXTENSIONS

This chapter discussed the synthesis of the LC model and choice models for pick any/J and (partial or incomplete) ranking data by combining the LC model with the choice models by Luce (1959) and Coombs (1964), respectively. The implementation of Coombs' unfolding approach has the additional advantage of yielding graphical displays that summarize the data concisely and are easy to interpret.

The selection of these choice models was motivated by the applications reported in this chapter. Because of their parsimonious structure, it is clear that in other applications it may prove necessary to apply more complex choice models that address the multidimensional structure and/or similarity relationships of the items. For example, Böckenholt and Böckenholt (1991) investigated preferences for magazines in a pick any/J study and found that the following r-dimensional unfolding model provided a significantly better fit than its unidimensional counterpart:

$$\pi_{Y_j|X(t)}(1) = 1 - \Phi\left[\tau_j + \sum_{h=1}^{r}(\beta_{th} - \delta_{jh})^2\right],$$

where the normal cumulative distribution function is denoted by Φ and τ_j is an item-specific threshold value [see Equation (2)]. In particular, when the number of items is large, it is likely that the perceptual space of the items is multidimensional. In addition to the distinction between unidimensional and multidimensional representations of the items, it is important to take into account perceived similarity relationships among the items on the individual level. Thurstonian choice models are well suited to represent these effects. Applications of Thurstonian models in a constrained LC framework are reported by DeSoete and Carroll (1992) for paired comparison data and by Böckenholt (1993) for ranking data.

Because many decision problems are faced not once but repeatedly (e.g., choice of a residential location, selection of a travel mode for a trip to work, purchase of consumer brands within a product class), it is also of interest to consider extensions of the LC approach for the investigation of intertemporal choices. Applications by Böckenholt and Langeheine (1996) and Böckenholt (1998) demonstrate that time-dependent LC models yield valuable information about decision-making processes and their inherently dynamic nature that is not available when modeling individual choices at a particular point in time. A crucial element of this work is that it takes into account the fact that individuals may differ both in their preferences and in the way they change their preferences over time. As a result, it is possible to test a rich set of hypotheses about the loci of change in latent preferences and systematic variations in the relationship between perception and choice over time (Loewenstein and Elster, 1992).

A further powerful feature of the restricted LC approach is its ability to allow for covariates in an analysis of choice data (see Chapter 8). Individual-level covariates may provide additional information about the latent variables and can thus improve the accuracy in assigning respondents to their respective classes and aid in interpreting differences among the classes. The introduction of continuous covariates in the basic LC model was first proposed by Dayton and Macready (1988; see also van der Heijden, Dessens, and Böckenholt, 1996). Using the paired comparison model given by Equation (5), Dillon, Kumar, and de Borrero (1993) applied this approach in an analysis of a food preference study. Similarly, using model (7), Kamakura, Wedel, and Agrawal (1994) took into account covariate information in a conjoint measurement study. In both applications, the covariates provided important additional insights about the relationships between the individual- and aggregate-level data.

7. CONCLUSION

One major problem in the analysis of choice data is to account for heterogeneity caused by individual taste differences. It is well known in the field of preference analysis that individuals may perceive and evaluate the options among which they choose in very different ways. Latent class models are well suited to account for these taste differences by allowing for subpopulations with distinctly different preferences.

The basic strategy for developing a constrained LC model is to identify the fundamental elements of the choice process at the individual level, specify the parameters that can vary in the populations, and link the individual-level choice and LC model to describe the aggregate behavior. By this explicit formulation of the relationship between individual- and group-level data, it is possible to validate the various components of the model separately. The synthesis of individual-level choice models and LC models provides a general and flexible approach that is well suited for modeling preference differences. If past progress in this area is any indication, many new applications of this approach can be expected in the near future.

REFERENCES

Andrews, R. L., & Manrai, A. K. (1999). "MDS maps for product attributes and market response: an application to scanner panel data," *Marketing Science*, **18**, 584–604.

Bartholomew, D. J. (1998). "Scaling unobservable constructs in social science," *Applied Statistics*, **47**, 1–13.

Ben-Akiva, M., McFadden, D., Abe, M., Böckenholt, U., Bolduc, D., Gopinath, D., Morikawa, T., Ramaswamy, V., Rao, V., Revelt, D., & Steinberg, D. (1997). "Modeling methods for discrete choice analysis," *Marketing Letters*, **8**, 273–86.

Böckenholt, U. (1993). "Applications of Thurstonian models to ranking data." In M. Fliegner & J. Verducci (eds.), *Probability Models and Statistical Analyses for Ranking Data*. New York: Springer, pp. 157–72.

Böckenholt, U. (1998). "Modeling time-dependent preferences: drifts in ideal points." In M. Greenacre & J. Blasius (eds.), *Visualization of Categorical Data*. Hillsdale, NJ: Erlbaum, pp. 461–76.

Böckenholt, U. (2001). "Mixed-effects analyses of rank-ordered data," *Psychometrika*, **66**, 45–62.

Böckenholt, U., & Böckenholt, I. (1990). "Modeling individual differences in unfolding preference data," *Applied Psychological Measurement*, **14**, 257–69.

Böckenholt, U., & Böckenholt, I. (1991). "Constrained latent class analysis: simultaneous classification and scaling of discrete choice data," *Psychometrika*, **56**, 699–716.

Böckenholt, U., & Langeheine, R. (1996). "Latent change in recurrent choice data," *Psychometrika*, **61**, 285–302.

Borg, I., & Staufenbiel, T. (1993). *Theorien und Methoden der Skalierung*. Bern: Huber.

Clogg, C. C. (1988). "Latent class models for measuring." In R. Langeheine & J. Rost (eds.), *Latent Trait and Latent Class Models*. New York: Plenum, pp. 173–205.

Clogg, C. C. (1995). "Latent class models: recent developments and prospects for the future." In G. Arminger, C. C. Clogg, & M. E. Sobel (eds.), *Handbook of Statistical Modeling on the Social Sciences*. New York: Plenum, pp. 311–59.

Coombs, C. H. (1964). *A Theory of Data*. New York: Wiley.

Croon, M. A. (1989). "Latent class models for the analysis of rankings." In G. de Soete, H. Feger, & K. C. Klauer (eds.), *New Developments in Psychological Choice Modelling*. Amsterdam: Elsevier Science, pp. 99–121.

Croon, M. A. (1990). "Latent class analysis with ordered latent classes," *British Journal of Mathematical and Statistical Psychology*, **43**, 171–92.

Dayton, C. M., & Macready, G. B. (1988). "Concomitant variable latent class models," *Journal of the American Statistical Association*, **83**, 173–8.

DeSoete, G., & Carroll, D. (1992). "Probabilistic multidimensional models of pairwise choice data." In G. Ashby (ed.), *Probabilistic Multidimensional Models of Perception and Cognition*. Hillsdale, NJ: Erlbaum, pp. 61–88.

Dillon, W. R., Böckenholt, U., de Borrero, M. S., Bozdogan, H., DeSarbo, W., Gupta, S., Kamakura, W., Kumar, A., Ramaswamy, V., & Zenor, M. (1994).

"Issues in the estimation and application of latent structure models of choice," *Marketing Letters*, **5**, 323–34.

Dillon, W. R., Kumar, A., & de Borrero, M. (1993). "Capturing individual differences in paired comparisons: an extended BTL model incorporating descriptor variables," *Journal of Marketing Research*, **30**, 42–51.

Formann, A. K. (1988). "Latent class models for nonmonotone items," *Psychometrika*, **53**, 45–62.

Formann, A. K. (1992). "Linear logistic latent class analysis for polytomous data," *Journal of the American Statistical Association*, **87**, 476–86.

Formann, A. K., & Kohlmann, T. (1998). "Structural latent class models," *Sociological Methods and Research*, **26**, 530–65.

Hoijtink, H. (1990). "A latent trait model for dichotomous choice data," *Psychometrika*, **55**, 641–56.

Kamakura, W. A., Wedel, M., & Agrawal, J. (1994). "Concomitant variable latent class models for conjoint analysis," *International Journal for Research in Marketing*, **11**, 451–64.

Kim, J.-S., & Böckenholt, U. (2000). "Modeling stage-sequential change in ordered categorical responses," *Psychological Methods*, **5**, 380–400.

Klingemann, H. D. (1972). "Testing the left–right continuum in a sample of German voters," *Comparative Political Studies*, **5**, 93–106.

Lazarsfeld, P. F. (1950). "The logical and mathematical foundation of latent structure analysis." In E. A. Suchman, P. F. Lazarsfeld, S. A. Starr, & J. A. Clausen (eds)., *Studies in Social Psychology in World War II, Vol. 4. Measurement and Prediction*. Princeton NJ: Princeton University Press, pp. XXX–XXX.

Lenk, P. J., & DeSarbo, W. S. (2000). "Bayesian inference for finite mixtures of generalized linear models with random effects," *Psychometrika*, **65**, 93–119.

Lindsay, B., Clogg, C., & Grego, J. (1991). "Semiparametric estimation in the Rasch model and related exponential response models, including a simple model for item analysis," *Journal of the American Statistical Association*, **86**, 96–107.

Loewenstein, G. F., & Elster, J. (1992). *Choice Over Time*. New York: Russell Sage Foundation.

Luce, R. D. (1959). *Individual Choice Behavior*. New York: Wiley.

McAdams, S., Winsberg. S., Donnadieu, S., DeSoete, G., & Krimphoff, J. (1995). "Perceptual scaling of synthesized musical timbres – common dimensions, specificities, and latent subject classes," *Psychological Research*, **58**, 177–92.

Meiser, T., & Ohrt, B. (1996). "Modeling structure and chance in transitions: mixed latent partial Markov-chain models," *Journal of Educational and Behavioral Statistics*, **21**, 91–109.

Norporth, H. (1979). "The parties come to order! Dimensions of preferential choice in the West German electorate, 1971–1976," *The American Political Science Review*, **73**, 724–36.

Pappi, F. U. (1983). "Die links–rechts-Dimension des Deutschen Parteiensystems und die Parteipräferenz-Profile der Wählerschaft." In M. Kaase & H. D. Klingemann (eds.), *Wahlen und Politisches System, Analysen aus Anlass der Bundestagswahl 1980.* Opladen: Westdeutscher Verlag, pp. 422–41.

Proctor, C. H. (1970). "A probabilistic formulation and statistical analysis of Guttman scaling," *Psychometrika*, **35**, 73–8.

Rost, J., & Langeheine, R. (1997). *Applications of Latent Trait and Latent Class Models in the Social Sciences*. Münster: Waxmann Verlag.

van der Heijden, P. G. M., Dessens, J., & Böckenholt, U. (1996). "Estimating the concomitant variable latent class model with the EM algorithm," *Journal of Educational and Behavioral Statistics*, **21**, 215–29.

Wedel, M., & Kamakura, W. A. (1999). *Market Segmentation: Conceptual and Metodological Foundations*. Boston: Kluwer Academic.

NOTE

1. This research was partially supported by the National Science Foundation under Grant SBR-9409531. I am grateful to Ingwer Borg for helpful discussions on the party preference data set. Requests for reprints should be sent to Ulf Böckenholt, Department of Psychology, University of Illinois at Urbana-Champaign, Champaign, IL 61820.

SEVEN

Three-Parameter Linear Logistic Latent Class Analysis

Anton K. Formann and Thomas Kohlmann

1. INTRODUCTION

The three-parameter linear logistic latent class model (3P-LCM) combines three well-known concepts in psychometrics: First, the concept of a discrete distribution for the subjects' latent ability or attitude, whereby the latent distribution is expressed by internally homogeneous classes (Lazarsfeld, 1950; Lazarsfeld and Henry, 1968); second, the concept of logistic regression for binary data (Cox, 1970; Cox and Snell, 1989), which traces back observed proportions to a smaller number of explanatory variables; and third, Birnbaum's (1968) concept of item-response theory, which allows different discriminatory power for the items as well as guessing.

The first and the second concepts have already been brought together in linear logistic latent class analysis (Formann, 1982, 1985, 1989, 1995), leading, among others, to the latent class/Rasch model; this is a latent class-model in which the latent response probabilities are restricted according to Rasch's (1960) model. Subsequently, this model has been shown to almost always become equivalent to the conditional maximum-likelihood approach in the Rasch model, provided that the number of classes is at least half the number of items (de Leeuw and Verhelst, 1986; Follmann, 1988; Lindsay, Clogg, and Grego, 1991). The 3P-LCM introduced in the following paragraphs can therefore be considered a generalization of the latent class/Rasch model in that it is enriched by discrimination and guessing parameters.

2. UNCONSTRAINED LATENT CLASS ANALYSIS

Assume that the subjects S_ν, $\nu = 1, \ldots, N$, responded to the dichotomous items $I_i, i = 1, \ldots, k$. Let the response $a_{\nu i}$ of subject S_ν to item I_i as a realization of random variable $A_{\nu i}$ be coded 1 in the case of a positive (correct or affirmative) response and 0 in the case of a negative (incorrect or disapproving) response. With k items, there are 2^k possible response patterns. These can be represented by the set of vectors $\boldsymbol{a}_s = [a_{s1}, \ldots, a_{si}, \ldots, a_{sk}]$, $s = 1, \ldots, 2^k$, whose elements assume values of 1 or 0. The frequencies with which these response patterns are observed in a given sample are denoted by n_s, $s = 1, \ldots, 2^k$:

$$a_s = \begin{Bmatrix} 0 & 0 & \cdots & 0 \\ 1 & 0 & \cdots & 0 \\ 0 & 1 & \cdots & 0 \\ 1 & 1 & \cdots & 0 \\ & \vdots & & \\ 1 & 1 & \cdots & 1 \end{Bmatrix}; \quad n_s = \begin{Bmatrix} n_1 \\ n_2 \\ n_3 \\ n_4 \\ \cdots \\ n_{2^k} \end{Bmatrix}. \tag{1}$$

Analyzing these data, latent class analysis (LCA) intends to identify homogeneous subgroups (latent classes) of subjects that differ from each other by their response behavior with respect to the items under consideration. Denote the number of latent classes by m and their relative sizes by π_j, $j = 1, \ldots, m$, with

$$\sum_{j=1}^{m} \pi_j = 1, \qquad 0 < \pi_j \leq 1, \tag{2}$$

and the conditional probability of a positive response of subject S_ν to item I_i given membership in class C_j with

$$\pi(A_{\nu i} = 1 \mid S_\nu \in C_j) = \pi_{i|j}. \tag{3}$$

Then, assuming stochastic independence within classes, the conditional probability of response pattern $\boldsymbol{a}_s$ given class C_j equals the product of that pattern's individual response probabilities:

$$\pi(\boldsymbol{a}_s \mid C_j) = \prod_{i=1}^{k} \pi_{i|j}^{a_{si}} \left(1 - \pi_{i|j}\right)^{1-a_{si}}. \tag{4}$$

Weighting Equation (4) with the class sizes and summing up the classes results in the unconditional probability of response pattern $\boldsymbol{a}_s$,

$$\pi(\boldsymbol{a}_s) = \sum_{j=1}^{m} \pi_j \prod_{i=1}^{k} \pi_{i|j}^{a_{si}} \left(1 - \pi_{i|j}\right)^{1-a_{si}}. \tag{5}$$

For estimating the unknown parameters π_j and $\pi_{i|j}$, the maximum-likelihood (ML) method is most often used, starting from the log likelihood

$$\log L \propto \sum_{s=1}^{2^k} n_s \log \pi(\boldsymbol{a}_s). \tag{6}$$

Before proceeding further, note that the notation used in this section and throughout the rest of the paper differs in certain respects from the traditional notation as introduced, for example, in the seminal papers by Leo A. Goodman and Clifford C. Clogg. One reason is that we focus exclusively on dichotomous observed variables so that we can simplify the usual notation considerably. Furthermore, the equations for the linear logistic parameterization of LCA, which will be introduced in the next section, would appear rather complex if the traditional ways of notation were used. Nevertheless, all elements of our notation correspond exactly to respective elements in the traditional notation (some examples of the correspondence between the notational styles are $\pi_j \equiv \pi^X$; $\pi_{1|j} \equiv \pi^{A|X}$; $\pi_{2|j} \equiv \pi^{B|X}$; $\pi_{3|j} \equiv \pi^{C|X}$; $\pi_{4|j} \equiv \pi^{D|X}$; $\pi(\boldsymbol{a}_s \mid C_j) \equiv \pi^{ABCD|X}$; $\pi(\boldsymbol{a}_s) \equiv \pi^{ABCD}$).

3. LINEAR LOGISTIC LATENT CLASS ANALYSIS

The logistic version of LCA reparameterizes the latent response probabilities $\pi_{i|j}$ according to

$$\log\left[\pi_{i|j}/(1-\pi_{i|j})\right] = x_{ij}, \tag{7}$$

so that

$$\pi_{i|j} = \exp(x_{ij})/[1+\exp(x_{ij})]. \tag{8}$$

In addition, linear logistic LCA (LOGLCA) subjects the log odds x_{ij} to linear constraints of more elementary parameters $\lambda_t^{(x)}$, $t = 1, \ldots, r^{(x)}$, that is,

$$x_{ij} = \sum_{t=1}^{r^{(x)}} q_{ijt}^{(x)} \lambda_t^{(x)}. \tag{9}$$

The weights $\boldsymbol{Q}^{(x)} = [q_{ijt}^{(x)}], i = 1, \ldots, k, j = 1, \ldots, m, t = 1, \ldots, r^{(x)}$, are assumed to be given constants, so that the elementary parameters $\lambda^{(x)}$ are conceived as fixed effects. Note that $\boldsymbol{Q}^{(x)}$ restates the hypothesis that the researcher wants to test with respect to the latent response probabilities. Notice that this parameterization corresponds closely to the usual specification of logistic regression models in which the probability of a "success" is related to a set of categorical and/or continuous explanatory variables.

Using the linear logistic parameterization, unconstrained LCMs as described in Section 2 can be easily specified. Moreover, LOGLCA provides a very flexible and convenient framework for imposing parameter constraints that result in meaningful models for scaling dichotomously scored data. Some of these models will be discussed in the following sections.

4. RASCH AND BIRNBAUM MODELS ASSUMING CONTINUOUS LATENT TRAITS

The Rasch model (RM) is one of the most widely used "latent trait" methods for scaling dichotomous items. As the RM originated primarily in educational testing, it seems to be natural to introduce some of its basic features by referring to this field of application. It should be noted, however, that the RM and its extensions that we introduce here are not confined to educational testing, but can be used successfully in many other areas of data analysis.

The RM assumes that the probability of a correct response to a given test item depends on the examinee's "ability" and on the "easiness" of the item. The greater the examinee's ability (assumed to be a continuous quantity) and the easier the item, the higher the probability that the examinee will answer this item correctly. This dependence of the probability of a correct response can be depicted by using a graphical display that is called the Item Characteristic Curve (ICC). The RM assumes the ICC to be of a specific form: the logistic function. Two logistic ICCs are shown in Figure 1, one of which represents an easy item (Item I_1) and the other represents a more difficult item (Item I_2).

Denoting by θ_ν the (unobserved) ability of subject S_ν, and by β_i the easiness of item I_i, the probability $\pi_{\nu i}$ that subject S_ν answers item I_i correctly is given by

$$\pi_{\nu i} = \frac{\exp(\theta_\nu + \beta_i)}{1 + \exp(\theta_\nu + \beta_i)}, \qquad \nu = 1, \dots, N, \quad i = 1, \dots, k. \tag{10}$$

Consequently, all ICCs have exactly the same shape; they differ only with respect to their location on the horizontal axis. The location of each item is given by its easiness parameter β; hence, the RM is called the *one-parameter logistic model.*

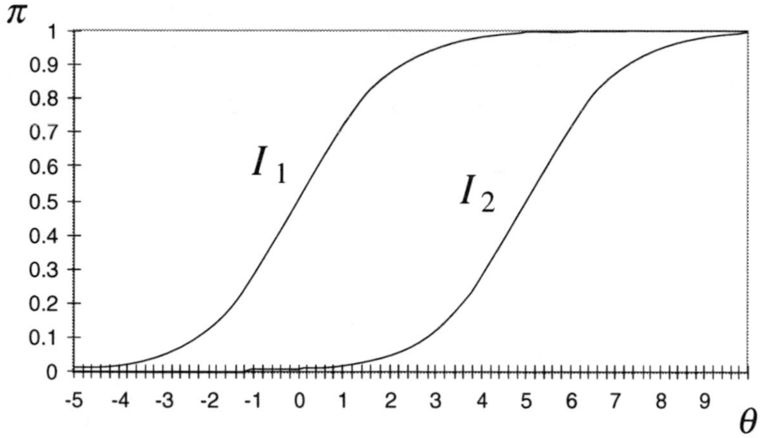

Figure 1. Logistic ICCs of two items with different easiness, showing the probability π of a correct response as a function of the ability θ.

With only one parameter describing the ICCs, the RM is a very parsimonious model. Yet, as a consequence of this parsimony, it sometimes does not fit the data very well. Therefore, several extensions of the RM have been proposed that allow the ICCs to differ in other respects than just their location on the horizontal axis. In these extensions, parameters are added to the RM that result in a two- or even three-parameter logistic model (sometimes called the *Birnbaum model*). The first of these extended models presented here is the model that includes "discrimination" parameters α_i:

$$\pi_{\nu i} = \frac{\exp[\alpha_i(\theta_\nu + \beta_i)]}{1 + \exp[\alpha_i(\theta_\nu + \beta_i)]}, \qquad \nu = 1, \ldots, N, \quad i = 1, \ldots, k. \tag{11}$$

By the inclusion of discrimination parameters, the items are allowed to differ also with respect to the steepness or slope of their ICCs. In Figure 2, ICCs for two items having different values of the α parameters are displayed. For a rough interpretation of the discrimination parameters, suppose there are two subjects, S_1 and S_2, who differ in their ability by a certain quantity $\Delta = \theta_2 - \theta_1$. Item I_1 having a steeper ICC will usually discriminate better between the two subjects than item I_2. This is because the difference in the probability of a correct response associated with a difference in ability of Δ will be more pronounced if the item has a steeper ICC (see Figure 2).

The second extension of the RM we present incorporates, "guessing" parameters γ_i. Whereas in the usual RM the ICC asymtotically

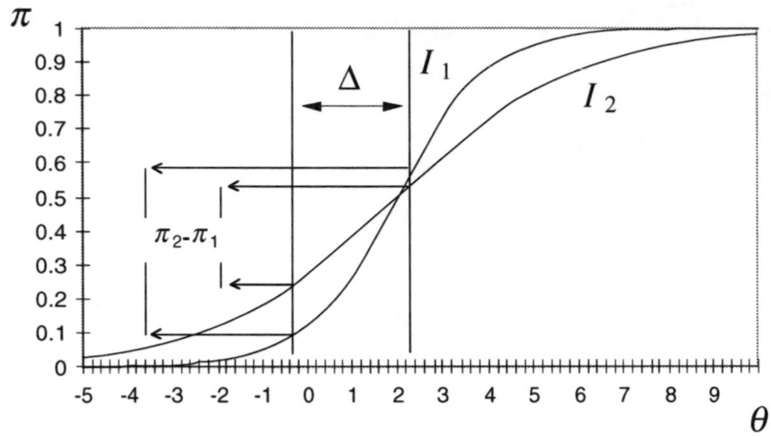

Figure 2. ICCs for two items with different discrimination. Item I_1 discriminates better between two subjects with different abilities (Δ) because the difference in probability of success ($\pi_2 - \pi_1$) expected under the model is greater for this item than for item I_2.

approaches 0 with decreasing ability, inclusion of a guessing parameter results in a lower bound of the ICC that may be greater than 0. Models of this type nicely refer to their origin in educational testing: an examinee with even a very low ability has a certain chance to chose the correct answer just by randomly selecting one of the given alternatives (or by using other "alternative" strategies, such as cheating). Thus, with decreasing ability, the ICC approaches a value of $\pi_{\nu i}$ that is greater than 0 and equal to the probability of success that can be achieved independently of ability simply by guessing. In Figure 3, three ICCs with different values of the guessing parameter (but equal easiness and discrimination) are shown; the model equation of the two-parameter model with guessing parameters is given by

$$\pi_{\nu i} = \frac{\gamma_i + \exp(\theta_\nu + \beta_i)}{1 + \exp(\theta_\nu + \beta_i)}, \qquad \nu = 1, \ldots, N, \quad i = 1, \ldots, k. \qquad (12)$$

Combining the two extensions of the RM introduced so far results in a three-parameter model that allows ICCs to differ with respect to both the steepness of their slopes and their lower bounds. The specification equation of the three-parameter logistic model is given by

$$\pi_{\nu i} = \frac{\gamma_i + \exp[\alpha_i(\theta_\nu + \beta_i)]}{1 + \exp[\alpha_i(\theta_\nu + \beta_i)]}, \qquad \nu = 1, \ldots, N, \quad i = 1, \ldots, k. \qquad (13)$$

This concludes our brief introduction to one-, two-, and three-parameter logistic models for dichotomously scored items. Additional

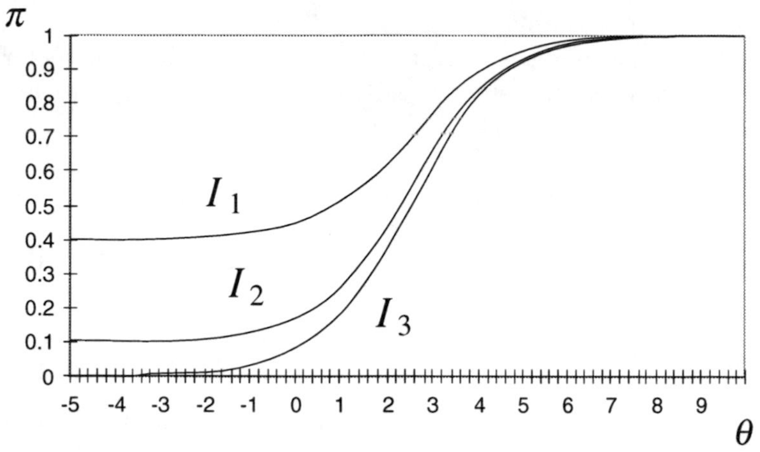

Figure 3. Three ICCs with different guessing parameters ($\gamma_1 = 0.4, \gamma_2 = 0.1, \gamma_3 = 0.0$).

information may be found, for example, in Baker (1992) and in the volumes edited by Fischer and Molenaar (1995), and by van der Linden and Hambleton (1997).

5. LATENT CLASS RASCH AND BIRNBAUM MODELS

In all models described in the previous section, the parameters θ_ν refer to the examinees' abilities, which are located somewhere on the latent continuum. Suppose now that the examinees' abilities are not spread continuously over the whole range of values on the latent ability continuum, but that N subjects belong to, say, m mutually exclusive and exhaustive latent classes $C_1, \ldots, C_j, \ldots, C_m$, each of which is located at some ability level $\theta_1, \ldots, \theta_j, \ldots, \theta_m$. In this case, all subjects in a given class C_j are assumed to have the same ability θ_j. This assumption of *located latent classes* can be applied to all the logistic models for continuous latent traits we introduced earlier, thus resulting in latent class counterparts of the RM and the two- and three-parameter models. If the linear logistic parameterization of LCA in Equations (8) and (9) is compared with the RM and Birnbaum models in Equations (10)–(13), we can easily see that LOGLCA appears to be particularly suitable for the formal specification of latent class variants of these models. It will be shown later how even a generalization of the models for continuous latent traits can be obtained by defining them in terms of LOGLCA.

Assuming that the position of each item I_i and of each class C_j on the underlying latent continuum is described by a single location parameter, β_i and θ_j, respectively, the latent class/Rasch model (LC/RM; Formann, 1985; see also Clogg, 1988, 1995) is given by

$$\pi_{i|j} = \frac{\exp(\theta_j + \beta_i)}{1 + \exp(\theta_j + \beta_i)}, \qquad i = 1, \ldots, k, \; j = 1, \ldots, m. \qquad (14)$$

and the latent class/three-parameter Birnbaum model (LC/3P-BM) is given by

$$\pi_{i|j} = \frac{\gamma_i + \exp[\alpha_i(\theta_j + \beta_i)]}{1 + \exp[\alpha_i(\theta_j + \beta_i)]}, \qquad i = 1, \ldots, k, \; j = 1, \ldots, m. \qquad (15)$$

Thus, for $\gamma_i = 0, i = 1, \ldots, k$, the LC/3P-BM becomes the latent class/two-parameter Birnbaum model (LC/2P-BM) with differential discrimination; for both $\gamma_i = 0$ and $\alpha_i = 1, i = 1, \ldots, k$, it reduces to the LC/RM, and for $\alpha_i = 1$ but $\gamma_i > 0$ one gets the two-parameter model with guessing.

Coming back to the LC/3P-BM, note that two side conditions of the parameters,

$$\sum_{i=1}^{k} \beta_i = 0 \text{ or } \beta_l = 0, \quad l \in \{1, \ldots, k\}, \qquad (16)$$

$$\prod_{i=1}^{k} \alpha_i = 1 \text{ or } \alpha_l = 1, \quad l \in \{1, \ldots, k\}, \qquad (17)$$

reduce their number, and that in addition, the restrictions

$$0 \leq \gamma_i \leq 1, \quad i = 1, \ldots, k, \qquad 0 \leq \alpha_i, \quad i = 1, \ldots, k, \qquad (18)$$

must be fullfilled (these restrictions also apply to the continuous latent trait models of Section 4). With an eye toward parameter estimation, we find that restriction (18) suggests we transform γ_i according to the logistic function and that we transform α_i according to the exponential function.

Let the two parameters be denoted by γ_i^* and α_i^*, respectively; then $\gamma_i^* = \log[\gamma_i/(1 - \gamma_i)]$ and $\alpha_i^* = \log(\alpha_i), i = 1, \ldots, k$, that is

$$\gamma_i = \exp(\gamma_i^*)/[1 + \exp(\gamma_i^*)], \qquad (19)$$

$$\alpha_i = \exp(\alpha_i^*), \qquad (20)$$

do hold. With these reparameterizations it is now easy to state the linear logistic generalization of the LC/3P-BM. This three-parameter linear logistic latent class model (3P-LCM) assumes for the latent response probabilities

$$\pi_{i|j} = \frac{z_{ij} + \exp(y_{ij}x_{ij})}{1 + \exp(y_{ij}x_{ij})}, \qquad i = 1, \ldots, k, \quad j = 1, \ldots, m \qquad (21)$$

with the following constraints:

1. For the guessing parameters z_{ij}, as a generalization of Equation (19),

$$z_{ij} = \frac{\exp\left[\sum_{i=1}^{r^{(z)}} q_{ijt}^{(z)} \lambda_t^{(z)}\right]}{1 + \exp\left[\sum_{i=1}^{r^{(z)}} q_{ijt}^{(z)} \lambda_t^{(z)}\right]}, \qquad i = 1, \ldots, k, \quad j = 1, \ldots, m. \qquad (22)$$

2. For the discrimination parameters y_{ij}, as a generalization of Equation (20),

$$y_{ij} = \exp\left[\sum_{i=1}^{r^{(y)}} q_{ijt}^{(y)} \lambda_t^{(y)}\right], \qquad i = 1, \ldots, k, \quad j = 1, \ldots, m. \qquad (23)$$

3. For the location parameters x_{ij}, the linear constraints of LOGLCA are as stated by Equation (9).

Altogether, three structural matrices of weights, $\boldsymbol{Q}^{(x)}$, $\boldsymbol{Q}^{(y)}$, and $\boldsymbol{Q}^{(z)}$, determine from which unknown parameters $\lambda^{(x)}$, $\lambda^{(y)}$, and $\lambda^{(z)}$ to what extent the latent response probabilities depend. Whereas with respect to the location parameters x_{ij} their linear decomposition makes it possible to test structural hypotheses (e.g., the hypothesis $x_{ij} = \theta_j + \beta_i$ of the LC/RM; or more general hypotheses concerning β_i, e.g., that β_i equals the weighted sum of operation difficulties; cf. Formann, 1982, and the linear logistic test model, LLTM, of Fischer, 1973, 1983), the linear decompositions of y_{ij} and z_{ij} are motivated by another reason: they allow a test for the equality of subsets of y_{ij} and z_{ij} by setting weights of the respective matrices, $\boldsymbol{Q}^{(y)}$ and $\boldsymbol{Q}^{(z)}$, appropriately. For example, to test the hypothesis that across items and classes guessing is always the same, $\boldsymbol{Q}^{(z)}$ reduces to the column unit vector of length *km*; and to test the hypothesis that the Items I_1 and I_2 have the same discrimination, which moreover does not change with classes, the constraints $y_{11} = y_{21} = \cdots = y_{1m} = y_{2m} = \lambda_i^{(y)}$, $t \in \{1, \ldots, r^{(y)}\}$ must be imposed.

Note that all of the three types of parameters $\lambda^{(x)}$, $\lambda^{(y)}$, and $\lambda^{(z)}$ are not indexed by j (classes) and i (items). In principle, the 3P-LCM, as a generalization of the latent class/Birnbaum model, therefore allows the testing of a huge variety of submodels because restrictions can be set across items and/or classes. In reality, however, identifiability conditions (see Section 7) as well as substantive reasons reduce this variety to a smaller number of typical submodels. Because, in general, structural hypotheses concerning guessing and discrimination can hardly be formulated a priori, real linear constraints of y_{ij} and z_{ij} seem to be of little relevance. Similarly, some types of equality restrictions of y_{ij} and z_{ij} seem to be of more interest – namely, those in which guessing and discrimination depend on the items, but not on the classes (cf. Birnbaum's two- and three-parameter logistic models) – than others. Nevertheless, the generally formulated 3P-LCM gives us a chance to deal with all of the submodels mentioned previously as special cases regarding parameter estimation, identifiability of parameters, and goodness of fit. In summary, the 3P-LCM may handle the following types of latent class models for dichotomous data: (i) unconstrained LCA; (ii) LCA with fixed class sizes and/or fixed latent response probabilities; (iii) LCA with equated class sizes and/or equated latent response probabilities; (iv) LOGLCA (no guessing, equal discrimination), including the special case of LC/RM; (v) LC/RM with guessing; (vi) LC/2P-BM (different discrimination, no guessing); (vii) LC/3P-BM (different discrimination plus guessing); (viii) special cases of (iv), (v), (vi), and (vii) resulting from equality restrictions for some parameters; (ix) mixtures of some models listed here; and (x) LCA with arbitrary constraints as specified by Equations (9) and (21)–(24), equating certain parameters to each other or to prespecified values.

The two following Sections (6 and 7) cover technical aspects of parameter estimation, assessment of model fit, and problems of identifiability of parameters. As some of these topics are slightly more specialized, they may be omitted at a first reading.

6. PARAMETER ESTIMATION, GOODNESS OF FIT

The log-likelihood of the sample data, as given by Equation (6), is a function of the unknown parameters $\lambda^{(x)}, \lambda^{(y)}$, and $\lambda^{(z)}$, and further depends on the unknown class sizes π_j, $j = 1, \ldots, m$. As before in the case of the other matrices of weights and their associated parameters, weights $\boldsymbol{Q}^{(w)}$ and parameters $\lambda^{(w)}$ are introduced. The elements of $\boldsymbol{Q}^{(w)} = [(q_{jt}^{(w)})]$, $j = 1, \ldots, m, t = 1, \ldots, r^{(w)}$ are assumed to be given constants expressing the hypothesis to be tested, now with respect to the class sizes.

For the class sizes, π_j, the logistic representations

$$\pi_j = \exp\left[\sum_{t=1}^{r^{(w)}} q_{jt}^{(w)} \lambda_t^{(w)}\right] \Bigg/ \left\{\sum_{l=1}^{m} \exp\left[\sum_{t=1}^{r^{(w)}} q_{lt}^{(w)} \lambda_t^{(w)}\right]\right\}, \qquad j = 1, \ldots, m \tag{24}$$

are introduced, especially to open the possibility of imposing equality restrictions for the π_j. With the additional restrictions of Equation (24) for the class sizes, the log-likelihood of the 3P-LCM becomes

$$\log L \propto \sum_{s}^{2^k} n_s \log \pi_s(\lambda), \tag{25}$$

with $\lambda = [\lambda^{(w)}; \lambda^{(x)}, \lambda^{(y)}, \lambda^{(z)}]$ and π_s instead of $\pi(\boldsymbol{a}_s)$.

Differentiating Equation (25) and setting it equal to 0 gives the ML equations

$$\begin{aligned}
\frac{\partial \log L}{\partial \lambda_t^{(w)}} &= \sum_{s=1}^{2^k} \frac{n_s}{\pi_s} \frac{\partial \pi_s}{\partial \lambda_t^{(w)}} = 0, \quad t = 1, \ldots, r^{(w)}, \\
\frac{\partial \log L}{\partial \lambda_t^{(x)}} &= \sum_{s=1}^{2^k} \frac{n_s}{\pi_s} \frac{\partial \pi_s}{\partial \lambda_t^{(x)}} = 0, \quad t = 1, \ldots, r^{(x)}, \\
\frac{\partial \log L}{\partial \lambda_t^{(y)}} &= \sum_{s=1}^{2^k} \frac{n_s}{\pi_s} \frac{\partial \pi_s}{\partial \lambda_t^{(y)}} = 0, \quad t = 1, \ldots, r^{(y)}, \\
\frac{\partial \log L}{\partial \lambda_t^{(z)}} &= \sum_{s=1}^{2^k} \frac{n_s}{\pi_s} \frac{\partial \pi_s}{\partial \lambda_t^{(z)}} = 0, \quad t = 1, \ldots, r^{(z)},
\end{aligned} \tag{26}$$

with

$$\begin{aligned}
\frac{\partial \pi_s}{\partial \lambda_t^{(w)}} &= \sum_{j=1}^{m} \pi_j \pi(s \mid j) \left[q_{jt}^{(w)} - \sum_{l=1}^{m} w_l q_{lt}^{(w)}\right], \\
\frac{\partial \pi_s}{\partial \lambda_t^{(x)}} &= \sum_{j=1}^{m} \pi_j \pi(s \mid j) \sum_{i=1}^{k} q_{ijt}^{(x)} y_{ij} \pi_{i|j}^{*} (a_{si} - \pi_{i|j}) / \pi_{i|j}, \\
\frac{\partial \pi_s}{\partial \lambda_t^{(y)}} &= \sum_{j=1}^{m} \pi_j \pi(s \mid j) \sum_{i=1}^{k} q_{ijt}^{(y)} x_{ij} y_{ij} \pi_{i|j}^{*} (a_{si} - \pi_{i|j}) / \pi_{i|j}, \\
\frac{\partial \pi_s}{\partial \lambda_t^{(z)}} &= \sum_{j=1}^{m} \pi_j \pi(s \mid j) \sum_{i=1}^{k} q_{ijt}^{(z)} z_{ij} (a_{si} - \pi_{i|j}) / \pi_{i|j},
\end{aligned} \tag{27}$$

where

$\pi(s \mid j) = \pi(\boldsymbol{a}_s \mid C_j)$, $j = 1, \ldots, m$, $s = 1, \ldots, 2^k$, according to Equation (4);

π_j, $j = 1, \ldots, m$, according to Equation (24) as functions of $\lambda^{(w)}$;

x_{ij}, $i = 1, \ldots, k$, $j = 1, \ldots, m$, according to Equation (9) as functions of $\lambda^{(x)}$;

y_{ij}, $i = 1, \ldots, k$, $j = 1, \ldots, m$, according to Equation (23) as functions of $\lambda^{(y)}$;

z_{ij}, $i = 1, \ldots, k$, $j = 1, \ldots, m$, according to Equation (22) as functions of $\lambda^{(z)}$;

$\pi_{i|j}$, $i = 1, \ldots, k$, $j = 1, \ldots, m$, according to Equation (21) as functions of x_{ij}, y_{ij}, and z_{ij}; and

$\pi^*_{i|j} = \exp(y_{ij}x_{ij})/[1 + \exp(y_{ij}x_{ij})]$, $i = 1, \ldots, k$, $j = 1, \ldots, m$, which are the latent response probabilities without guessing.

As a way to estimate the parameters of the 3P-LCM, a FORTRAN program available for parameter estimation in LOGLCA for dichotomous data (source code published in Formann, 1984, pp. 215–52) was extended by inserting new parts for the guessing and discrimination parameters. This program uses a simple gradient method, a description of which can be found in Fischer (1974, pp. 252–5). In earlier applications, this gradient method was found to be rather stable with respect to convergence, but sometimes it is very slow in terms of the number of iterations. It shares this property with the expectation–maximization (EM) algorithm (Dempster, Laird, and Rubin, 1977; Wu, 1983), as well as the further property to require no second partial derivatives of the log likelihood. Parameters of the 3P-LCM can also be estimated by using a GAUSS program for derivative-free optimization, which is available from the second author; some of the submodels of the 3P-LCM, in particular the LC/RM, can be estimated with Vermunt's (1997) program LEM.

The constraints are realized in the aforementioned FORTRAN program in a more general form compared with the model Equations (9) and (21)–(24) to also allow fixation of the model parameters and of the original parameters π_j and $\pi_{i|j}$. Both fixations play an important role for hypothesis testing. Of special interest are fixations of the original parameters $\pi_{i|j}$ to the boundary 0 or 1, arising, for example, in cases when one or more $\lambda^{(x)}$ or $\lambda^{(y)}$ parameters diverge during their estimation (terminal solutions for one or more latent response probabilities), and in cases in which more than one group of persons should be analyzed simultaneously by multisample LCA (Clogg and Goodman, 1984). Fixations of parameters, especially of z (guessing) and y (discrimination), allow to specify restricted models such as the LC/RM with guessing (equal discrimination, all $y_{ij} = 1$) and the LC/2P-BM (no guessing, all $z_{ij} = 0$).

Providing all types of constraints for the class sizes and the latent response probabilities gives the modified specification equations of the 3P-LCM.

1. For the class sizes,

$$\pi_j = \begin{cases} \pi_j^* & \text{if } \pi_j \text{ should be fixed to the predetermined value } \pi_j^* \\ \dfrac{\left(1 - \sum_l \pi_l^*\right) \exp\left[\sum_t q_{jt}^{(w)} \lambda_t^{(w)}\right]}{\sum_l \exp\left[\sum_t q_{lt}^{(w)} \lambda_t^{(w)}\right]} & \text{otherwise} \end{cases}, \tag{28}$$

   with $\pi_j^* = 0$ if π_j should not be fixed.
2. For the latent response probabilities,

$$\pi_{i|j} = \begin{cases} \pi_{i|j}^* & \text{if } \pi_{i|j} \text{ should be fixed to the predetermined value } \pi_{i|j}^* \\ \dfrac{z_{ij} + \exp(y_{ij} x_{ij})}{1 + \exp(y_{ij} x_{ij})} & \text{otherwise} \end{cases}, \tag{29}$$

   with

$$z_{ij} = \begin{cases} z_{ij}^* & \text{if } z_{ij} \text{ should be fixed to the predetermined value } z_{ij}^* \\ \dfrac{\exp\left[\sum_t q_{ijt}^{(z)} \lambda_{it}^{(z)}\right]}{1 + \exp\left[\sum_t q_{ijt}^{(z)} \lambda_{it}^{(z)}\right]} & \text{otherwise} \end{cases},$$

$$y_{ij} = \begin{cases} y_{ij}^* & \text{if } y_{ij} \text{ should be fixed to the predetermined value } y_{ij}^* \\ \exp\left[\sum_t q_{ijt}^{(y)} \lambda_{it}^{(y)}\right] & \text{otherwise} \end{cases}.$$

Having estimated the parameters of a model that was locally identifiably (see Section 7), the acceptability of this model can be evaluated by using likelihood-ratio (G^2) or Pearson (X^2) goodness-of-fit statistics. Both G^2 and X^2 are asymptotically X^2 distributed with degrees of freedom df,

$$\text{df} = 2^k - r - 1, \tag{30}$$

with $r = r^{(w)} + r^{(x)} + r^{(y)} + r^{(z)}$ being the number of estimated parameters. In the case in which two competing models M_1 and M_2 assuming r_1

and r_2 parameters, respectively, with $r_1 > r_2$ and M_2 nested in M_1, should be compared, this may be done provided that M_1 holds by means of the likelihood-difference statistic $^*G^2 = G^2(M_2) - G^2(M_1)$ associated with $\mathrm{df} = r_1 - r_2$.

For more details concerning goodness-of-fit tests in latent class models, see Goodman (1974); Haberman (1979, p. 562); and, especially for testing the number of classes (i.e., the number of components in a mixture), Titterington et al. (1985, Chapters 1.2.2 and 5.4); and McLachlan and Basford (1988, Chapters 1.10 and 3.5).

7. NOTES ON IDENTIFIABILITY

Looking at specification Equation (21) makes it evident that the 3P-LCM is overparameterized as compared with its unconstrained latent class analog having the same number of classes, m. Unconstrained LCA has km latent response probabilities; thus, the total number $r^* = r^{(x)} + r^{(y)} + r^{(z)}$ of parameters $\lambda^{(x)}$, $\lambda^{(y)}$, and $\lambda^{(z)}$ describing these latent response probabilities must be (*Condition 1*)

$$r^* \leq \begin{cases} km & \text{in the case that the } m\text{-class unconstrained} \\ & \text{latent class model is identifiable, or} \\ u & \text{in the case that only } u \text{ of the } km \text{ latent} \\ & \text{response probabilities of the } m\text{-class} \\ & \text{unconstrained latent class model are} \\ & \text{identifiable.} \end{cases} \tag{31}$$

Introducing constraints given by Equation (9) for x, Equation (23) for y, and Equation (22) for z clearly leads to the trivial rank conditions (*Condition 2*) for the corresponding matrices of weights $\boldsymbol{Q}^{(x)}$, $\boldsymbol{Q}^{(y)}$, and $\boldsymbol{Q}^{(z)}$,

$$\mathrm{Rank}\left[\boldsymbol{Q}^{(x)}\right] = \boldsymbol{r}^{(x)}, \quad \mathrm{Rank}\left[\boldsymbol{Q}^{(y)}\right] = \boldsymbol{r}^{(y)}, \quad \mathrm{Rank}\left[\boldsymbol{Q}^{(z)}\right] = \boldsymbol{r}^{(z)}, \tag{32}$$

with $\boldsymbol{r}^{(x)}$, $\boldsymbol{r}^{(y)}$, and $\boldsymbol{r}^{(z)}$ being the number of $\lambda^{(x)}$, $\lambda^{(y)}$, and $\lambda^{(z)}$ parameters, respectively. Analogously, for the class sizes

$$\mathrm{Rank}\left[\boldsymbol{Q}^{(w)}\right] = r^{(w)} \tag{33}$$

must hold.

However, *Conditions 1* and *2* cannot guarantee that all of the $r = r^{(w)} + r^{(x)} + r^{(y)} + r^{(z)}\lambda$ parameters are identifiable. With the parameters having been estimated, in addition, their identifiability has to be checked by means of the Jacobian $\hat{\boldsymbol{J}}$,

$$\hat{\boldsymbol{J}} = \left\{ \left[\frac{\partial \pi_s(\hat{\lambda})}{\partial \lambda_i} \right] \right\}, \qquad s = 1, \ldots, 2^k, \; i = 1, \ldots, r, \tag{34}$$

with $\hat{\lambda} = [\hat{\lambda}^{(w)}; \hat{\lambda}^{(x)}, \hat{\lambda}^{(y)}, \hat{\lambda}^{(z)}]$. In the case that $\hat{\boldsymbol{J}}$ is of full rank r,

$$\text{Rank}\ (\hat{\boldsymbol{J}}) = \boldsymbol{r}. \tag{35}$$

(*Condition 3*) all parameters are locally identifiable, and the obtained solution of the ML equations corresponds to a (local) maximum of the likelihood (cf. McHugh, 1956; Goodman, 1974; Formann, 1985). Apart from the disadvantage that only local but not global identifiability can be checked numerically a posteriori but not in advance, one must be prepared for two further problems when applying LCMs: first, a solution to the ML equations may fail to maximize the likelihood, and second, multiple maxima of the likelihood may occur (Haberman, 1977, p. 1135; for an example of multiple maxima, see Formann, 1994).

For some information to be gained regarding identifiable submodels of the 3P-LCM, large synthetic data sets ($N = 1{,}000{,}000$) were generated. For each data set, $k = 6$ items and $m = 3$ classes as well as a certain submodel [analogous to the LC/3P-BM; see Equation (15)] with given parameters were assumed (see Table 1). Then, the response pattern probabilities $\pi_s, s = 1, \ldots, 2^k$ were calculated according to Equation (5), together with Equations (9) and (21)–(24). Setting $n_s = N\pi_s, s = 1, \ldots, 2^k$ resulted in the observed frequencies of the response patterns, so that each data set is completely conformed to the model used for its generation. In the following paragraphs, each model was applied to "its" data set.

The check for local identifiability by means of the Jacobian Equation (34) led to the results also contained in Table 1. In all cases, $k - 1$ item parameters β_i and m class parameters θ_j are identifiable [$\lambda^{(x)}$ parameters of the 3P-LCM; cf. Equation (9)]; in addition, $k - 1$ discrimination parameters α_i [$\lambda^{(y)}$ parameters of the 3P-LCM; cf. Equation (23)] are identifiable if the discrimination is assumed to be item specific, and $m - 1$ discrimination parameters α_j if the discrimination is assumed to be class specific. The results for the guessing parameters [$\lambda^{(z)}$ parameters of the 3P-LCM; cf. Equation (22)] are more complicated. If guessing is assumed to be class and item independent, then the single guessing parameter γ is identifiable. If guessing is assumed to be class specific, then only $m - 1$ of the guessing parameters γ_j are identifiable when the discrimination is assumed to be constant; but all m guessing parameters γ_j are identifiable when the discrimination is assumed to be different across classes or items. If guessing

Table 1. Synthetic Data ($k = 6$ items, $m = 3$ classes) – Submodels of 3P-LCM (Line 1) with π_j, β_i, θ_j, and Their Identifiability Conditions

	Discrimination		
Guessing	**Equal**	**Class Specific** $[\alpha^* = (0.5, 0, -0.5)]$	**Item Specific** $[\alpha^* = (-1.25, -0.75, -0.25, 0.25, 0.75, 1.25)]$
No	$\frac{\exp(\theta_j + \beta_i)}{1 + \exp(\theta_j + \beta_i)}$ —/— (LC/RM)	$\frac{\exp[\alpha_j(\theta_j + \beta_i)]}{1 + \exp[\alpha_j(\theta_j + \beta_i)]}$ —/2	$\frac{\exp[\alpha_i(\theta_j + \beta_i)]}{1 + \exp[\alpha_i(\theta_j + \beta_i)]}$ —/5 (LC/2P-BM)
Class and item independent $[\gamma = 0.1]$	$\frac{\gamma + \exp(\theta_j + \beta_i)}{1 + \exp(\theta_j + \beta_i)}$ 1/—	$\frac{\gamma + \exp[\alpha_j(\theta_j + \beta_i)]}{1 + \exp[\alpha_j(\theta_j + \beta_i)]}$ 1(0)/2	$\frac{\gamma + \exp[\alpha_i(\theta_j + \beta_i)]}{1 + \exp[\alpha_i(\theta_j + \beta_i)]}$ 1/5
Class specific $[\gamma = (0.3, 0.2, 0.1)]$	$\frac{\gamma_j + \exp(\theta_j + \beta_i)}{1 + \exp(\theta_j + \beta_i)}$ 2/—	$\frac{\gamma_j + \exp[\alpha_j(\theta_j + \beta_i)]}{1 + \exp[\alpha_j(\theta_j + \beta_i)]}$ 3(2)/2	$\frac{\gamma_j + \exp[\alpha_i(\theta_j + \beta_i)]}{1 + \exp[\alpha_i(\theta_j + \beta_i)]}$ 3(2)/5
Item specific $[\gamma = (0.3, 0.25, 0.2, 0.15, 0.1, 0.05)]$	$\frac{\gamma_i + \exp(\theta_j + \beta_i)}{1 + \exp(\theta_j + \beta_i)}$ 5/—	$\frac{\gamma_i + \exp[\alpha_j(\theta_j + \beta_i)]}{1 + \exp[\alpha_j(\theta_j + \beta_i)]}$ 6(5)/2	$\frac{\gamma_i + \exp[\alpha_i(\theta_j + \beta_i)]}{1 + \exp[\alpha_i(\theta_j + \beta_i)]}$ 5/5 (LC/3P-BM)

Note: $\pi_j = (0.2, 0.3, 0.5); \beta_i = (-5, -3, -2.5, -1.5, -0.5, 0); \theta_j = (-0.5, 1, 2)$; identification conditions are shown on line 2 (number of identifiable guessing/discrimination parameters).

is assumed to be item specific, then $k-1$ of the guessing parameters γ_i are identifiable in the case that the discrimination is assumed to be constant or item specific, but all k guessing parameters γ_i are identifiable when the discrimination is assumed to be class specific.

In some cases, however, the "practical" identifiability for the guessing parameters differs from their theoretical (local) identifiability. This discrepancy became visible when models with the smallest guessing parameter γ fixed to zero were used: the likelihood of the restricted model was nearly indistinguishable from the likelihood of the respective full model, even if in some cases the parameter estimates of the restricted model were slightly biased. The practical identifiability conditions (cf. Table 1; number of practically identifiable guessing parameters in parentheses, if applicable), established by using this principle, are therefore as follows: $k-1$ guessing parameters γ_i are identifiable if guessing is assumed to be item specific, and $m-1$ guessing parameter γ_j if guessing is assumed to be class specific; the one guessing parameter γ is identifiable if guessing is assumed to be item and class independent, except in the case that at the same time the discrimination is assumed to be class specific.

For the normalization of the item parameters β_i, see Equation (16); for the normalization of item-dependent discrimination parameters, see Equation (17), which analogously applies also to the case of class-dependent discrimination parameters; the guessing parameters must be normalized in that their smallest is set equal to 0, that is, $\min_i \gamma_i = 0$ or $\min_j \gamma_j = 0$ [through $z_{ij} = z^*_{ij} = 0$; cf. Equation (29)].

As a further control of practical identifiability, the synthetic data generated under the assumption of the LC/3P-BM were analyzed by using some misspecified, that is, simpler models. The results of these analyses are summarized in Table 2, showing that perfect fit is attained for the three-class unconstrained LCM (which is as general as the three-class LC/3P-BM), and clearly also for the LC/3P-BM itself, which is denoted by M1 in Table 2. A nearly perfect fit shows for M2 (class-specific guessing, item-specific discrimination), and nonsignificant test statistics result further for M3 (item-specific discrimination, guessing independent of items and classes) and M4 (class-specific discrimination, item-specific guessing). A poor fit is found for the remaining models M5 (item-specific guessing only; LC/RM + guessing), M6 (item-specific discrimination only; LC/2P-BM), and M7 (LC/RM).

Note that these results refer to a sample size of $N = 1{,}000{,}000$. As a way to give an impression regarding the more realistic sample size of $N = 1{,}000$, the goodness-of-fit statistics simply have been divided by 1,000

Table 2. Synthetic Data ($k = 6$ items, $m = 3$ classes) Generated under the Three-Class LC/3P-BM(=M1) – Likelihood-Ratio-Statistic (G^2) for Some Models

		No. of Parameters			G^2			
Model	**Classes**	**Guessing**	**Discrimination**	$\pi_{i\|j}$ **or** $\theta_j + \beta_i$	$N = 1{,}000{,}000$	$N = 1{,}000$	**df**	$X_{0.95}^2$
Unrestricted	1	—	—	6	402,377.66	402.38	57	75.62
	2	—	—	12	1,080.64	1.08	50	67.50
	3	—	—	18	0.00	0.00	43	59.30
M1	3	5	5	3 + 5	0.00	0.00	43	59.30
M2	3	2	5	3 + 5	0.61	0.00	46	62.83
M3	3	1	5	3 + 5	42.27	0.04	47	64.00
M4	3	5	2	3 + 5	6.38	0.01	46	62.83
M5	3	5	—	3 + 5	227.26	0.23	48	65.17
M6	3	—	5	3 + 5	5,794.41	5.79	48	65.17
M7	3	—	—	3 + 5	235,038.16	235.04	53	70.99

to be reinterpreted (see Table 2). Somewhat surprisingly, now the test statistics also allow one to accept nearly all simpler models as a sufficient description of the data, for example, even the two-class unconstrained latent class model. Only M7 (LC/RM) and the one-class unconstrained model (i.e., the independence model) must be rejected. From this it must be concluded that only rather massive misspecifications could be detected by the chi-squared goodness-of-fit tests, or, in other words, that rather large samples are needed when models of the present type must be differentiated. For LCMs in general, this recommendation is not new; see, for example, Espeland and Handelman (1989).

8. EXAMPLE: GRADED PAIN STATUS

Back pain is a common health problem in many industrialized countries, and population surveys regularly report considerable variation in the severity of this condition, ranging from short episodes of minor discomfort to chronic and severe pain (Raspe and Kohlmann, 1994). As no objective clinical tests for the assessment of pain and its severity exist, epidemiologic and clinical studies must rely on subjective reports of symptoms when they use scoring systems for grading the severity of back pain. These scoring systems usually include multiple indicators that are combined in certain ways to reflect a subject's graded pain status (for an example, see von Korff et al., 1992).

In an epidemiologic mail survey conducted in Lübeck, Germany (Michel et al., 1997), data on intensity of back pain (low/high intensity; I_1), pain-related disability in activities of daily living (absent/present; I_2), duration of current back pain episode ($\leq$3 months/$>$3 months; I_3), presence of pain in back-related regions of the body (neck, shoulder, hip; no/yes; I_4), and presence of pain in other regions of the body (head, face, arms, etc.; no/yes; I_5) were collected. Pain intensity and disability are well-accepted criteria for grading pain severity; data on pain in other regions of the body were included in this study to account for the concept that often "back pain is more than pain in the back." Persistence of pain for more than 3 months has been selected as a criterion because it represents the cut-off point between acute and chronic pain as defined by the International Association for the Study of Pain (IASP; see Merskey and Bogduk, 1994). However, the relevance of this criterion has been questioned because it appears to be too simplistic from a substantive point of view and does not sufficiently address the complexity of the process leading to chronic pain.

The analysis presented next investigates whether these five indicators may be used to develop a measurement model in which pain severity is treated as an unobserved variable composed of a certain number of classes located on a latent continuum. In this respect, it is of special interest how variable I_3, the IASP criterion for chronic pain, fits into this measurement model.

Unconstrained LCMs and models with located classes and items (with and without discrimination or guessing parameters) were estimated for data from 527 respondents (age 25–74 years) with a current episode of back pain (Table 3). Goodness-of-fit statistics show that the unconstrained two-class model just fails to fit the data, whereas the three-class model is well acceptable (Table 4). Thus, the three-class model was chosen as a starting point for estimating restricted models by using the linear logistic parameterization.

With only two exceptions, the latent response probabilities of the unconstrained three-class model suggest a monotonic ordering of the latent classes (Table 5), but the three-class LC/RM does not fit. After an inspection of the univariate marginals, it might be suspected that I_4 (back-related pain) and/or I_5 (other pain), because of their high prevalence, are subject to guessing. However, this is not guessing in the sense of educational testing, but rather is caused here by approaching a lower bound greater than 0 as a result of the high rate with which coexisting pain occurred. Although estimates of the parameters of the LC/RM with guessing by using normalization $\gamma_3 = 0$ indicate that guessing might play a role in answering $I_2(\gamma_2 = 0.101)$ and $I_4(\gamma_4 = 0.264)$, goodness-of-fit statistics lead to the rejection of this model.

Nevertheless, including item-specific discrimination parameters, that is, estimating the LC/2P-BM (normalization $\alpha_1 = 1$), results in an acceptable fit that is close to that of the unconstrained three-class model. In contrast to the unconstrained model, however, the LC/2P-BM provides a scoring scheme that locates respondents on a latent continuum. From the estimates of item easiness parameters β_i (normalization $\beta_5 = 0$), it can be seen that their order coincides with the observed item marginals. The locations of classes on the latent continuum (θ_j) are not strictly equidistant as $\theta_3 - \theta_2 > \theta_2 - \theta_1$. Interestingly, the doubtful relevance of the IASP criterion for chronic pain is demonstrated by the estimates of the discrimination parameters, with $I_3(\alpha_3 = 0.476)$ being the least discriminative indicator (for a graphical display of this result, see Figure 4).

From the results for the LC/2P-BM, further constraints have been imposed upon the model parameters. Because of the similarity of the

Table 3. Empirical Example: Response Patterns, Their Observed and Expected Frequencies under Models A and B, and Marginal Distributions

Response Pattern					Frequencies			Response Pattern					Frequencies		
						Expected								Expected	
I_1	I_2	I_3	I_4	I_5	Observed	A	B	I_1	I_2	I_3	I_4	I_5	Observed	A	B
0	0	0	0	0	38	37.76	38.33	1	0	0	0	0	4	7.67	5.57
0	0	0	0	1	18	17.82	18.84	1	0	0	0	1	3	3.91	9.09
0	0	0	1	0	44	44.08	42.72	1	0	0	1	0	15	11.35	11.03
0	0	0	1	1	74	73.69	68.79	1	0	0	1	1	43	40.55	47.82
0	0	1	0	0	5	6.95	6.87	1	0	1	0	0	2	1.56	1.27
0	0	1	0	1	6	6.72	6.04	1	0	1	0	1	4	2.94	3.37
0	0	1	1	0	11	8.12	9.71	1	0	1	1	0	2	3.62	3.38
0	0	1	1	1	28	27.77	25.47	1	0	1	1	1	28	30.50	26.68
0	1	0	0	0	10	7.05	7.03	1	1	0	0	0	3	1.75	1.50
0	1	0	0	1	10	9.05	8.12	1	1	0	0	1	6	5.47	4.90
0	1	0	1	0	5	8.23	11.42	1	1	0	1	0	5	5.39	4.48
0	1	0	1	1	36	37.43	36.97	1	1	0	1	1	54	56.80	51.38
0	1	1	0	0	2	1.30	1.46	1	1	1	0	0	1	0.53	0.43
0	1	1	0	1	3	3.41	2.88	1	1	1	0	1	3	4.12	2.28
0	1	1	1	0	0	1.52	3.25	1	1	1	1	0	3	3.14	1.55
0	1	1	1	1	15	14.10	17.09	1	1	1	1	1	46	42.72	47.26

Note: Univariate percentages are 42.1, 38.3, 30.2, 77.6, and 71.5 for I_1, I_2, I_3, I_4, and I_5, respectively. Model A is the unconstrained three-class model; Model B is the LC/2P-BM; marginal distributions are for the five observed variables.

Table 4. Empirical Example: Goodness-of-fit Statistics for Some Models

		No. of Parameters						
Model	**Classes**	**Guessing**	**Discrimination**	$\pi_{i\|j}$ **or** $\theta_j + \beta_i$	X^2	G^2	**df**	$X^2_{0.95}$
Unrestricted LCA	1	—	—	5	256.41	199.18	26	38.88
	2	—	—	10	32.42	36.48	20	31.41
	3	—	—	15	13.14	14.78	18[a]	28.87
LC/RM	3	—	—	$3+4$	40.71	45.12	22	33.92
LC/RM + guessing	3	4	—	$3+4$	35.95	40.33	18	28.87
LC/2P-BM	3	—	4	$3+4$	22.37	26.94	18	28.87
LC/2P-BM*	3	—	2	$1+2$	27.80	30.66	22	33.92

Note: X^2 = Pearson chi-squared statistic; G^2 = likelihood-ratio statistic. The LC/2P-BM* has restrictions $\alpha_1 = \alpha_2 = \alpha_4 = \alpha$ and $\theta_3 - \theta_2 = \theta_2 - \theta_1 = \Delta$.

[a] df is corrected because of four boundary solutions.

Table 5. Empirical Example: Class Sizes and Latent Response Probabilities for the Unconstrained Three-Class Model and the Three-Class LC/2P-BM, and Parameter Estimates for the LC/2P-BM

			Latent Response Prob. of Item					
Model	**Class**	**Class Size**	I_1	I_2	I_3	I_4	I_5	θ_j
Unconstrained LCA	C_1	0.261	0.164	0.157	0.156	0.539	0^a	
	C_2	0.361	0^a	0.337	0.274	0.801	1^a	
	C_3	0.378	1^a	0.584	0.429	0.912	0.938	
LC/2P-BM	C_1	0.200	0.096	0.134	0.140	0.450	0.101	−0.989
	C_2	0.585	0.363	0.333	0.265	0.812	0.821	0.692
	C_3	0.215	0.882	0.752	0.552	0.982	0.999	3.267
β_i			−1.253	−1.682	−2.832	0.788	0^b	
α_i			1^b	0.700	0.476	0.988	2.204	

[a] Boundary solution.
[b] normalized.

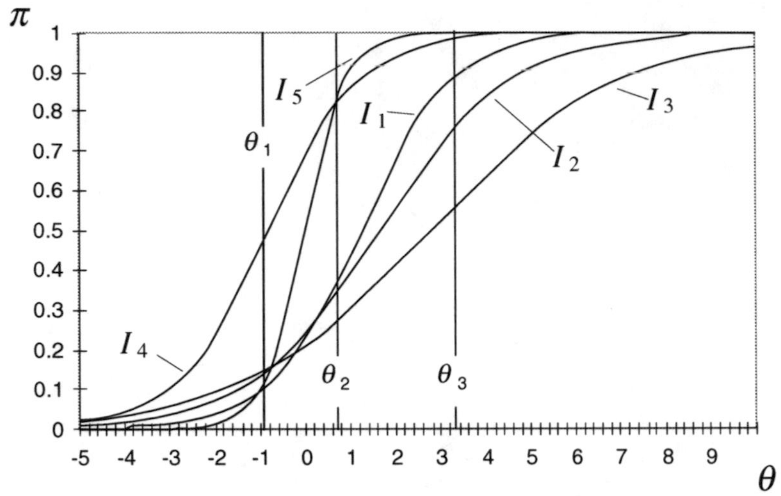

Figure 4. Fitting the LC/2P-BM to the pain data: Intersections of the vertical lines representing the ability levels θ_j with the five ICCs mark the latent response probabilities.

discrimination parameters for variables I_1, I_2, and I_4, equality of the respective parameters ($\alpha_1 = \alpha_2 = \alpha_4 = \alpha$) and, in addition, equidistance of latent class scores ($\theta_3 - \theta_2 = \theta_2 - \theta_1 = \Delta$) was specified. When both sets of constraints are imposed simulaneously, the resulting goodness-of-fit statistics indicate an acceptable fit (see LC/2P-BM* in Table 4). As the restricted model LC/2P-BM* is nested in LC/2P-BM, the acceptability of the constraints can also be evaluated by means of the likelihood-difference statistic $^*G^2$.

From these findings, it may be concluded that a three-class Birnbaum model provides an adequate model for scaling the five indicators of severity of back pain. On the latent level, the model assigns the respondents to three located classes representing different grades of severity. All five criteria discriminate between the classes, but duration of the current back pain episode contributes much less to the assignment than the other four criteria. From a substantive point of view, the latter result is of special importance. Although temporal aspects of pain such as persistence over a certain period of time have often been suggested as main indicators for evaluating the severity of pain, our results demonstrate that a widely accepted criterion (persistence > 3 months) discriminates poorly between various grades of severity. Persistence for longer than 3 months does fit in the scaling model for measuring severity of pain and may therefore be used as a valid indicator of this attribute. However, the other items included in the model provide more information about the location of a subject on the latent severity scale and would be reasonably retained in a reduced set of indicators. Grapical displays such as that in Figure 4 can be used to communicate such results even to groups with limited knowledge of the underlying statistical concepts.

In order for our results to be used in further analyses, each respondent could be assigned unequivocally to one class based on the modal probabilities associated with the observed response pattern (see, e.g., Formann, 1995, p. 241). These classes, along with the quantitative information about their location, could then be entered, for example, into the statistical analysis of group differences, association with other variables, or change over time. Another attractive option in this respect would be to include substantive hypotheses directly into the LCM. For this purpose, "structural latent class models" (Formann and Kohlmann, 1998) are available that can be specified by using the same linear logistic framework presented in this chapter.

9. FINAL REMARKS

The 3P-LCM assumes a specific sort of guessing. Let us consider as a simple special case the LC/RM plus guessing (γ), with the ICCs defined by

$$\pi_{i|j} = \frac{\gamma + \exp(\theta_j + \beta_i)}{1 + \exp(\theta_j + \beta_i)}. \tag{36}$$

For increasing θ_j (ability or attitude of class C_j) or increasing β_i (easiness or attractivity of item I_i), the effect of the guessing parameter γ on the latent probability of positive reactions, $\pi_{i|j}$, decreases. That is, in terms of the latent response probabilities, guessing is not assumed to be independent of classes (persons) and items, but becomes less relevant for large θ_j and β_i.

Other types of guessing are realized in the latent distance models (Lazarsfeld, 1950; Lazarsfeld and Henry, 1968; Proctor, 1970) and in Goodman's (1975) model with unscalable subjects. In the latent distance models, all subjects irrespective of their class membership have the same probability (item specific or not) for showing response errors. These are defined in relation to Guttman's (1950) deterministic scaling model: "intrusions" – which can be interpreted to be caused by guessing – occur when a positive reaction is observed instead of a negative reaction, which was expected according to the perfect deterministic scale; and "omissions" occur when a negative response is unexpectedly observed. In contrast to the latent distance models and the 3P-LCM, in which one and the same person may guess or not, Goodman's (1975) model with unscalable subjects postulates two basically differently reacting types of persons: the first conforming to Guttman's perfect scale, and the second being unscalable – possibly as a result of guessing. The combination of both approaches, the latent distance model with unscalable subjects (Dayton and Macready, 1980), allows both violations of the perfect scale: unscalable persons who show an unsystematic response behavior as well as response errors of the scalable persons.

The older scaling models of the latent class type mentioned previously do not explicitly model different discrimination of items (or classes = persons). Whereas Guttman's deterministic model and Goodman's model with unscalable persons assume that all items perfectly discriminate, in the latent distance models the item discrimination is in some sense confounded with guessing: the item discrimination, as the deviation from the perfect scale, is expressed by response errors – large response errors lead to a "flat" ICC; but the one type of response errors, the intrusions, can also be attributed to guessing. In contrast to that, the three-parameter Birnbaum model and the 3P-LCM try to separate guessing and discrimination.

REFERENCES

Baker, F. B. (1992). *Item Response Theory*. New York: Dekker.

Birnbaum, A. (1968). "Some latent trait models and their use in inferring an examinee's ability." In, F. M. Lord & M. R. Novick (eds.), *Statistical Theories of Mental Test Scores*. Reading, MA: Addison–Wesley, pp. 395–479.

Clogg, C. C. (1988). "Latent class models for measuring." In R. Langeheine & J. Rost (eds.), *Latent Trait and Latent Class Models*. New York: Plenum, pp. 173–206.

Clogg, C. C. (1995). "Latent class models." In G. Arminger, C. C. Clogg & M. E. Sobel (eds.), *Handbook of Statistical Modeling for the Social and Behavioral Sciences*. New York: Plenum, pp. 311–59.

Clogg, C. C., & Goodman, L. A. (1984). "Latent structure analysis of a set of multidimensional contingency tables," *Journal of the American Statistical Association*, **79**, 762–71.

Cox, D. R. (1970). *The Analysis of Binary Data*. London: Methuen.

Cox, D. R., & Snell, E. J. (1989). *Analysis of Binary Data*. London: Chapman & Hall.

Dayton, C. M., & Macready, G. B. (1980). "A scaling model with response errors and intrinsically unscalable respondents," *Psychometrika*, **45**, 343–56.

de Leeuw, J., & Verhelst, N. (1986). "Maximum likelihood estimation in generalized Rasch models," *Journal of Educational Statistics*, **11**, 183–96.

Dempster, A. P., Laird, N. M., & Rubin, D. B. (1977). "Maximum likelihood from incomplete data via the EM algorithm," *Journal of the Royal Statistical Society, B*, **39**, 1–38.

Espeland, M. A., & Handelman, S. A. (1989). "Using latent class models to characterize and assess relative error in discrete measurements," *Biometrics*, **45**, 587–99.

Fischer, G. H. (1973). "The linear logistic test model as an instrument in educational research," *Acta Psychologica*, **37**, 359–74.

Fischer, G. H. (1974). *Einführung in die Theorie Psychologischer Tests.* [*Introduction to Mental Test Theory.*] Bern: Huber.

Fischer, G. H. (1983). "Logistic latent trait models with linear constraints," *Psychometrika*, **48**, 3–26.

Fischer, G. H., & Molenaar, I. W., Eds. (1995). *Rasch Models – Foundations, Recent Developments, and Applications*. New York: Springer.

Follmann, D. A. (1988). "Consistent estimation in the Rasch model based on nonparametric margins," *Psychometrika*, **53**, 553–62.

Formann, A. K. (1982). "Linear logistic latent class analysis," *Biometrical Journal*, **24**, 171–90.

Formann, A. K. (1984). *Die Latent-Class-Analyse* [*Latent Class Analysis*]. Weinheim: Beltz.

Formann, A. K. (1985). "Constrained latent class models: theory and applications," *British Journal of Mathematical and Statistical Psychology*, **38**, 87–111.

Formann, A. K. (1989). "Constrained latent class models: some further applications," *British Journal of Mathematical and Statistical Psychology*, **42**, 37–54.

Formann, A. K. (1994). "Measurement errors in caries diagnosis: some further latent class models," *Biometrics*, **50**, 865–71.

Formann, A. K. (1995). "Linear logistic latent class analysis and the Rasch model." In, G. H. Fischer & I. W. Molenaar (eds.), *Rasch Models – Foundations, Recent Developments, and Applications*. New York: Springer, pp. 239–55.

Formann, A. K., & Kohlmann, T. (1998). "Structural latent class models," *Sociological Methods & Research*, **26**, 530–65.

Goodman, L. A. (1974). "Exploratory latent structure analysis using both identifiable and unidentifiable models," *Biometrika*, **61**, 215–31.

Goodman, L. A. (1975). "A new model for scaling response patterns: an application of the quasi-independence concept," *Journal of the American Statistical Association*, **70**, 755–68.

Guttman, L. (1950). "The basis for scalogram analysis." In S. A. Stouffer, L. Guttman, E. A. Suchman, P. F. Lazarsfeld, S. A. Star & J. A. Clausen (eds.), *Studies in Social Psychology in World War II. Vol. IV: Measurement and Prediction*. Princeton, NJ: Princeton University Press, pp. 60–90.

Haberman, S. J. (1977). "Product models for frequency tables derived by indirect observation," *The Annals of Statistics*, **5**, 1124–47.

Haberman, S. J. (1979). *Analysis of Qualitative Data*. (Vol. **2**: *New Developments*). New York: Academic.

Lazarsfeld, P. F. (1950). "The logical and mathematical foundation of latent structure analysis." In S. A. Stouffer, L. Guttman, E. A. Suchman, P. F. Lazarsfeld, S. A. Star, & J. A. Clausen (eds.), *Studies in Social Psychology in World War II*

Vol. IV: Measurement and Prediction. Princeton, NJ: Princeton University Press, pp. 362–412.

Lazarsfeld, P. F., & Henry, N. W. (1968). *Latent Structure Analysis*. Boston: Houghton Mifflin.

Lindsay, B., Clogg, C. C., & Grego, J. (1991). "Semiparametric estimation in the Rasch model and related exponential response models, including a simple latent class model for item analysis," *Journal of the American Statistical Association*, **86**, 96–107.

McHugh, R. B. (1956). "Efficient estimation and local identification in latent class analysis," *Psychometrika*, **21**, 331–47.

McLachlan, G. J., & Basford, K. E. (1988). *Mixture Models*. New York: Dekker.

Merskey, H., & Bogduk, N. (1994). *Classification of Chronic Pain*. Seattle: IASP Press.

Michel, A., Kohlmann, T., & Raspe, H. (1997). "The association between clinical findings on physical examination and self-reported severity in back pain," *Spine*, **22**, 296–304.

Proctor, C. H. (1970). "A probabilistic formulation and statistical analysis of guttman scaling," *Psychometrika*, **35**, 73–8.

Rasch, G. (1960). *Probabilistic Models for Some Intelligence and Attainment Tests*. Copenhagen: Nielsen & Lydiche.

Raspe, H., & Kohlmann, T. (1994). "Disorders characterised by pain: a methodological review of population surveys," *Journal of Epidemiology and Community Health*, **48**, 531–7.

Titterington, D. M., Smith, A. F. M., & Makov, U. E. (1985). *Statistical Analysis of Finite Mixture Distributions*. Chichester: Wiley.

van der Linden, W., & Hambleton, R. K. (1997). *Handbook of Modern Item Response Theory*. New York: Springer.

Vermunt, J. K. (1997). "LEM: a general program for the analysis of categorical data. User's manual," Tilburg, The Netherlands: Tilburg University.

Von Korff, M., Ormel, J., Keefe, F. J., & Dworkin, S. F. (1992). "Grading the severity of chronic pain," *Pain*, **50**, 133–49.

Wu, C. F. J. (1983). "On the convergence properties of the EM algorithm," *The Annals of Statistics*, **11**, 95–103.

CAUSAL ANALYSIS AND DYNAMIC MODELS

EIGHT

Use of Categorical and Continuous Covariates in Latent Class Analysis

C. Mitchell Dayton and George B. Macready

1. INTRODUCTION

In this chapter, we consider latent class models that include concomitant variables. As in an analysis of variance, concomitant variables may be classified as blocking variables or as continuous covariates. For example, cases may be grouped by sex in order to investigate the equivalence of latent class proportions and/or conditional probabilities for male and female respondents. Sex would play the role of a blocking, or stratification, variable, and standard methods of estimation and hypothesis testing for latent class models can be easily extended to situations of this type (Clogg and Goodman, 1984, 1985, 1986). In contrast, latent class parameters can be conditioned on continuous concomitant variables in a manner analogous to the analysis of covariance. For example, in a two-class model, the proportion of cases in the first latent class, say, might be represented as a function of a respondent characteristic such as scholastic aptitude (e.g., SAT score). Modeling a concomitant variable of this type is more complicated than modeling a blocking variable because a specific functional form must be assumed for the relationship between the covariate and latent class proportion (Dayton and Macready, 1988a, 1988b). In an analysis of covariance, the functional form is ordinarily assumed to be linear, but this is clearly unsatisfactory for representing relations when the variable being modeled is a probability.

In the ensuing sections of this chapter, we present and illustrate theory related to categorical blocking variables and continuous covariates. In Section 2, a basic latent class model without concomitant variables is defined and extended to include blocking variables. In addition, the problem of comparing hierarchic and nonhierarchic models is reviewed. An exemplary analysis for the blocking model is presented in Section 3.

The basic model is extended to include covariates in Section 4 and illustrated in Section 5.

2. MODELS WITH BLOCKING VARIABLES

In an ordinary T-class latent class model, there are J manifest categorical variables that are assumed to be independent conditional upon latent class membership. For example, if T latent classes are posited to exist for three manifest variables A, B, and C, then responses to these three variables are assumed to be independent within each of the latent classes. Thus, conditional upon membership in class t of latent variable X, the probability for a particular observed response vector $[i, j, k]'$ for variables A, B, and C is

$$\pi_{ijkt}^{\overline{ABC}X} = \pi_{it}^{\bar{A}X}\pi_{jt}^{\bar{B}X}\pi_{kt}^{\bar{C}X}. \tag{1}$$

In many applications, the manifest variables are dichotomous with responses $i, j, k = \{1, 0\}$, although this is not a necessary condition. Assuming latent class proportions, π_t^X for $t = 1, \ldots, T$, we find that the unconditional probability for the response vector, $[i, j, k]'$, is

$$\pi_{ijk}^{ABC} = \sum_{t=1}^{T} \pi_t^X \pi_{it}^{\bar{A}X}\pi_{jt}^{\bar{B}X}\pi_{kt}^{\bar{C}X} \tag{2}$$

Assuming that certain conditions are met, we find that an estimation of the parameters in the model represented by Equation (2) can be carried out by the methods of maximum-likelihood estimation (MLE). Let f_{ijk} be the observed frequency of cases for the response vector $[i, j, k]'$. Then, the likelihood associated with that response is $(\pi_{ijk}^{ABC})^{f_{ijk}}$, and the log-likelihood for the entire sample of size $N = \sum_{i,j,k} f_{ijk}$ is given by

$$\lambda = \sum_{i,j,k} f_{ijk} \log_e \left(\pi_{ijk}^{ABC}\right), \tag{3}$$

where summation is over the set of possible response patterns (e.g., for three dichotomous manifest variables, there are $2^3 = 8$ patterns). Maximization of λ with respect to the conditional probabilities, $\pi_{it}^{\bar{A}X}$, $\pi_{jt}^{\bar{B}X}$, and $\pi_{kt}^{\bar{C}X}$, and latent class proportions, π_t^X, can be accomplished by using full-information iterative MLE methods or computationally simpler iterative procedures based on missing-data methods.

Assume that the cases are grouped within S levels of a manifest blocking variable, G. Typical blocking variables include sex, grade level in school, and socioeconomic status. A general approach to incorporating this information into latent class modeling is to condition the parameters of the model on the known levels of the blocking variable. In effect, the model in Equation (2) is written separately for each blocking level. Thus, a latent class proportion may be denoted as $\pi_{st}^{G\bar{X}}$, representing the size of the tth latent class within the sth level of the blocking variable. Similarly, item conditional probabilities may be written so as to take into account the blocking variable. For example, $\pi_{ist}^{\bar{A}GX}$ represents the probability for response i to variable A for a case at blocking level s given membership in latent class t. Then, Equation (1) can be rewritten to take into account the blocking variable:

$$\pi_{ijkst}^{\overline{ABC}GX} = \pi_{ist}^{\bar{A}GX}\pi_{jst}^{\bar{B}GX}\pi_{kst}^{\bar{C}GX}. \tag{4}$$

Assuming the number of latent classes, T, per blocking level is constant, we find that the model summing over latent classes becomes

$$\pi_{ijks}^{\overline{ABC}G} = \sum_{t=1}^{T} \pi_{st}^{G\bar{X}}\pi_{ist}^{\bar{A}GX}\pi_{jst}^{\bar{B}GX}\pi_{kst}^{\bar{C}GX}. \tag{5}$$

If the observed proportion of cases at blocking level s is P_s, with $\sum_{s=l}^{S} P_s = 1$, then there exists S restrictions of the form $\sum_{t=1}^{T} \pi_{st}^{GX} = P_s$ as well as the overall restriction $\sum_{s=1}^{S}\sum_{t=1}^{T} \pi_{st}^{GX} = 1$, where $\pi_{st}^{GX} = P_s\pi_{st}^{G\bar{X}}$ is the joint probability of membership in the tth latent class within the sth block. Note that in addition to the requirement that the latent class proportions sum to 1 overall, they must also sum to the observed proportions of cases within the fixed blocking levels. As usual, the conditional probabilities are subject to conventional restrictions for probabilities. As examples, if A is dichotomous, then $\sum_{i=0}^{1} \pi_{ist}^{\bar{A}GX} = 1$ for all s and t and $\sum_{t=1}^{T} \pi_{st}^{G\bar{X}} = 1$ for all s.

Often the purpose for introducing blocking variables into latent class analysis is to assess the homogeneity of the latent class structure with respect to the levels of the blocking variable(s). For example, Davey and Macready (1990) grouped students by reading ability into "good" and "poor" readers and investigated the issue of whether the latent class structure for various reading tasks was comparable for these two groups of readers. This process of exploratory model fitting is often divided into two distinct phases. In the first phase, an effort is undertaken to discover interpretable models within each blocking level. Then, in the second phase, the equivalence, or lack of equivalence, of these models is investigated. With respect to conditional probabilities for manifest variables, general equality constraints across levels of a blocking variable would be of the form (where ∀ means "for all")

$$\pi_{ist}^{\bar{A}GX} = \pi_{is't}^{\bar{A}GX} \quad \forall i,\ s \neq s',\ t,$$
$$\pi_{jst}^{\bar{B}GX} = \pi_{js't}^{\bar{B}GX} \quad \forall j,\ s \neq s',\ t,$$
$$\pi_{kst}^{\bar{C}GX} = \pi_{ks't}^{\bar{C}GX} \quad \forall k,\ s \neq s',\ t. \tag{6}$$

In addition, equality constraints can be imposed on latent class proportions, and these are of the form

$$\pi_{st}^{G\bar{X}} = \pi_{s't}^{G\bar{X}} \forall s \neq s',\ t. \tag{7}$$

However, these latter constraints have limited interpretability unless it is assumed that the conditional probabilities for the manifest variables are homogeneous. That is, if the latent classes have different interpretations at various levels of a blocking variable, then it may not be meaningful to ask whether the sizes of these classes are equivalent across blocking levels. Thus, in practice, constrained models are typically assessed in sequence, with homogeneity of conditional probabilities being imposed first; then, if it is assumed that homogeneity is warranted, equality of latent class proportions is assessed.

3. EXEMPLARY ANALYSIS WITH BLOCKING VARIABLES

A data set comprising four items dealing with role conflict, presented by Stouffer and Toby (1951), has been the subject of several previous analyses (e.g., Goodman, 1974, 1975; Clogg and Goodman, 1985, 1986; Dayton and Macready, 1980; Dayton, 1999). These items describe four different situations with responses coded as 1 if they represent particularistic values or as 0 if they represent universalistic values. The items were presented in different forms to three independent groups of respondents. The first group received the items in a form in which the situations referred to themselves (i.e., "Ego faces dilemma"), the second group in a form in which the situations referred to a stranger (i.e., "Smith faces dilemma"), and the third group in a form in which the situations referred to a close friend (i.e., "Ego's Friend faces dilemma"). Frequencies for the $2^4 = 16$ response patterns for the three groups of respondents are shown in Table 1. Clogg and Goodman (1986) present extensive results based on a Guttman scale with the item sequence D → C → B → A, but include data for only

Table 1. Stouffer–Toby Data

Item				Frequency		
A	**B**	**C**	**D**	**1 (Ego)**	**2 (Smith)**	**3 (Friend)**
0	0	0	0	42	37	35
0	0	0	1	23	31	17
0	0	1	0	6	6	9
0	0	1	1	25	15	26
0	1	0	0	6	5	3
0	1	0	1	24	29	27
0	1	1	0	7	6	3
0	1	1	1	38	25	32
1	0	0	0	1	2	3
1	0	0	1	4	4	5
1	0	1	0	1	3	2
1	0	1	1	6	4	5
1	1	0	0	2	3	0
1	1	0	1	9	23	20
1	1	1	0	2	3	3
1	1	1	1	20	20	26
Total				216	216	216

the first and third groups (i.e., Ego and Ego's Friend face dilemma). In the present analyses, we compare scaling models across all three groups and consider models representing structures that are more complex than a Guttman scale.

For all three groups, there arc relatively large frequencies for the seven response patterns $[0, 0, 0, 0]'$, $[0, 0, 0, 1]'$, $[0, 0, 1, 1]'$, $[0, 1, 0, 1]'$, $[0, 1, 1, 1]'$, $[1, 1, 0, 1]'$, and $[1, 1, 1, 1]'$, although the pattern $[1, 1, 0, 1]'$ is less distinctive for the first group, in which Ego faces dilemma. It is apparent from the consistently large frequency for the fourth response pattern, $[0, 1, 0, 1]'$, for all three groups that a simple linear hierarchy (i.e., Guttman scale) can represent only a reasonable model for these data if certain of the item conditional probabilities permit relatively large "errors." Specifically, for example, the pattern $[0, 1, 0, 1]'$ could arise from a permissible vector $[0, 1, 1, 1]'$ corresponding to the fifth pattern if the conditional probability $\pi_{05}^{\bar{C}X}$ were suitably large. In fact, for two of the scaling models fit by Clogg and Goodman (1986), this conditional probability was estimated to be 0.21 and 0.36, respectively. A hierarchic structure that is consistent with the set of seven response patterns is a joint scale comprising the union of the two sequences D → B → A and D → C [see Figure 1(a)]. If the pattern [1, 1, 0, 1]′ is omitted, the remaining six response patterns are consistent with a joint scale comprising the union of the two sequences D → B → A and D → C → A [see Figure 1(b)]. Results for a series of scaling models based on the joint scales are summarized in Table 2.

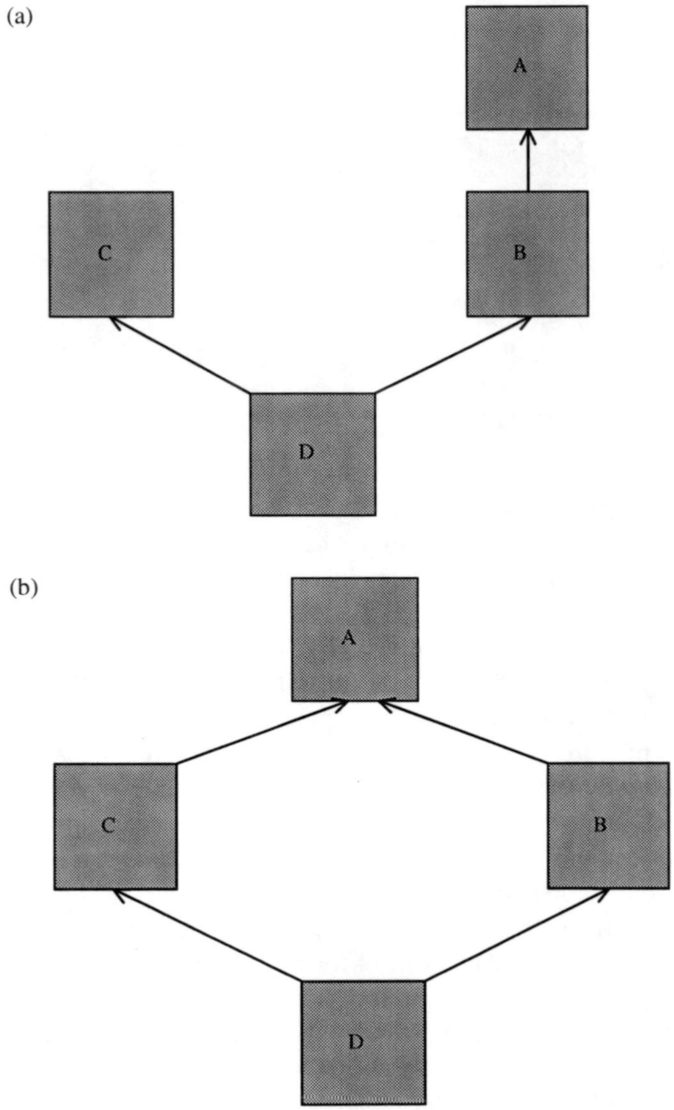

Figure 1. (a) First and (b) second hierarchies.

Table 2. Chi-Square and AIC Values for Scaling Models Applied to Stouffer–Toby Data

Model	Likelihood-Ratio Chi Square	df	*p* Value	AIC	AIC Rank	Description
M_1	17.69	21	.6685	−24.31	9	7 classes [Fig. 1(a)]; intrusion–omission (I–O) errors per group
M_1^*	21.07	25	.6887	−28.93	5	M_1 with equal I–O errors across groups
M_1^{**}	39.03	37	.3786	−34.97	1	M_1 with equal I–O errors and equal latent class proportions across groups
M_1^{***}	33.37	33	.4493	−32.63	3	M_1 with equal latent class proportions across groups
M_2^{**}	41.89	38	.3058	−34.11	2	M_1^{**} with equal intrusion, omission error rates (Proctor model)
M_3	22.11	22	.4533	−21.89	10	6 classes [Fig. 1(b)] in group 1;7 classes [Fig. 1(a)] in groups 2 & 3; I–O errors per group
M_3^*	25.77	26	.4758	−26.23	6	M_3 with equal I–O errors across groups
M_4	58.19	24	.0001	10.19	15	6 classes [Fig. 1(b)]; I–O errors per group
M_4^*	64.31	28	.0001	8.31	14	M_4 with equal I–O errors across groups
M_G	5.15	12	.9528	−18.85	11	8 classes [Fig. 1(a) + intrinsically unscalable class]; Goodman model per group
M_G^*	25.80	26	.4741	−26.20	7	M_G with equal latent class proportions across groups
M_{1G}	4.01	6	.6753	−7.99	13	M_G with I–O errors per groups (Dayton–Macready model)
M_{1G}^a	6.56	10	.7662	−13.44	12	M_{1G} with equal I–O errors across groups
$M_{1G}^{a^*}$	16.28	24	.8777	−31.72	4	M_{1G} with equal latent class proportions across groups
$M_{2G}^{a^*}$	24.35	25	.4992	−25.65	8	M_{1G} with equal I–O errors and equal latent class proportions across groups

Model M_1 is an intrusion–omission error model (Dayton and Macready, 1976) incorporating the restrictions

$$\begin{aligned}\pi_{11}^{\bar{A}X} &= \pi_{11}^{\bar{B}X} = \pi_{11}^{\bar{C}X} = \pi_{11}^{\bar{D}X} = \pi_{12}^{\bar{A}X} = \pi_{12}^{\bar{B}X} = \pi_{12}^{\bar{C}X} \\ &= \pi_{14}^{\bar{A}X} = \pi_{14}^{\bar{C}X} = \pi_{15}^{\bar{A}X} = \pi_{16}^{\bar{C}X} = \alpha_1 \\ \pi_{02}^{\bar{D}X} &= \pi_{03}^{\bar{C}X} = \pi_{03}^{\bar{D}X} = \pi_{04}^{\bar{B}X} = \pi_{04}^{\bar{D}X} = \pi_{05}^{\bar{B}X} = \pi_{05}^{\bar{C}X} \\ &= \pi_{05}^{\bar{D}X} = \pi_{06}^{\bar{A}X} = \pi_{06}^{\bar{B}X} = \pi_{06}^{\bar{D}X} = \pi_{07}^{\bar{A}X} = \pi_{07}^{\bar{B}X} \\ &= \pi_{07}^{\bar{C}X} = \pi_{07}^{\bar{D}X} = \alpha_0 \end{aligned} \tag{8}$$

within each of the groups for the seven-latent-class model based on permissible response patterns corresponding to the seven response patterns with large frequencies and consistent with Figure 1(a) (within groups, the restricted values are denoted α_{Is} and α_{Os}). There are six independent latent class proportions and two error rates to estimate per group, with the resulting degrees of freedom for the model being $48 - 3(9) = 21$. The model fits the data quite well, as evidenced by the likelihood-ratio chi-square (G^2) value of 17.69. Three constrained forms of this model were investigated; the first two conform to the recommended sequence presented in the previous section. Model M_1^* does not restrict the sizes of the latent classes but imposes the condition of equal error rates across groups. Using the notation of Equation (8), we see that these constraints are equivalent to imposing the following four restrictions on the errors:$\alpha_{Is} = \alpha_I, \alpha_{Os} = \alpha_s, s = 1, \ldots, 3$. Thus, this model has 25 degrees of freedom (i.e., 4 more degrees of freedom than model M_1). Model M_1^{**} is a model of complete homogeneity that restricts both the latent class proportions and error rates to be equal across groups. This model is equivalent to simultaneously imposing the restrictions for models M_1^* and the additional 12 restrictions on the latent class proportions: $\pi_{1t}^{G\bar{X}} = \pi_{2t}^{G\bar{X}} = \pi_{3t}^{G\bar{X}}, t = 1, \ldots, 7$. Thus, model M_1^{**} has 37 degrees of freedom (i.e., $12 + 4 = 16$ more degrees of freedom than model M_1). For completeness, we also fit a model M_1^{***} that retains separate intrusion–omission error rates for the three groups but restricts the sizes of the latent classes to be equal across groups. This model has 33 degrees of freedom (i.e., 12 more degrees of freedom than model M_1^*). All three of these restricted models provide adequate fit to the frequencies, as evidenced by their chi-square goodness-of-fit statistics. Although each of these restricted models can be compared with the heterogeneous model, M_1, by means of chi-square difference statistics, models M_1^* and M_1^{***} are not, themselves, hierarchically related and cannot be compared by this approach. However, the four competing models can be evaluated simultaneously by using an Akaike min (AIC) strategy. As shown in Table 2, the model of complete homogeneity, M_1^{**}, is selected on this basis.

The remaining models in Table 2 represent attempts to improve on the model of complete homogeneity with intrusion and omission errors. The simpler Proctor model, M_2^{**}, is obtained by further imposing the restriction $\alpha_I = \alpha_O$ on model M_1^{**}. Although this model fits the frequencies in terms of its chi-square value, the increase in chi square is too large to justify this additional restriction as evidenced by an AIC value that is somewhat larger than for model M_1^{**}.

Two additional models were derived from the consideration that the response vector, $[1, 1, 0, 1]'$, occurs distinctly less frequently in the first group than in the remaining two groups. This suggests fitting models with permissible vectors based on the two sequences D → B → A and D → C → A [see Figure 1(b)] for the first group, but with permissible vectors based on the two sequences D → B → A and D → C [see Figure 1(a)] for the second and third groups. Model M_3 is a heterogeneous model similar to model M_1, but with only six latent classes for the first group. Although this model fits the data, its AIC value is considerably larger than that for the preferred model based on seven latent classes per group (i.e., model M_1^{**}). We also consider a constrained form of this model, M_3^*, that does not restrict the sizes of the latent classes but does impose the condition of equal error rates across groups. Once again, an adequate fit is attained although the AIC value does not suggest that this model is preferred to model M_1^{**}. The six-class model based on the two sequences D → B → A and D → C → A was fit to all three groups in an unrestricted, or heterogeneous, form in model M_4 and with equal error rates across groups in model M_4^*. As expected, neither of these models provided satisfactory fit to the data.

The final set of models considered for the Stouffer–Toby data utilizes the notion of intrinsically unscalable classes as proposed by Goodman (1975). For each group, model M_G fits eight latent classes, with the first seven classes corresponding to those for model M_1 and the final class representing intrinsically unscalable respondents. This model is heterogeneous and is equivalent to fitting a Goodman intrinsically unscalable class model separately to the three groups and pooling the fit statistics. There are 11 parameters to estimate for each group and this, in combination with the requirement that expected frequencies sum to the group total, results in fit statistics based on 4 degrees of freedom per group, or 12 degrees of freedom overall. The likelihood ratio G^2 value, 5.15, represents excellent fit to the data, although this fit is at the expense of estimating a substantial proportion of respondents as being unscalable. In any case, because of the unparsimonious nature of this model, its AIC value is substantially larger than that of, for example, model M_1^{**}. The form of this model with latent class sizes constrained to equality across the groups, model M_G^*, also provides an adequate fit with a much smaller AIC value than the unconstrained model. Dayton and Macready (1980) proposed including intrusion and omission errors for the latent classes corresponding to permissible response patterns in the Goodman model. The unconstrained form of this model, M_{1G}, is the most highly parameterized model considered for this data set and has only 6 degrees of freedom for assessing fit. Although the chi-square fit statistic is satisfactory, the associated AIC value is among the worst of the models being compared. A restricted form of this model, M_{1G}^a, with equal intrusion–omission error rates across groups, provides satisfactory fit, but, still, a relatively large AIC value. A further restricted model, with latent class sizes constrained to equality across the groups, model $M_{1G}^{a^*}$, in addition to providing adequate fit, attains a much smaller AIC value, being the fourth best among the models under consideration. The final model imposes the additional restriction of equality for the intrusion and omission error rates and is denoted model $M_{2G}^{a^*}$. Once again, this model provides adequate fit to the data.

For the three groups comprising the Stouffer–Toby data set, the relatively simple, homogeneous seven-class model with separate intrusion and omission error rates (i.e., model M_1^{**}) results in the smallest AIC value among the set of 15 models considered. Furthermore, imposing the restriction of equal intrusion and omission error rates (i.e., model M_2^{**}) results in only a slightly larger AIC value and has a chi-square fit statistic that differs by a nonsignificant amount (i.e., $41.89 - 39.03 = 2.86$, with 1 degree of freedom; $p = .0908$). Parameter estimates for these two models are displayed in Table 3. The row labeled "Scaled proportion" represents the proportion of respondents actually observed for the permissible response patterns after scaling these proportions to sum to 1. For comparison, parameter estimates are presented for the heterogeneous version of the Goodman model, model M_G (Table 4), and for the constrained version, model $M_{1G}^{a^*}$ (Table 5). Although each of these models provides good fit to the observed frequencies, a substantial proportion of respondents are classified as unscalable. As shown in Table 4 for model

Table 3. Estimated Parameters for Models M_1^{**} and M_2^{**}

	Permissible Pattern						
Parameter	**[0,0,0,0]**	**[0,0,0,1]**	**[0,0,1,1]**	**[0,1,0,1]**	**[0,1,1,1]**	**[1,1,0,1]**	**[1,1,1,1]**
Model M_1^{**}							
LC proportion[a]	0.20	0.10	0.12	0.12	0.20	0.09	0.17
Intrusion error[a]	0.06						
Omission error[a]	0.14						
Model M_2^{**}							
LC proportion[a]	0.24	0.11	0.12	0.13	0.18	0.09	0.13
Intrusion error[a,b]	0.10						
Omission error[a,b]	0.10						
Scaled proportion	0.21	0.13	0.12	0.15	0.17	0.10	0.12

[a] The latent class proportions and intrusion–omission error rates are constrained to be equal across groups.
[b] The intrusion and omission error rates are constrained to be equal.

M_G, for example, only approximately one-third of the respondents are estimated in the scalable classes for Group 1 (Ego). In addition, conditional probabilities for particularistic values for the intrinsically unscalable class show a very large range from 0.20 to 0.80. In contrast, model M_1^{**} shows reasonable proportions in each of the seven latent classes representing scalable types with modest values for intrusion and omission errors (i.e., rates of 0.06 and 0.14, respectively).

Latent class analyses with blocking variables can be extended to situations with two or more blocking variables. Consider, for example, modeling the Stouffer–Toby data set if information were available concerning sex of respondent. There would be six groups resulting from the combinations of item form and respondent sex. In addition to unconstrained models, a variety of models with homogeneity and partial homogeneity restrictions could be explored (e.g., conditional probabilities for items can be constrained to equality within item forms but allowed to be heterogeneous with respect to respondent sex; similar restrictions could be considered for latent class proportions).

4. COVARIATE MODELS

In this section, we consider latent class models in which the latent class proportion is assumed to be functionally dependent upon one or more concomitant variables (Dayton and Macready, 1988a, 1988b). These models are important for two fundamental reasons. First, as with models

Table 4. Estimated Parameters for Model M_G

	Permissible Pattern							
Parameter	**[0,0,0,0]**	**[0,0,0,1]**	**[0,0,1,1]**	**[0,1,0,1]**	**[0,1,1,1]**	**[1,1,0,1]**	**[1,1,1,1]**	**[IUC]**
LC prop.								
Group 1 (Ego)	0.17	0.03	0.02	0.00	0.04	0.01	0.07	0.67
Group 2 (Smith)	0.13	0.02	0.05	0.10	0.12	0.08	0.11	0.39
Group 3 (Friend)	0.15	0.11	0.03	0.10	0.07	0.09	0.07	0.38
	$\pi_{18}^{\bar{A}X}$	$\pi_{18}^{\bar{B}X}$	$\pi_{18}^{\bar{C}X}$	$\pi_{18}^{\bar{D}X}$				
IUC probs.								
Group 1 (Ego)	0.20	0.58	0.56	0.80				
Group 2 (Smith)	0.27	0.30	0.55	0.64				
Group 3 (Friend)	0.33	0.52	0.54	0.60				

Note: The intrinsically unscalable class is shown as [IUC].

Table 5. Estimated Parameters for Model M_{1G}^{a*}

	Permissible Pattern							
Parameter	**[0,0,0,0]**	**[0,0,0,1]**	**[0,0,1,1]**	**[0,1,0,1]**	**[0,1,1,1]**	**[1,1,0,1]**	**[1,1,1,1]**	**[IUC]**
LC Proportion[a]	0.18	0.09	0.12	0.01	0.09	0.00	0.10	0.42
	$\pi_{18}^{\bar{A}X}$	$\pi_{18}^{\bar{B}X}$	$\pi_{18}^{\bar{C}X}$	$\pi_{18}^{\bar{D}X}$				
IUC Probs.								
Group 1 (Ego)	0.29	0.80	0.60	0.91				
Group 2 (Smith)	0.46	0.87	0.51	1.00				
Group 3 (Friend)	0.44	0.86	0.36	0.91				
Instrusion Error[a]	0.04							
Omission Error[a]	0.25							

Note: The intrinsically unscalable class is shown as [IUC].

[a] The latent class sizes and intrusion–omission error rates are constrained to be equal across groups.

incorporating blocks, hypotheses can be tested concerning the relation between the latent class proportions and the covariate(s). Second, it may be possible to estimate and fit covariate models in situations that have inadequate degrees of freedom for fitting basic models or models with blocking. In fact, in certain situations, covariate models can be fitted when there is only a single dichotomous manifest variable.

For the hth respondent, let the responses to three manifest variables, A, B, and C, be Y_{h1}, Y_{h2}, and Y_{h3}, and assume that these are augmented by values for m known covariates, $\mathbf{Z}_h = [Z_{h1}, Z_{h2}, \ldots, Z_{hm}]'$. Thus, the complete response vector for the hth respondent is

$$\mathbf{v_h} = [Y_{h1}, Y_{h2}, Y_{h3}, Z_{h1}, Z_{h2}, \ldots, Z_{hm}]'. \tag{9}$$

The covariates are assumed to have properties similar to those of independent variables in fixed-effects multiple regression models. That is, they may be continuous or categorical in nature and take on fixed, known values. If a covariate is categorical and exists at more than two levels, then 0/1 dummy variables, or other coded variables, may be used to represent the categories in the model. Also, product variables representing interactions between two categorical covariates or one categorical and one continuous covariate may be introduced into the models.

An approach to incorporating covariates into two-class models, suggested by Dayton and Macready (1988a), is to modify the basic latent class model of Equation (2) by positing a submodel for the proportion in the first latent class, π_1^X. In general, the submodel may be written in the form

$$\pi_{1h}^{\bar{X}Z} = g(Z_h \mid \beta), \tag{10}$$

where $\pi_{1h}^{\bar{X}Z}$ is the proportion in the first latent class conditional on the covariates, $\mathbf{Z}_h$, $g(\bullet)$ is some monotone function that must be specified, and β is a vector of parameters characterizing the function.

Because $\pi_{1h}^{\bar{X}Z}$ is defined on the 0,1 interval, the function, $g(\bullet)$, must map $\mathbf{Z}_h$ onto this interval. An especially simple functional form is the one-parameter cumulative exponential,

$$\pi_{1h}^{\bar{X}Z} = 1 - \exp(-Z_h/\beta_0). \tag{11}$$

However, a functional form that is more widely used in psychometrics is the logistic:

$$\pi_{1h}^{\bar{X}Z} = e^{\beta_0+\sum_{m=1}^{M}\beta_m Z_{hm}} \Big/ \left(1+e^{\beta_0+\sum_{m=1}^{M}\beta_m Z_{hm}}\right) = 1\Big/\left(1+e^{-\beta_0-\sum_{m=1}^{M}\beta_m Z_{hm}}\right). \tag{12}$$

The logistic function has the interpretation that the log odds in favor of the first latent class are a linear function of the covariates; that is,

$$\log_e \frac{\pi_{1h}^{\bar{X}Z}}{1-\pi_{1h}^{\bar{X}Z}} = \beta_0 + \sum_{m=1}^{M} \beta_m Z_{hm}. \tag{13}$$

Note that the model in Equation (12) is formally equivalent to that for logistic regression analysis (Hosmer and Lemeshow, 1989), except that in latent class analysis the dependent variable is latent rather than manifest.

The logistic covariate latent class model can be extended to the case of three or more latent classes by the introduction of the notion of continuation ratios (Fienberg, 1980). Continuation ratios are defined in such a manner that the monotonicity of the relation between the covariates, $\mathbf{Z}_h$, and the latent class proportions, $\pi_{1h}^{\bar{X}Z}$, is preserved. For example, with three classes, the following model would be appropriate:

$$\begin{aligned}
\pi_{1h}^{\bar{X}Z} &= g(Z_h \mid \beta_1) = 1/\left(1 + e^{-\beta_{10} - \sum_{m=1}^{M} \beta_{1m} Z_{hm}}\right),\\
\pi_{2h}^{\bar{X}Z} &= \left(1 - \pi_{1h}^{\bar{X}Z}\right) g(Z_h \mid \beta_2)\\
&= \left\{\frac{1}{1 + e^{\beta_{10} + \sum_{m=1}^{M} \beta_{1m} Z_{hm}}}\right\}\left\{\frac{1}{1 + e^{-\beta_{20} - \sum_{m=1}^{M} \beta_{2m} Z_{hm}}}\right\},\\
\pi_{3h}^{\bar{X}Z} &= 1 - \pi_{1h}^{\bar{X}Z} - \pi_{2h}^{\bar{X}Z}.
\end{aligned} \tag{14}$$

Although a more elaborate exemplary analysis for a covariate model is presented in the next section, a relatively simple example from the research literature may be useful in understanding the model. Dayton and Scheers (1997) report results from a survey of undergraduate college students. Agree/disagree responses were available from six survey items concerning academic dishonesty (e.g., "I have cheated by obtaining a copy of an exam ahead of time") and a two-class model provided reasonable fit to the data. The two latent classes were interpretable as representing "cheaters" and "noncheaters," respectively. With the use of grade point average (GPA, self-reported) as a covariate, the following logistic function was estimated: $\pi_{1h}^{\bar{X}Z} = 1/(1 + e^{-9.96+3.30 Z_h})$. This function shows the expected inverse relation between GPA and membership in the class representing cheaters. For example, substituting values of GPA equal to 2.5, 3.0, 3.5, and 4.0 into the function yields expected proportions in the "cheater" class of 0.80, 0.44, 0.13, and 0.03, respectively.

One advantage of both blocking and covariate latent class models is that an evaluation of models based on very few response variables is possible. For example, with two dichotomous variables, there are only four distinct response patterns. The most complex latent class model that can be fit would be for two classes with constant intrusion–omission error rates for the two classes (i.e., a Proctor model), and this model has no degrees of freedom for assessing fit (i.e., it always fits the set of frequencies perfectly). However, if, in addition to the two variables, there were a concomitant variable at, say, three levels, there is now a total of 12 response patterns and more complex modeling is possible. In contrast, the researcher should be aware of the fact that, in these circumstances, there will be restrictions on the complexity of the models that can be assessed and, hence, some generality will be lost with respect to exploring a wide range of models. For the example just cited, a linear hierarchy (i.e., Guttman scale) with intrusion–omission errors could be fit only if it were assumed to be homogeneous across the three levels of the concomitant variable. The example in the next section illustrates some analytical possibilities.

5. EXEMPLARY ANALYSIS WITH COVARIATES

The Second International Mathematics Study (SIMS) was based on data from 20 participating countries during the time period 1977 to 1984 and was one of the comparative studies conducted by IEA, the International Association for the Evaluation of Educational Achievement (Pelgrum et al., 1986). Students involved in the study were approximately 13 years of age and would, in the United States, represent an eighth-grade population. A core set of mathematics items in five-option multiple choice format, in the content areas of arithmetic, algebra, geometry, statistics, and measurement, was taken by all participating students. Among the items in the geometry area were two that dealt directly with the Pythagorean theorem. One of these items presented the lengths of the two sides of a right triangle (i.e., $a = 5, b = 12$) and asked for the length of the hypotenuse (i.e., 13). The second item presented similar information (i.e., $a = 3, b = 4$) and asked the student to recognize the alternative that correctly stated the theorem (i.e., $x^2 = 3^2 + 4^2$). As a way to illustrate the covariate latent class model, the score for the separate algebra area was used as a covariate and the outcome variable was the response vector of 0,1 scores for the two Pythagorean theorem items. The algebra score ranged from 0 through 6, and the response patterns for the two Pythagorean theorem items are shown for 1,241 respondents for the distinct algebra scores in Table 6.

The underlying notion for these analyses was that successful performance on the two Pythagorean theorem items was expected to be an

Table 6. Geometry Item Responses

	Pythagorean Items		
Algebra Score	**1**	**2**	**Frequency**
0	0	0	61
	0	1	24
	1	0	9
	1	1	6
1	0	0	92
	0	1	50
	1	0	28
	1	1	17
2	0	0	107
	0	1	60
	1	0	30
	1	1	37
3	0	0	75
	0	1	44
	1	0	32
	1	1	71
4	0	0	56
	0	1	30
	1	0	44
	1	1	102
5	0	0	28
	0	1	17
	1	0	20
	1	1	112
6	0	0	5
	0	1	2
	1	0	9
	1	1	73
Total			1,241

increasing function of the ability to complete algebra problems because, in addition to knowledge of the theorem per se, correct solutions to the two items involved some fundamental notions of algebraic manipulation. It was also assumed that performance on the two Pythagorean theorem items may be viewed as mastery or nonmastery because the two items are essentially identical in content. That is, it was posited that there existed two types of students, or latent classes, in this population: those who had mastered the basic concepts and manipulations involved in the Pythagorean theorem and those who had not done so. These two groups correspond to the permissible response vectors, $[1, 1]'$ and $[0, 0]'$, respectively.

Table 7. Chi-Square and AIC Values for Covariate Models

Model	Likelihood-Ratio Chi Square	df	*p* Value	AIC	AIC Rank
M_E	83.34	16	0.0000	51.34	5
M_L	17.03	15	0.3171	−12.97	1
M_L^*	45.77	17	0.0002	11.77	3
M_B	12.21	10	0.2712	−7.79	2
M_B^*	40.89	12	0.0001	16.89	4

The first three results summarized in Table 7 are for covariate latent class models. The first model, M_E, assumes a one-parameter cumulative exponential function [Equation (11)] for the relation between membership in the mastery latent class and the covariate, algebra score. Because there is a total of $7 \times 4 = 28$ distinct response patterns with a fixed total frequency for each score group, there are, potentially, 21 degrees of freedom for model fitting. For the exponential model, there are two conditional probabilities associated with each latent class and one exponential parameter to estimate. Thus, this model has $21 - 5 = 16$ degrees of freedom, but it fits the data poorly as evidenced by the chi-square fit statistic of 83.34. The second model, M_L, that was considered posited a logistic relation [Equation (12)] between membership in the mastery latent class and the covariate. Model M_L has one more parameter than the exponential model and, thus, has 15 degrees of freedom for assessing fit. This model provides very good fit to the data with a chi-square value of 17.03. A restricted form of this model, M_L^*, was based on the assumption of equal conditional probabilities across the two items within each latent class. That is, the two restrictions, $\pi_{11}^{\bar{A}X} = \pi_{11}^{\bar{B}X}$ and $\pi_{12}^{\bar{A}X} = \pi_{12}^{\bar{B}X}$, were imposed. This model did not fit the data well, as indicated by the chi-square value of 45.77 based on 17 degrees of freedom.

The final two models summarized in Table 7 treat algebra scores as a blocking variable. Because there are only seven distinct algebra scores, there are, at most, 6 degrees of freedom for purposes of modeling the regression of latent class membership on algebra score. The regression component of the exponential covariate model, M_E, was based on only 1 degree of freedom whereas the logistic covariate model, M_L was based on 2 degrees of freedom. Of course, more complex models could be posited and fit to the data. Model M_B is equivalent to fitting the most complex regression model possible with respect to the regression of latent class membership on algebra score. That is, there were two latent classes at each level of algebra score, but conditional probabilities for the two items were assumed to be homogeneous across the latent classes. In absolute terms, this model fits the data quite well as evidenced by a chi-square value of 12.21 based on 10 degrees of freedom. However, it not a very parsimonious model and yields an AIC value that is larger than that of the unrestricted logistic regression model, M_L. The final model presented in Table 7, model M_B^*, restricts conditional probabilities to be equal within latent classes for the two items. As for the comparable model using the logistic regression model, this model does not provide good fit to the data.

Both in terms of chi-square fit statistics and associated AIC values, the preferred covariate latent class model for the Pythagorean theorem items is based on a logistic regression of latent class membership on algebra score with unrestricted conditional probabilities for the items (model M_L). The estimated values for the conditional probabilities for a 1, or correct, response in the first latent class were 0.99 and 0.85, respectively, for the two items. This suggests that interpreting this class as representing mastery of concept related to the Pythagorean theory is not unreasonable. The comparable estimated values in the second latent class were 0.20 and 0.34, respectively, for the two items. Because the items were in a five-choice multiple choice format, it appears reasonable to interpret this class as representing nonmastery. Estimated log odds for the mastery class, as a function of algebra score, Z, were $\log_e \hat{\pi}_{1h}^{\bar{X}Z}/(1-\hat{\pi}_{1h}^{\bar{X}Z}) = -3.91 + 0.99Z$. The corresponding logistic regression is shown in Figure 2. As expected, performance on the Pythagorean theorem items is strongly related to algebra skill, with the odds being nearly 8:1 in favor of mastery for a student with the highest algebra score, 6, and nearly 50:1 against

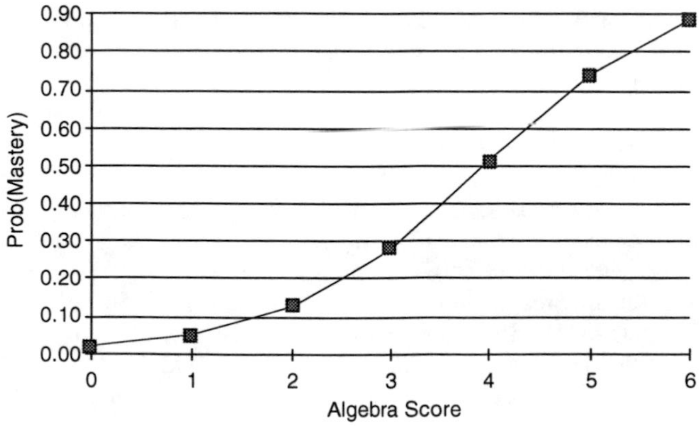

Figure 2. Mastery as a function of algebra score.

mastery for a student with an algebra score of 0. Note that, with respect to mastery, these two extreme groups show an odds ratio of approximately 380:1.

6. CONCLUSION

A natural extension of latent class modeling involves conditioning membership in any given latent class on known covariates. If no particular functional form for the relation between latent class membership and the covariates is specified, then it is often appropriate to treat the covariates as blocking variables and utilize the methods of simultaneous latent class analysis as presented by Clogg and Goodman (1984, 1985, 1986). In contrast, if the covariates are measured along some continuum, then more parsimonious models can be established that posit, say, an exponential or logistic relation (Dayton and Macready, 1988a, 1988b). As illustrated in Section 5, an additional benefit of imposing functional restrictions on the relation is that identified latent class models can be established when the number of response variables is very small (e.g., two in the exemplary analysis).

REFERENCES

Clogg, C. C., & Goodman, L. A. (1984). "Latent structure analysis of a set of multi-dimensional contingency tables," *Journal of the American Statistical Association*, **79**, 762–71.

Clogg, C. C., & Goodman, L. A. (1985). "Simultaneous latent structure analysis in several groups." In N. B. Tuma (ed.), *Sociological Methodology, 1985*. San Francisco: Jossey–Bass, pp. XXX–XXX.

Clogg, C. C., & Goodman, L. A. (1986). "On scaling models applied to data from several groups," *Psychometrika*, **51**, 123–35.

Davey, B., & Macready, G. B. (1990). "An exploration of latent structure underlying multiple-choice reading comprehension items for good and poor readers," *Applied Measurement in Education*, **3**, 209–29.

Dayton, C. M. (1999). *Latent Class Scaling Analysis*. Sage University Papers Series on Quantitative Applications in the Social Sciences, 07-126. Thousand Oaks, CA: Sage.

Dayton, C. M., & Macready, G. B. (1976). "A probabilistic model for validation of behavioral hierarchies," *Psychometrika*, **41**, 189–204.

Dayton, C. M., & Macready, G. B. (1980). "A scaling model with response errors and intrinsically unscalable respondents," *Psychometrika*, **45**, 343–56.

Dayton, C. M., & Macready, G. B. (1988a). "Concomitant-variable latent class models," *Journal of the American Statistical Association*, **83**, 173–8.

Dayton, C. M., & Macready, G. B. (1988b). "A latent class covariate model with applications to criterion-referenced testing." In R. Langeheine & J. Rost (eds.), *Latent Class and Latent Trait Models*. New York: Plenum, pp. XXX–XXX.

Dayton, C. M., & Scheers, N. J. (1997). "Latent class analysis of survey data dealing with academic dishonesty." In J. Rost & R. Langeheine (eds.), *Application of Latent Trait and Latent Class Models in the Social Sciences*. Munster, New York: Waxmann, pp. XXX–XXX.

Fienberg, S. E. (1980). *The Analysis of Cross-Classified Categorical Data*, 2nd ed. Cambridge, MA: MIT Press.

Goodman, L. A. (1974). "Exploratory latent structure analysis using both identifiable and unidentifiable models," *Biometrika*, **61**, 215–31.

Goodman, L. A. (1975). "A new model for scaling response patterns: an application of the quasi-independence concept," *Journal of the American Statistical Association*, **70**, 755–68.

Hosmer, D. W., & Lemeshow, S. (1989). *Applied Logistic Regression*. New York: Wiley.

Pelgrum, W. J., Eggen, T. J. H. M., & Plomp, T. (1986). "The implemented and attained mathematics curriculum: a comparison of eighteen countries," Research Report, Twente University of Technology.

Stouffer, S. A., & Toby, J. (1951). Role conflict and personality. *American Journal of Sociology*, **56**, 395–406.

NINE

Directed Loglinear Modeling with Latent Variables

Causal Models for Categorical Data with Nonsystematic and Systematic Measurement Errors

Jacques A. Hagenaars

1. INTRODUCTION

The founding fathers of Latent Class Analysis (LCA) have explained LCA mainly as a categorical latent structure model, that is, as a particular scaling model (Lazarsfeld, 1950; Lazarsfeld and Henry, 1969; Goodman, 1974a, 1974b; Haberman, 1974, 1979; Clogg, 1981a, 1995). LCA is used to measure the true scores on underlying, not directly observed, categorical variables by means of a set of observed categorical variables that function as the indicators of these underlying, latent variables. The indicators are not perfect measures of the latent variables, but they are subjected to "random," nonsystematic errors that lead to probabilistic relationships between the latent variables and the indicators. The standard LCA assumption of local independence specifies that the errors are conditionally independent of each other, when one controls for the latent variables. Throughout this chapter, the local independence assumption will be denoted as the independent classification error (ICE) assumption. Interpreting particular conditional response probabilities in terms of probabilities of classification errors (also termed *misclassifications*) is most appropriate when the scores on an indicator directly depend on just one latent variable and when there is a one-to-one correspondence between the categories of the indicator and the latent variable (Sutcliffe, 1965a, 1965b; Hagenaars, 1990; Kuha and Skinner, 1997). (For LCA models in which the latent variable has just a few categories and the indicators have many more, see Clogg, 1981b; also see Chapter 4 in this volume).

As the chapters in this volume show, the basic LCA model can be extended in many ways and into many directions. Here three extensions will be discussed. First, it will be shown in Section 2 how LCA (measurement) models can take the ordered nature of the categories of the indicators into account. The latent class models described have much in common with (discretized) latent trait models that are often used in psychological research (Heinen, 1996; Bartholomew and Knott, 1999; Van der Linden and Hambleton, 1997; Chapter 7 in this volume). Second, in Section 3, it will be discussed how to set up models in which the causes and consequences of the latent variables are investigated. In other words, causal models for categorical data will be defined in which some variables are latent, some variables act as the indicators of the latent variables, and the remaining (observed) variables influence or are influenced by the latent variables (Hagenaars, 1993, 1998; Vermunt, 1997a).[1] In the models presented in Sections 2 and 3, the standard ICE assumption is made and, in this way, the strengths of the causal relationships are estimated by taking random, nonsystematic errors into account. However, applying models in which the misclassifications are systematic and dependent upon each other is also possible and often desirable (Hagenaars, 1988). This will be the topic of Section 4.

2. LATENT CLASS MODELS FOR ORDERED DATA

A. Standard Two-Latent-Variable Model

The starting point for the analyses presented in this section is the data in Table 1, taken from the cross-national Political Action Study (Barnes and Kaase, 1979; Jennings and van Deth, 1990). The data in Table 1 are from the United States only and refer to five of the basic political concepts and indices that play a central role in this study, namely, System Responsiveness, Ideological Level, Repression Potential, Protest Approval, and Conventional Participation. LCA is used to discover whether these five basic indices can be reduced to even more fundamental political orientations.

From previous analyses (Hagenaars, 1993) using dichotomized rather than trichotomized indicators, it appeared that a reduction to two basic, latent orientations is possible, that is, System Involvement (Y) and Protest Tolerance (Z). (Their precise meaning will be further outlined when the results in Table 3 are discussed). The standard two-latent-variable LCA model can be denoted as $\{YZ, YA, YB, ZC, ZD, YE\}$; here the usual shorthand notation for hierarchical loglinear models is used. The model is rendered in Figure 1 and in Equation (1) in terms of (conditional response) probabilities and loglinear parameters. All parameters

Table 1. Indices of Political Orientation

A	*B*	*C*	*D*	*E*	Freq.	*A*	*B*	*C*	*D*	*E*	Freq.	*A*	*B*	*C*	*D*	*E*	Freq.	*A*	*B*	*C*	*D*	*E*	Freq.	*A*	*B*	*C*	*D*	*E*	Freq.	*A*	*B*	*C*	*D*	*E*	Freq.
1	1	1	1	1	0	1	2	1	1	1	8	2	1	1	1	1	0	2	2	1	1	1	9	3	1	1	1	1	0	3	2	1	1	1	2
1	1	1	1	2	0	1	2	1	1	2	3	2	1	1	1	2	2	2	2	1	1	2	3	3	1	1	1	2	2	3	2	1	1	2	4
1	1	1	1	3	0	1	2	1	1	3	2	2	1	1	1	3	1	2	2	1	1	3	2	3	1	1	1	3	1	3	2	1	1	3	2
1	1	1	2	1	0	1	2	1	2	1	12	2	1	1	2	1	4	2	2	1	2	1	23	3	1	1	2	1	2	3	2	1	2	1	17
1	1	1	2	2	3	1	2	1	2	2	13	2	1	1	2	2	7	2	2	1	2	2	29	3	1	1	2	2	9	3	2	1	2	2	16
1	1	1	2	3	1	1	2	1	2	3	5	2	1	1	2	3	9	2	2	1	2	3	14	3	1	1	2	3	20	3	2	1	2	3	16
1	1	1	3	1	3	1	2	1	3	1	18	2	1	1	3	1	2	2	2	1	3	1	17	3	1	1	3	1	3	3	2	1	3	1	7
1	1	1	3	2	4	1	2	1	3	2	12	2	1	1	3	2	6	2	2	1	3	2	23	3	1	1	3	2	4	3	2	1	3	2	12
1	1	1	3	3	7	1	2	1	3	3	13	2	1	1	3	3	15	2	2	1	3	3	12	3	1	1	3	3	15	3	2	1	3	3	19
1	1	2	1	1	0	1	2	2	1	1	24	2	1	2	1	1	1	2	2	2	1	1	20	3	1	2	1	1	0	3	2	2	1	1	7
1	1	2	1	2	1	1	2	2	1	2	5	2	1	2	1	2	4	2	2	2	1	2	5	3	1	2	1	2	1	3	2	2	1	2	10
1	1	2	1	3	1	1	2	2	1	3	3	2	1	2	1	3	1	2	2	2	1	3	2	3	1	2	1	3	4	3	2	2	1	3	5
1	1	2	2	1	3	1	2	2	2	1	19	2	1	2	2	1	4	2	2	2	2	1	38	3	1	2	2	1	2	3	2	2	2	1	15
1	1	2	2	2	7	1	2	2	2	2	15	2	1	2	2	2	8	2	2	2	2	2	31	3	1	2	2	2	6	3	2	2	2	2	30
1	1	2	2	3	6	1	2	2	2	3	9	2	1	2	2	3	3	2	2	2	2	3	17	3	1	2	2	3	19	3	2	2	2	3	31
1	1	2	3	1	0	1	2	2	3	1	3	2	1	2	3	1	1	2	2	2	3	1	9	3	1	2	3	1	3	3	2	2	3	1	6
1	1	2	3	2	2	1	2	2	3	2	7	2	1	2	3	2	5	2	2	2	3	2	15	3	1	2	3	2	6	3	2	2	3	2	16
1	1	2	3	3	3	1	2	2	3	3	8	2	1	2	3	3	1	2	2	2	3	3	7	3	1	2	3	3	4	3	2	2	3	3	9
1	1	3	1	1	3	1	2	3	1	1	32	2	1	3	1	1	3	2	2	3	1	1	57	3	1	3	1	1	2	3	2	3	1	1	8
1	1	3	1	2	1	1	2	3	1	2	14	2	1	3	1	2	2	2	2	3	1	2	41	3	1	3	1	2	3	3	2	3	1	2	19
1	1	3	1	3	1	1	2	3	1	3	8	2	1	3	1	3	2	2	2	3	1	3	20	3	1	3	1	3	6	3	2	3	1	3	15
1	1	3	2	1	0	1	2	3	2	1	25	2	1	3	2	1	1	2	2	3	2	1	38	3	1	3	2	1	1	3	2	3	2	1	21

1	1	3	2	2	4	1	2	3	2	2	16	2	1	3	2	2	6	2	2	3	2	2	29	3	1	3	2	2	9	3	2	3	2	2	27
1	1	3	2	3	1	1	2	3	2	3	10	2	1	3	2	3	4	2	2	3	2	3	24	3	1	3	2	3	15	3	2	3	2	3	26
1	1	3	3	1	0	1	2	3	3	1	3	2	1	3	3	1	0	2	2	3	3	1	6	3	1	3	3	1	1	3	2	3	3	1	5
1	1	3	3	2	2	1	2	3	3	2	4	2	1	3	3	2	3	2	2	3	3	2	3	3	1	3	3	2	4	3	2	3	3	2	7
1	1	3	3	3	0	1	2	3	3	3	1	2	1	3	3	3	1	2	2	3	3	3	4	3	1	3	3	3	2	3	2	3	3	3	2

Total 1437

Notes: Variable A: System Responsiveness is an index based on agreement with three statements, such as, "I don't think that public officials care much about what people like me think." Agreement with these items points to a low opinion of the system's responsiveness (1. Low: original scores 1.00–1.99 or missing; 2. Medium: 2.00–2.59; 3. High: 2.60–4.00). Variable *B:* Ideological Level indicates whether the respondent used left and right ideological concepts when describing political parties [1. (Near)Ideologues] or whether they had a idiosyncratic or false understanding of the nature of political parties [2. Nonideologues (or missing)]. Variable *C:* Repression Potential is an index based on approval of four actions, such as, "The police using force against demonstrators"; disapproval of these actions points to a low repression potential (1. Low: original scores 0, 1; 2. Medium: 2, missing; 3. High: 3, 4). Variable *D:* Protest Approval is based on approval of seven protest actions ranging from "Signing a petition" to "Occupying buildings or factories" and "Blocking traffic"; disapproval of these actions points to a low tolerance level (1. Low: original scores 0, 1, missing; 2. Medium: 2, 3; 3. High: 4–7). Variable *E:* Conventional Participation is an index based on how often the respondent engages in conventional political activities ranging from "Reading about politics in the newspapers" to "Attending a political meeting or rally"; not taking part in these activities indicates a low level of political participation (1. Low: original scores 0, 1, missing; 2. Medium: 2, 3; 3. High: 4–7). Note that the (few) "missing scores" have been assigned to substantive categories.

Source: The Political Action Study; 1981; see text.

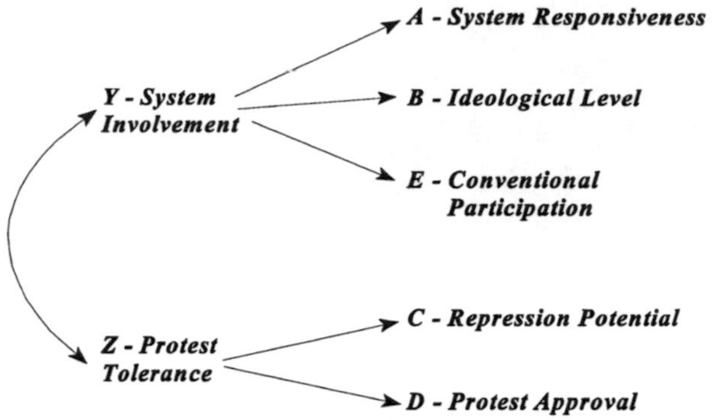

Figure 1. Two-latent-variable model for Table 1.

are restricted to the usual identifying restrictions; the (conditional) probabilities do not exceed the boundaries zero or one and they sum to one where appropriate; the loglinear parameters add up to zero whenever they are summed over any of their subscripts.

$$\pi_{yzabcde}^{YZABCDE} = \pi_{yz}^{YZ}\pi_{ay}^{A|Y}\pi_{by}^{B|Y}\pi_{cz}^{C|Z}\pi_{dz}^{D|Z}\pi_{ey}^{E|Y},$$

$$\ln \pi_{yzabcde}^{YZABCDE} = \lambda + \lambda_y^Y + \lambda_z^Z + \lambda_a^A + \lambda_b^B + \lambda_c^C + \lambda_d^D + \lambda_e^E$$
$$+ \lambda_{yz}^{YZ} + \lambda_{ya}^{YA} + \lambda_{yb}^{YB} + \lambda_{zc}^{ZC} + \lambda_{zd}^{ZD} + \lambda_{ye}^{YE}. \quad (1)$$

The test statistics for the model in Equation (1) in which both latent variables Y and Z have three categories are presented in Table 2, Model 1.[2]

Table 2. Test Results for Two-Latent-Variable Models Applied to Table 1

Model	BIC	X^2	L^2	df	$p(L^2)$
1. $\{YZ, YA, YB, ZC, ZD, YE\}$	−769.50	135.43	146.56	126	0.102
2. All relationships linear × linear	−837.22	185.43	195.16	142	0.002
3. As Model 2, but YZ: nominal × nominal	−827.42	173.86	183.16	139	0.007
4. As Model 1, but latent–manifest: interval × nominal YZ: nominal × nominal	−821.92	154.73	159.57	135	0.073
5. As Model 1, but (YB), ZC, ZD: lin. × lin. YA, YE: int. × nom. YZ: nom. × nom.	−835.13	155.36	160.90	137	0.080
6. As Model 5, but YA: lin. × lin.	−822.19	172.89	181.10	138	0.008
7. As Model 5, but YE: lin. × lin.	−837.70	156.78	165.60	138	0.055

Note: The latent variables Y and Z each have three categories for all models. See the text for exact descriptions of the models.

There is no reason to reject Model 1. From the estimated conditional response probabilities (not shown here, but cf. Table 3), it was concluded that latent variables Y and Z could indeed be interpreted as System Involvement and Protest Tolerance, respectively, with categories Low, Medium, and High.

B. Models for Ordered Data: Linear × Linear Association Models

Model 1 is a strictly nominal level latent class model that contains many parameters. Given the ordered nature of the categories of the indicators A through E, it was investigated whether more restricted models would fit the data. First, it was tested whether all relationships in Figure 1 might be linear: Uniform association models (also called linear × linear association models or logistic latent class models) were defined in which the restrictions presented in Equations (2) and (3) were imposed on the loglinear two-variable parameters of Equation (1) (Haberman, 1979; Goodman, 1984; Clogg and Shihadeh, 1994; Heinen, 1996; McCutcheon, 1993, 1996). The symbols Y_y and Z_z in Equations (2) and (3) denote the scores assigned to the categories of the latent variables Y and Z, respectively. As is often done, the scores are restricted to sum to zero and have equal unit distance (here, for three categories: −1, 0, and 1); A_a, B_b, C_c, D_d, and E_e are the category scores of the observed indicators, restricted in the same way as Y_y and Z_z; λ_{YZ}, λ_{YA}, λ_{YB}, λ_{ZC}, λ_{ZD}, and λ_{YE} are parameters to be estimated showing the strength of the relationship between the variables concerned (i.e., the strength of the uniform association):

$$\lambda_{yz}^{YZ} = (\lambda_{YZ})(Y_y)(Z_z), \tag{2}$$

$$\lambda_{ya}^{YA} = (\lambda_{YA})(Y_y)(A_a),$$
$$\lambda_{yb}^{YB} = (\lambda_{YB})(Y_y)(B_b),$$
$$\lambda_{zc}^{ZC} = (\lambda_{ZC})(Z_z)(C_c),$$
$$\lambda_{zd}^{ZD} = (\lambda_{ZD})(Z_z)(D_d),$$
$$\lambda_{ye}^{YE} = (\lambda_{YE})(Y_y)(E_e). \tag{3}$$

It is doubtful whether the linear × linear latent class model with three latent classes for Y and Z fits the data (see Table 2, Model 2). According to the Bayesian information criterion (BIC) statistic (Raftery, 1993), Model 2 has to be preferred above Model 1, showing BIC's preferential treatment of parsimonious models. However, in terms of L^2, Model 2 does not fit the data. Moreover, the conditional L^2 test of Model 2 against Model 1, using the differences in L^2 between Models 1 and 2, also leads to

Table 3. Estimated Parameters for Model 7 in Table 2, Applied to Data in Table 1

		Probabilities			Loglinear Parameters			Standard Errors $\hat{\sigma}_\lambda$			
		Y			*Y*			*Y*			
		1	**2**	**3**	**1**	**2**	**3**	**1**	**2**	**3**	
			$\hat{\pi}_{zy}^{Z\mid Y}$			$\hat{\lambda}_{yz}^{YZ}$					
Z	1	0.818	0.295	0.147	1.263	−0.316	−0.947	0.652	0.380	0.354	Wald = 8.69
	2	0.132	0.518	0.575	−0.594	0.212	0.382	1.168	0.674	0.511	df = 4
	3	0.049	0.187	0.278	−0.669	0.104	0.565	1.139	0.677	0.476	$p = 0.069$
			$\hat{\pi}_{ay}^{A\mid Y}$			$\hat{\lambda}_{ya}^{YA}$					
A	1	0.370	0.275	0.144	−0.534	0	−0.534	0.117			Wald = 38.57
	2	0.551	0.471	0.283	0.394	0	−0.394	0.096			df = 2
	3	0.079	0.254	0.573	−0.927	0	0.927	0.149			$p = 0.000$
			$\hat{\pi}_{by}^{B\mid Y}$			$\hat{\lambda}_{yb}^{YB}$					
B	1	0.011	0.085	0.435	−1.055	0	1.055	0.222			
	2	0.989	0.915	0.565	1.055	0	−1.055				
			$\hat{\pi}_{ey}^{E\mid Y}$			$\hat{\lambda}_{ye}^{YE}$					
E	1	0.794	0.405	0.073	1.492	0	−1.492	0.246			
	2	0.190	0.430	0.343	0	0	0				
	3	0.016	0.165	0.585	−1.492	0	1.492				

		Z				Z				Z		
		1	2	3		1	2	3		1	2	3
			$\hat{\pi}_{cz}^{C\|Z}$				$\hat{\lambda}_{zc}^{ZC}$					
C	1	0.082	0.308	0.635		−1.016	0	1.016		0.190		
	2	0.272	0.372	0.278		0	0	0				
	3	0.646	0.320	0.087		1.016	0	−1.016				
			$\hat{\pi}_{dz}^{D\|Z}$				$\hat{\lambda}_{zd}^{ZD}$					
D	1	0.759	0.036	0.000		4.314	0	−4.314		1.468		
	2	0.240	0.846	0.087		0	0	0				
	3	0.001	0.118	0.913		−4.314	0	4.314				
$\hat{\pi}_y^Y$		0.149	0.480	0.371	$\hat{\pi}_z^Z$	0.318	0.481	0.200	$\hat{\alpha}_a$	0.468	0.345	−0.813

rejecting Model 2 in favor of Model 1 ($L^2_{2/1} = 195.16 - 146.56 = 48.60$; $df_{2/1} = 142 - 126 = 16$; $p = 0.000$). One way to try to improve the fit of the linear × linear model is to enlarge the number of latent classes, that is, the number of categories of the two latent variables, to four, five, and so on. However, this did not result in models that fit the data better than Model 2.

Then it was investigated whether the restriction in Equation (2) on the relationship between the two latent variables Y and Z might be the source of the lack of fit of Model 2. Even if the two latent variables may be regarded as ordered variables with categories Low, Medium, and High, it does not automatically follow that their relationship must be linear. One might argue that the relationship is nonlinear because people with both high and with low involvement in the political system have reasons to tolerate political protests, but for different reasons. Model 2 was relaxed accordingly by deleting the restriction in Equation (2), resulting in Model 3 of Table 2. Still, Model 3 did not fit the data in terms of L^2, and according to BIC, Model 3 has to be rejected in favor of Model 2. In contrast, the outcomes of the conditional $L^2_{2/3}$ test led to preferring Model 3 above Model 2 ($p = .007$). Given these "inconsistent" test outcomes in combination with the lack of a solid, a priori reason for assuming a linear relationship between Y and Z, it was decided not to impose the restriction of Equation (2) any longer.

C. Models for Ordered Data: Linear × Nominal Association Models

The reason for the lack of fit of Models 2 and 3 might be that the restrictions in Equation (3) are too strict, in the sense that fixed scores had to be assigned to the categories of the indicators. With that, one assumes that the indicators have been measured at the interval level, an assumption that may not be valid here. After all, when answering the original questions, the respondents had to choose among categories such as "often, sometimes, rarely, never" or "approve very much, approve, disapprove, disapprove very much." Assigning interval level scores, such as "1, 2, 3, 4," to these categories, as was done originally, may be arbitrary. Moreover, the categories of the original indices in the Political Action Study obtained by summing the "arbitrary" original item scores were collapsed into three more-or-less uniformly distributed categories (see Note in Table 1). Therefore, assigning interval level scores "−1, 0, 1" to the categories of the collapsed indices A, C, D, and E may indeed be arbitrary.

It is probably more realistic to assume that the relationships between the latent variables and the indicators are linear, that the latent variables Y and Z may be regarded as equally spaced interval level variables, but that the scores on the indicators are not equidistant.

A way to deal with possibly unequally spaced scores is to treat the category scores as unknowns, that is, as extra parameters to be estimated. Within the loglinear framework, these extra parameters can be included by means of a column or row association model (Goodman, 1984; Clogg and Shihadeh, 1994; Luijkx and Hagenaars, 1990). In terms of the column association model, the scores on the indicators acting as the column variables are estimated, whereas the latent variables (the row variables) are still regarded as interval level variables with fixed scores. This column association model implies the following restrictions on those loglinear parameters in Equation (1) that represent the relationships between the latent and manifest variables:

$$\begin{aligned}
\lambda_{ya}^{YA} &= (\lambda_{YA})(Y_y)(\alpha_a),\\
\lambda_{yb}^{YB} &= (\lambda_{YB})(Y_y)(\beta_b),\\
\lambda_{zc}^{ZC} &= (\lambda_{ZC})(Z_z)(\gamma_c),\\
\lambda_{zd}^{ZD} &= (\lambda_{ZD})(Z_z)(\delta_d),\\
\lambda_{ye}^{YE} &= (\lambda_{YE})(Y_y)(\epsilon_e).
\end{aligned} \tag{4}$$

Equations (3) and (4) are very similar to each other; the only difference is that the fixed scores A_a, B_b, C_c, D_d, and E_e have been replaced by α_a, β_b, γ_c, δ_d, and ϵ_e, respectively, and are now parameters whose values have to be estimated. Two extra identifying restrictions are needed. The restrictions $\sum_a \alpha_a = 0$ and $\sum_a \alpha_\alpha^2 = 1$, and similar restrictions for β_b, γ_c, δ_d, and ϵ_e, will be used here. Furthermore, it may be expected that the more the manifest response categories reflect the same intensity of the underlying latent attitude, the closer their estimated scores will be. Also, given the outcomes for Model 1 and the ordered nature of the categories of the indicators and the latent variables, the estimates of the consecutive category scores α_a (and so on) will most probably be ordered as well. Yet this not necessarily true and, therefore, this column association model will be called an *interval–nominal*, rather than an interval–ordinal, LCA: The latent (row) variables are considered interval level variables, but the manifest (column) variables are essentially treated as nominal level variables. This may become more clear when Equation (4) is rewritten in more familiar loglinear form without the estimated scores:

$$\begin{aligned}
\lambda_{ya}^{YA} &= (\lambda_{YA_a})(Y_y),\\
\lambda_{yb}^{YB} &= (\lambda_{YB_b})(Y_y),\\
\lambda_{zc}^{ZC} &= (\lambda_{ZC_c})(Z_z),\\
\lambda_{zd}^{ZD} &= (\lambda_{ZD_d})(Z_z),\\
\lambda_{ye}^{YE} &= (\lambda_{YE_e})(Y_y).
\end{aligned} \tag{5}$$

where (λ_{YA_a}) in Equation (5) equals $(\lambda_{YA})(\alpha_a)$ in Equation (4), and so on, and the scores Y_y and Z_z are fixed.[3]

The relevant test statistics for the model in Figure 1 with two trichotomous latent variables Y and Z in which the restrictions in Equations (4), or (5), are imposed are provided in Table 2, Model 4. Model 4 as such need not be rejected, given the value of L^2. It also has to be preferred above Models 1, 2, and 3, when one compares the L^2's of the first four models. In contrast, the more parsimonious Model 2 still has to be chosen based on the BIC.

Inspection of the estimates of category scores γ_c and δ_d in Model 4 (not shown here) suggested that the categories of C and D were equally spaced. Therefore, linear $\times$ linear models, as in Equation (3), were reintroduced for the relations between Z and its indicators C and D. The resulting Model 5 fitted the data well according to all (conditional) test outcomes. Regarding the relations between Y and its indicators A, B, and E, it should first be noted that the (linear) restriction in Equation (3) concerning B is by definition identical to the restriction for B in Equation (4) or (5) because B is a dichotomous variable. Furthermore, the estimates of the successive scores α_a were ordered but the distances between the successive scores of A were definitely not equal (compare Table 3). Not surprisingly, reimposing the restriction of Equation (3) regarding the relationship $Y-A$ yields a poorly fitting model (Model 6 in Table 2). In constrast, the estimates of ϵ_e, pointed toward E being a more-or-less equally spaced variable. Reintroduction of the linear $\times$ linear model of Equation (3) for the relationship $Y-E$ resulted in Model 7. The (conditional) L^2 test statistics for Model 7 yielded borderline results. In terms of BIC, Model 7 is the best model and at least as well fitting as the most parsimonious model, Model 2. Although the test outcomes are not completely unambiguous, the fact that the parameter estimates for Models 5 and 7 were the same for all practical purposes led to choosing the more parsimonious model. Model 7, in which no restrictions are imposed on the relationship between the latent variables Y and Z, in which the relation Y–A is regarded as interval–nominal and all other relationships between the latent and manifest variables as linear × linear, was chosen as the final model to be used in all subsequent analyses.[4]

D. Interpretation of the Outcomes

The parameter estimates for the final Model 7 have been presented in Table 3. Latent variable Y, System Involvement, is measured by means of three indicators representing an evaluative (A), a cognitive (B), and an action (E) component of involvement in the political system. As follows from the meanings of the categories of the indicators (see Note to Table 1) and from the relationships between the indicators and the latent variables (see later), category 1 of Y, containing 15% of the respondents (see last line of Table 3), consists of people who have "low system involvement"; category 3 (37%) consists of the "highly involved"; and category 2 (48%) occupies a medium involvement position. Variable Y has strong and statistically significant linear relations with indicators B and E (see $\hat{\lambda}_{11}^{YB} = -1.055$ and $\hat{\lambda}_{11}^{YE} = 1.492$): the more one is involved in the political system, the larger is the probability of being an Ideologue, that is, the more one is able to describe the political system in appropriate political terms (B), and the more one engages in regular political activities (E). To give a more precise indication of the strengths of these relationships, remember that odds ratios form the cornerstones of loglinear analysis and that, for example, the local log odds ratio for table YB, involving cell 11, can be found by adding and subtracting the loglinear parameter estimates in the following way: cell 11 + cell 22 − cell 12 − cell 21, here $-1.055 + 0 - 1.055 - 0 = -2.110$. The odds ratio itself then equals $e^{-2.110} = 0.121$ and its inverse is $1/0.1212 = 8.248$. This leads to the conclusion that the odds of being an Ideologue ($B = 1$) rather than a Nonideologue ($B = 2$) are more than eight times larger for people who are moderately involved in the system ($Y = 2$) than for those who have a low involvement ($Y = 1$). It further follows from the linearly restricted parameter estimates for Y–B in Table 3 that the step from medium involvement ($Y = 2$) to high involvement ($Y = 2$) has a similar impact on the odds of being an Ideologue. Considering the extremes, one sees that having a high rather than a low involvement increases the odds of being an Ideologue by a factor $8.248^2 = 68.033$, a very strong effect indeed. The effect of Y on E is even stronger. A comparison of the pertinent parameter estimates for the extreme cells (cell 11 + cell 33 − cell 13 − cell 31) eventually leads to an odds ratio of $e^{(4\times 1.492)} = e^{5.968} = 390.723$: the odds of having a high level of political participation ($E = 3$) rather than a low level ($E = 1$) are almost 400 times larger for high-involvement people ($Y = 3$) than for low-involvement ones ($Y = 1$). To put these figures in perspective, one should also consider the confidence interval of the estimates. From the standard errors presented in Table 3, the 95% confidence interval for the loglinear parameter estimates can be determined, and from these the confidence limits for the (extreme) odds ratios. The confidence interval of the "extreme" odds ratio of 68.033 for the relation between Y and B has a lower limit of 11.936 and an upper limit of 387.796 (68.033 being in the "middle" of this interval as determined by the geometric mean), whereas the confidence interval of the "extreme" odds ratio of 390.723 for the relation between Y and E lies between 56.790 and 2,688.235. Computing these intervals may prevent the researcher from attaching too rigidly a numerical precision to the conclusions and, for instance, concluding too hastily that a particular effect is much larger than another. Also in this little example, a formal test is needed before one would conclude that the effect of Y on E is much larger than that of Y on B. However, the conclusion that E and B are strongly related to Y in the (substantive) sense indicated previously is inescapable.

The (statistically significant) relationship between Y and indicator A shows that the more the respondent is involved in the system, the more she or he evaluates the political system as responsive. However, Y–A is not a linear relationship. As follows from inspection of the relevant loglinear parameters $\hat{\lambda}_{ya}^{YA}$ (e.g., 0.534, 0.394, and −0.927) or, equivalently, from the estimated scores $\hat{\alpha}_a$ (in the lower right corner of Table 3: 0.468, 0.345, −0.813), the categories 1 and 2 of indicator A can almost be collapsed as far as their relationship with latent variable Y is concerned (a possibility that might be further formally tested).

Latent variable Z, labeled Protest Tolerance, is measured by Repression Potential (C) and by Protest Approval (D). The people in category 1 of Z (32%) are the least tolerant for political protest; those in category 3 (20%) are the most tolerant; and category 2 (48%) occupies the middle position. Protest Tolerance has a very strong and statistically significant linear relation with indicator D (e.g., 4.314): the higher the protest tolerance is, the more people approve of more "violent" forms of protest. The linear relation Z–C is a bit less strong, but definitely not weak (e.g., −1.016). It is statistically significant and in the expected direction: the more tolerant a respondent is, the less he or she likes government to repress protests.

Finally, there is the relation between the two latent variables Y and Z. The relationship is reasonably strong (e.g., for the diagonal cells: 1.263, 0.212, 0.565), monotonically increasing, but not exactly linear: the odds that someone has a low level of protest tolerance ($Z = 1$) rather than a nonlow level ($Z = 2$ or $Z = 3$) decrease with increasing system involvement (Y), but the odds of having a medium rather than a high level of protest tolerance hardly change when system involvement increases.[5] However, almost none of the parameters $\hat{\lambda}_{yz}^{YZ}$ are statistically significant, and according to the Wald test statistic for testing the hypothesis that all parameters are equal to zero, Y and Z might be independent of each other in the population ($p = .069$). In contrast, if Model 7 is estimated with the additional restriction that Y and Z are independent of each other, the test statistics are very negative: BIC $= -792.28$, $X^2 = 240.49$, $L^2 = 240.11$, df $= 142$, $p = 0.000$. It might be that the standard errors of the parameter estimates (and accordingly the Wald statistic) are poorly estimated. Drawing any definite conclusions about the relationship between the two latent variables is difficult. In the next section, in which some causes and consequences of having differing degrees of System Involvement and Protest Tolerance will be investigated, the relationship $Y-Z$ will be looked into again.

3. DLM: DIRECTED LOGLINEAR MODELS WITH LATENT VARIABLES

A. Causal Model with Latent Variables

Once a satisfactory (measurement) model for the data in Table 1 has been obtained, an obvious next step is to examine the causes and consequences of the latent variables. It will be investigated how, on the one hand, the respondents' scores on the two (latent) basic political orientations Y and Z are rooted in the respondents' social positions as determined by some important background characteristics and how, on the other hand, the latent scores influence their voting behavior.

The starting point is the causal model depicted in Figure 2 (partly borrowed from Hagenaars, 1993). Figure 2 contains two exogenous variables S, Sex, and G, Age (or Generation), which are not influenced by the other variables. According to the model in Figure 2, S and G are correlated without further specification of the (causal) nature of their relationship; hence, the curved two-headed arrow. One might have expected S and G to be uncorrelated, were it not for possible differential

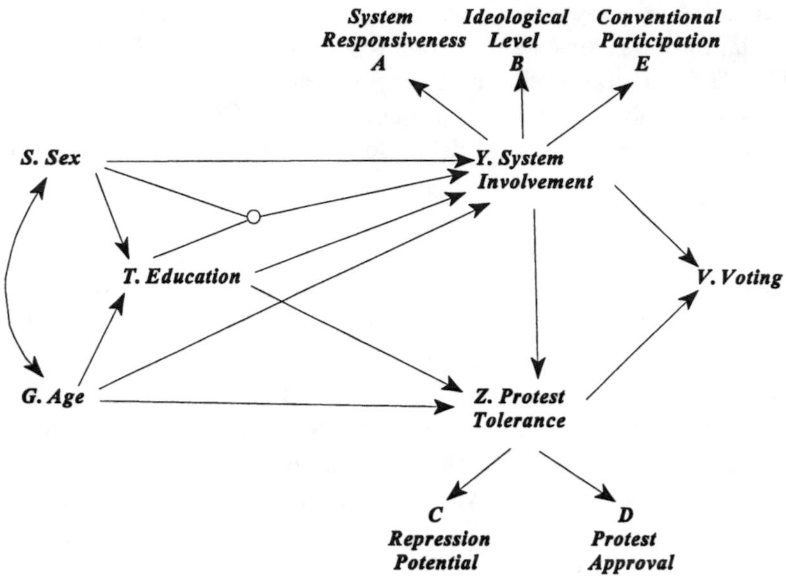

Figure 2. Causal model with latent variables (Model 0, Table 5).

nonresponse effects in the Political Action survey or for the differential mortality rates of men and women. Both exogenous variables directly influence *T*, Education (educational Training). The hypotheses are that later generations and men have better educational opportunities than earlier generations and women, respectively. These assumed direct causal (partial) relationships are represented in Figure 2 by directed arrows. If no arrows are drawn between two variables, there is no *direct* relationship between them, meaning that the two variables are (conditionally) independent of each other when one controls for the relevant intervening or antecedent variables (if any). Variables that are not directly connected may have a nonzero marginal association because of the presence of indirect effects or spurious associations.

Latent variable *Y*, System Involvement, is assumed to be directly influenced by *S*, *G*, and *T*. It is hypothesized that these direct effects are such that men, earlier generations (also being older), and higher-educated people are more involved in the political system than their respective counterparts. The node connecting the three variables *S*, *T*, and *Y* indicates a three-variable interaction effect among the three connected variables: the direct influence of education on system involvement is supposed to be different for men and women; more specifically, it is supposed to be stronger for men than for women. Four-variable interactions, not present here, would be represented by nodes connecting four variables and so on. If no nodes are drawn, no three- or more-variable interactions are assumed to exist.

Latent variable Y is considered a direct cause of latent variable Z, Protest Tolerance, rather than vice versa. Although it is somewhat arbitrary, system involvement is regarded as a more basic characteristic than protest approval. As was remarked earlier, the relationship between Y and Z might be linear or curvilinear. Protest Tolerance (Z) is also directly influenced by G and T: It is assumed that younger and higher-educated people are comparatively more tolerant toward protests than their respective counterparts. S does not directly influence Z: No differences between men and women with regard to protest tolerance are expected when one controls for G, T, and Y. Finally, there is the ultimate dependent variable V, Voting: V is only directly influenced by the two latent variables Y and Z. The expectation is that system involvement mainly relates to the decision to vote or not: the higher someone's system involvement is, the higher her or his probability of voting rather than choosing "Other" (mainly nonvoters). Republicans and Democrats are expected to show the same degree of system involvement. Protest Approval is supposedly more related to party choice: the higher the protest approval is, the more one is inclined to vote for the Democrats rather than the Republicans; the nonvoters (Other) probably occupy a middle position. It is further hypothesized and depicted in Figure 2 that the effects of the background variables S, T, and G on V are completely mediated through the two latent variables.

Besides these causal relationships that form the structural part of the causal model, there are the relationships within the measurement (LCA) part, connecting the latent variables to their indicators. The relationships between the latent variables and the indicators is assumed to be essentially the same as in Model 7 from the previous Section (see also Figure 1). As explained later, it is not trivial that variables S, G, T, and V are supposed to have no direct influences on indicators A through E.

B. The Modified Path Approach and Directed Loglinear Modeling

As Goodman has made clear, the causal model in Figure 2 cannot be evaluated by means of a single loglinear model, but a "modified path approach," that is, a particular system of loglinear or logit equations, is needed, one for each "next" dependent variable (Goodman, 1973). Application of the modified path approach poses some extra problems here because variables Y and Z are not observed; this difficulty will be dealt with later. The starting points of the modified path approach for the causal model in Figure 2 are the full cross-classification of all variables, that is, table $SGTYZVABCDE$ with entries $\pi_{sgtyzvabcde}^{SGTYZVABCDE}$ and the decomposition of this joint probability $\pi_{sgtyzvabcde}^{SGTYZVABCDE}$ into a series of "simpler" (conditional) probabilities in line with the causal order of the variables in Figure 2. The nature of this decomposition can be explained by means of a very simple example with only two variables P and Q. The joint probability π_{pq}^{PQ} can be decomposed as $\pi_{pq}^{PQ} = \pi_{p}^{P}\pi_{qp}^{Q|P}$ or as $\pi_{pq}^{PQ} = \pi_{q}^{Q}\pi_{pq}^{P|Q}$. Both decompositions follow from elementary rules of probability and are tautological, that is, are true by definition, and do not imply any restrictions on the joint distribution of P and Q (contrary to, e.g., the decomposition $\pi_{pq}^{PQ} = \pi_{p}^{P}\pi_{q}^{Q}$, which is valid if and only if P and Q are statistically independent of each other). Nevertheless, the choice between $\pi_{pq}^{PQ} = \pi_{p}^{P}\pi_{qp}^{Q|P}$ and $\pi_{pq}^{PQ} = \pi_{q}^{Q}\pi_{pq}^{P|Q}$ may not be arbitrary in a causal analysis. If the causal order is $P \rightarrow Q$, then the decomposition $\pi_{pq}^{PQ} = \pi_{p}^{P}\pi_{qp}^{Q|P}$ is applicable: One studies the marginal distribution of independent, exogenous variable P by means of π_{p}^{P}, and then one investigates by means of $\pi_{qp}^{Q|P}$ how the scores on Q vary with, in other words, depend on P. If, in contrast, P is considered determined by Q, that is, $Q \rightarrow P$, the decomposition $\pi_{pq}^{PQ} = \pi_{q}^{Q}\pi_{pq}^{P|Q}$ is relevant for analogous reasons.

Looking in this way at the causal model in Figure 2 and assuming $P \rightarrow Q$, we see that there is a set of variables that together form "P," namely, S, G, T, Y, Z, and V, and a set of "Q" variables, namely A, B, C, D, and E. Type P variables influence type Q variables, but not vice versa. This leads to the following (first) decomposition of the joint probability $\pi_{sgtyzvabcde}^{SGTYZVABCDE}$:

$$\pi_{sgtyzvabcde}^{SGTYZVABCDE} = \left(\pi_{sgtyzv}^{SGTYZV}\right)\left(\pi_{abcdesgtyzv}^{ABCDE|SGTYZV}\right). \tag{6}$$

The set of P variables consists here of the variables whose mutual causal connections we are interested in; the Q set of variables are the indicators A through E. The first element at the right side of Equation (6), π_{sgtyzv}^{SGTYZV}, represents the joint probability distribution of the proper causal variables. The set of (loglinear) equations that describes the probability distribution of $SGTYZV$ in table $SGTYZV$, that, in other words, describes the causal relationships among S, G, T, Y, Z, and V, is denoted as the structural part of the model. Because the set of P variables is not influenced by the set of Q variables, the relationships among the structural variables S, G, T, Y, and Z should be determined by means of loglinear models for marginal table $SGTYZ$ and not for the complete table $SGTYZABCDE$. Use of the latter table would result in (loglinear) direct effects among the structural variables that condition on the scores on the indicators A through E. This obviously does not make sense here from a causal point of view. The second element at the right side of Equation (6), $\pi_{abcdesgtyzv}^{ABCDE|SGTYZV}$, forms the measurement part that describes how the scores on indicators A through E are related to each other and how they depend on the (latent and manifest) structural variables S through V.

Further application of the principles of the modified path approach to the structural part in Equation (6) yields Equation (7):

$$\pi_{sgtyzvabcde}^{SGTYZVABCDE} = \left(\pi_{sg}^{SG}\pi_{tsg}^{T|SG}\pi_{ysgt}^{Y|SGT}\pi_{zsgty}^{Z|SGTY}\pi_{vsgtyz}^{V|SGTYZ}\right) \times \left(\pi_{abcdesgtyzv}^{ABCDE|SGTYZV}\right) \tag{7}$$

This particular decomposition or factorization of π_{sgtyzv}^{SGTYZV} mirrors the causal order among the structural variables in Figure 2: $(S, G)\rightarrow T\rightarrow Y\rightarrow Z\rightarrow V$, where (S, G) indicates the lack of causal order between S and G. Departing from the adage that what is "causally posterior" cannot influence what is "causally prior," essential to any causal model (whether the variables are continuous or not) is that one should investigate the relations between any two variables while controlling for all antecedent and intervening variables, but ignoring the variables that appear later in the causal chain. Because S and G are the two exogenous variables, which are not influenced by any of the other variables in the model, their (noncausal) relationship has to be investigated in marginal table SG with entries π_{sg}^{SG}, the first element at the right-hand side of Equation (7). The first endogenous variable is T. The influence of all prior variables, that is, of S and G, on T are determined in marginal table SGT, defining loglinear (logit) models for the conditional probabilities $\pi_{tsg}^{T|SG}$, the second element at the right-hand side of Equation (7). Following the same logic, the conditional probabilities $\pi_{ysgt}^{Y|SGT}$ in marginal table $SGTY$ must be used to determine the effects of all prior variables (S, G, and T) on Y, whereas the effects of S, G, T, and Y on Z are determined by means of the conditional probabilities $\pi_{zsgty}^{Z|SGTY}$ obtained from Table $SGTYZ$. The influences on the ultimate dependent variable V have to be investigated by means of $\pi_{vsgtyz}^{V|SGTYZ}$.

As explained earlier, Equations (6) and (7) are tautologies whose truths follow from elementary rules of probability and, as such, do not impose any restrictions on the data. Many other different tautological factorizations π_{sgtyzv}^{SGTYZV} (or $\pi_{sgtyzvabcde}^{SGTYZVABCDE}$) are possible, but only the one in Equation (7) is appropriate here because it points out which (independent) variables have to be considered as the possible causes of a particular dependent variable and, accordingly, which (marginal) tables or probabilities have to be used to determine the effects among the structural variables, given their assumed causal order. However, the causal model in Figure 2 is not a saturated model as it does imply certain restrictions on the data, restrictions that must be imposed by formulating the appropriate (nonsaturated) logit (or loglinear; see Note 5) model for each of the structural components in Equation (7) (Goodman, 1973).

Although leading to exactly the same estimates of the parameters and standard errors as the standard Goodman procedure, it may have practical advantages in terms of computing time and computer memory storage capacities, as Vermunt (1997a) has shown, to follow a somewhat modified Goodman procedure: Before defining the appropriate loglinear (sub)models, Equation (7) is simplified and made nontautological, applying principles and results from directed graphical modeling (see the references in Note 2). If the causal model in Figure 2 is valid in the population, it follows that particular variables are conditionally independent of each other. Variable Z depends directly only on G, T, and Y but is conditionally independent of S, given G, T, and Y. Therefore, conditional probability $\pi_{zsgty}^{Z|SGTY}$ in Equation (7) does not vary with subscript s and, if the model is true, it can be replaced by $\pi_{zgty}^{Z|GTY}$ without altering any essential characteristic of the model. Furthermore, variable V is conditionally independent of S, G, and T, given Y and Z. So, conditional probability $\pi_{vsgtyz}^{V|SGTYZ}$ can be replaced by $\pi_{vyz}^{V|YZ}$.

Similar simplifications and restrictions can be applied to the measurement part in Equation (7), that is, to conditional probability $\pi_{abcdesgtyzv}^{ABCDE|SGTYZV}$. First, the standard local (conditional) independence assumption has been made for the relationships among the indicators. Therefore, the joint conditional probability $\pi_{abcdesgtyzv}^{ABCDE|SGTYZV}$ can be replaced by the product of the marginal conditional probabilities for each indicator, that is, the product of $\pi_{asgtyzv}^{A|SGTYZV}$, $\pi_{bsgtyzv}^{B|SGTYZV}$, $\pi_{csgtyzv}^{C|SGTYZV}$, and so on. Second, according to the model in Figure 2, the score on each particular indicator only depends directly on the score on "its" latent variable, and therefore $\pi_{asgtyzv}^{A|SGTYZV}$ can be replaced by $\pi_{ay}^{A|Y}$, $\pi_{csgtyzv}^{C|SGTYZV}$ by $\pi_{cz}^{C|Z}$, and so on. Equation (7) can then be rewritten as follows:

$$\pi_{sgtyzvabcde}^{SGTYZVABCDE} = \left(\pi_{sg}^{SG}\pi_{tsg}^{T|SG}\pi_{ysgt}^{Y|SGT}\pi_{zgty}^{Z|GTY}\pi_{vyz}^{V|YZ}\right) \times\left(\pi_{ay}^{A|Y}\pi_{by}^{B|Y}\pi_{cz}^{C|Z}\pi_{dz}^{D|Z}\pi_{ey}^{E|Y}\right). \quad (8)$$

Remembering the difference between the tautological decomposition $\pi_{pq}^{PQ} = \pi_p^P\pi_{qp}^{Q|P}$ and the nontautological decomposition $\pi_{pq}^{PQ} = \pi_p^P\pi_q^Q$ mentioned above, we find that Equation (8) is clearly no longer a tautology but is only true if the postulated causal model is valid in the populations. In other words, Equation (8) imposes restrictions on the data that can be tested. These restrictions are constraints in terms of the occurrence of a (conditional) independence relationship between two variables.

However, not all restrictions intended by the causal diagram in Figure 2 and the substantive theory behind it can be represented as conditional independence restrictions. For example, the absence of the three-variable interaction between *S*, *G*, and *T* or particular linear restrictions on the relationships among variables cannot be expressed in terms of conditional independence. Therefore, an appropriate set of logit (or loglinear) models has to be defined – one model equation for each element at the right-hand side of Equation (8). An overview of the required set of loglinear models in the form of hierarchical models, ignoring possible nonhierarchical restrictions on the parameters, is provided in Table 4. (The entries $\hat{f}$ in the last column in Table 4 will be explained later.)

Forgetting for a moment that *Y* and *Z* are latent variables, we find it is straightforward to obtain in this way the maximum-likelihood estimates for all effects in Figure 2, as well as the maximum-likelihood estimates $\hat{\pi}$ for each of the right side elements of Equation (8), or (7). Substituting these right-hand side maximum-likelihood estimates $\hat{\pi}$ in Equation (8), or (7), we can compute the maximum-likelihood estimate of the joint probability $\pi_{sgtyzvabcde}^{SGTYZVABCDE}$ on the left side of Equation (8). The empirical validity of the complete causal model can then be evaluated by comparing $\hat{F}_{sgtyzvabcde}^{SGTYZVABCDE} (= N\hat{\pi}_{sgtyzvabcde}^{SGTYZVABCDE})$ with the observed frequencies $f_{sgtyzvabcde}^{SGTYZVABCDE}$ by means of standard chi-squared tests or otherwise. However, *Y* and *Z* are latent variables here and some complications arise.

Table 4. System of Loglinear Equations for Causal Model in Figure 2 and Equation (8)

(Conditional) Probabilities	Table	Loglinear Model	Estimated Observed Frequencies $\hat{f}$
Structural part			
π_{sg}^{SG}	SG	$\{SG\}$	$\hat{f}_{sg+++++++++}^{SGTYZVABCDE}\left(=f_{sg}^{SG}\right)$
$\pi_{tsg}^{T\|SG}$	SGT	$\{SG, ST, GT\}$	$\hat{f}_{sgt++++++++}^{SGTYZVABCDE}\left(=f_{sgt}^{SGT}\right)$
$\pi_{ysgt}^{Y\|SGT}$	$SGTY$	$\{SGT, STY, GY\}$	$\hat{f}_{sgty+++++++}^{SGTYZVABCDE}$
$\pi_{zgty}^{Z\|GTY}$	$GTYZ$	$\{GTY, GZ, TZ, YZ\}$	$\hat{f}_{+gtyz++++++}^{SGTYZVABCDE}$
$\pi_{vyz}^{V\|YZ}$	YZV	$\{YZ, YV, ZV\}$	$\hat{f}_{+++yzv+++++}^{SGTYZVABCDE}$
Measurement part			
$\pi_{ay}^{A\|Y}$	YA	$\{YA\}$	$\hat{f}_{+++y++a++++}^{SGTYZVABCDE}$
$\pi_{by}^{B\|Y}$	YB	$\{YB\}$	$\hat{f}_{+++y+++b+++}^{SGTYZVABCDE}$
$\pi_{ey}^{E\|Y}$	YE	$\{YE\}$	$\hat{f}_{+++y++++++e}^{SGTYZVABCDE}$
$\pi_{cz}^{C\|Z}$	ZC	$\{ZC\}$	$\hat{f}_{++++z++++c++}^{SGTYZVABCDE}$
$\pi_{dz}^{D\|Z}$	ZD	$\{ZD\}$	$\hat{f}_{++++z++++d+}^{SGTYZVABCDE}$

C. The Modified LISREL Approach and DLM with Latent Variables

The main problem in the approach sketched earlier is that there does not exist an observed table $SGTYZABCDE$ with observed cell entries $f_{sgtyzvabcde}^{SGTYZVABCDE}$ that can be used to estimate all relevant (causal) effects and obtain the estimates $\hat{\pi}_{sgtyzvabcde}^{SGTYZVABCDE}$. Because Y and Z are latent, not directly observed variables, only table $SGTABCDE$ is observed with cell entries $f_{sgtvabcde}^{SGTVABCDE}$.

Nevertheless, it is possible to find the maximum-likelihood estimates for (directed, causal) loglinear models with latent variables by applying methods that are very similar to the ones used for obtaining the maximum-likelihood estimates for ordinary latent class models. Among the several methods and algorithms that can be used to estimate the parameters of this "modified LISREL model" (Hagenaars, 1990), the expectation–maximization (EM) algorithm occupies a central place and will be very briefly and roughly explained here (Hagenaars, 1993; Vermunt, 1997a). As usual, in the E step of the EM algorithm, the entries $\hat{f}_{sgtyzvabcde}^{SGTYZVABCDE}$ of the complete "observed" cross-classification table $SGTYZVABCDE$ are estimated, conditional upon the observed frequencies $f_{sgtvabcde}^{SGTVABCDE}$ and using the estimated model parameters or the estimates of $F_{sgtyzvabcde}^{SGTYZVABCDE}$ obtained so far. In the M step, this estimated complete "observed" table is treated as if it were an ordinary observed table to obtain better estimates of the model parameters (or of $F_{sgtyzvabcde}^{SGTYZVABCDE}$), which in their turn are used in the E step to improve the estimates $\hat{f}$, and so forth. The estimates $\hat{F}^{SGTVABCDE}_{sgtvabcde}$, obtained by collapsing the final estimates $\hat{F}^{SGTYZVABCDE}_{sgtyzvabcde}$ with regard to the latent variables Y and Z, can be used to test the model by comparing them with the observed frequencies $f^{SGTVABCDE}_{sgtvabcde}$.

In this way, this is not different from the standard application of the EM algorithm to obtain the parameter estimates of the ordinary latent class model, certainly not with regard to the E step and the succession of E and M steps. What is more complicated here in comparison to the standard latent class model is what has to be done within the M step, namely, the application of Goodman's modified path approach or DLM described above (see also Table 4).

This modified LISREL approach or DLM with latent variables (Hagenaars, 1998; Vermunt, 1996) works only in this simple, direct way if the parameters belonging to one (sub)model equation can be estimated apart from the other equations. This is, for instance, not true when equality restrictions are imposed on parameters that belong to different (sub)models (e.g., the effect of T on Y is the same as the effect of T on Z) or when the causal model is nonrecursive (Y influences Z, but Z also directly or indirectly influences Y). Such complexities will not be dealt with here; for an exposition of how to handle these and other complications, see Cox and Wermuth (1996), Hagenaars (1998), Koster (1996), Lauritzen (1996), and Vermunt (1996, 1997a).[6]

D. Application: Selecting Models

The approach described earlier has been applied to the United States data from the Political Action Study mentioned before, using the variables (and their causal order) from Figure 2. The observed table $SGTVABCDE$ (for obvious reasons not presented here) has 8,748 cells, whereas the number of observations is only 1,437. Thus this table is very sparse and test results should be treated very cautiously. First, it was tested how well Model 0, that is, the causal model in Figure 2 [Equation (8) and Table 4] fitted the data. With regard to the measurement part of Model 0, it is assumed here and in all forthcoming models for these data that the scores on the indicators A through E only depend on the latent variables and in the ways specified in Model 7, Table 2. In other words, models $\{YA\}$, $\{YB\}$, $\{ZC\}$, and so on in Table 4 do not refer to the saturated model but rather to particular linearly restricted models. The test statistics for Model 0 are reported in Table 5. Pearson X^2- and the (log)likelihood ratio L^2 test yield

Table 5. Test Results for Causal Models for Variables in Figure 2, Table *SGTYZVABCDE*

Model	BIC	X^2	L^2	df
0. Model in Figure 2	−59,345.75	9,206.03	3,753.30	8,679
1. {*SGTYZV*, *AY*, *BY*, *EY*, *CZ*, *DZ*}	−56,577.63	7,078.79	3,380.65	8,247
2. As Model 1, but for Table *SG*{*S*, *G*}	−56,575.31	7,122.13	3,397.50	8,249
3. As Model 1, but for Table *SGT*{*SG*, *ST*, *GT*}	−56,598.84	7,046.27	3,388.52	8,251
4. As Model 3, but for Table *SGTY*{*SGT*, *SY*, *GY*, *TY*}	−56,756.92	7,112.56	3,404.92	8,275
5. As Model 4, but *TY*: lin. × lin.	−56,776.01	7,104.53	3,407.64	8,278
6. As Model 5, but for Table *SGTYZ* {*SGTY*, *SZ*, *GZ*, *TZ*, *YZ*}	−57,358.68	7,670.22	3,493.83	8,370
7. As Model 6, but *GZ* lin. × lin. *TZ* lin. × lin.	−57,395.55	7,704.82	3,500.59	8,376
8. As Model 7, but *YZ* interval × nom.	−57,405.58	7,745.73	3,505.11	8,378
9. As Model 8, but for Table *SGTYZV* {*SGTYZ*, *SV*, *GV*, *TV*, *YV*, *ZV*}	−59,384.19	9,333.33	3,736.67	8,682
10. As Model 8, but for Table *SGTYZV* {*SGTYZ*, *SGV*, *STV*, *SYV*, *SZV*, *GTV*, *GYV*, *GZV*, *TYV*, *TZV*, *YZV*}	−59,030.89	9,094.00	3,624.67	8,618
11. As Model 8, but for Table *GTYZV* {*GTYZ*, *GV*, *TV*, *YV*, *ZV*}	−59,397.43	9,341.08	3,737.96	8,684
12. As Model 8, but for Table *GTYZV* {*GTYZ*, *GTV*, *GYV*, *GZV*, *TYV*, *TZV*, *YZV*}	−59,142.99	8,713.17	3,643.43	8,636
13. As Model 11, but *GV*, *TV*, *YV*, all lin × nom.	−59,432.74	9,249.90	3,746.28	8,690
14. Final Model: Table *SG*, Model 1; Table *SGT*, Model 3; Table *SGTY*, Model 5 + *SY* lin. × lin.; Table *SGTYZ*, Model 8 + *YZ* lin. × lin.; Table *SGTYZV*, Model 13	−59,444.36	9,309.80	3,749.21	8,692

Note: The measurement part is always restricted in the manner of Model 7, Table 2; see text for exact descriptions of the models.

widely diverging results; the former leads to a *p* value of 0.00 and the latter to one of 1.00. It is hard or rather impossible to tell from these outcomes whether Model 0 as such fits the data. A better approach might be to compare models and choose among a set of theoretically plausible models. The strategy in search of alternative models will be to relax in a stepwise manner particular restrictions in the causal, structural part of the model but to retain the restrictions in the measurement part. Therefore, the following decomposition will be used as the starting point (after imposing the extra restrictions on the measurement part):

$$\pi_{sgtyzvabcde}^{SGTYZVABCDE} = \left(\pi_{sg}^{SG}\pi_{tsg}^{T|SG}\pi_{ysgt}^{Y|SGT}\pi_{zsgty}^{Z|SGTY}\pi_{vsgtyz}^{V|SGTYZ}\right) \times\left(\pi_{ay}^{A|Y}\pi_{by}^{B|Y}\pi_{cz}^{C|Z}\pi_{dz}^{D|Z}\pi_{ey}^{E|Y}\right). \tag{9}$$

Before trying to impose restrictions on the structural part of the model, we must first investigate whether it is true that the relations between the indicators (A through E) and the external variables (S, G, T, and V) are completely mediated through the latent variables Y and Z. This is an important measurement assumption and crucial for our interpretation of the latent variables Y and Z, as it amounts to the assumption of absence of differential item functioning; for example, item A does not depend directly on S and as such does not function differently for men and women (Van de Vijver and Leung, 1997; Camilli and Shephard, 1994; Hagenaars, 1988, 1993, pp. 38, 51). By applying saturated models to all elements of the structural part of Equation (9) but the (linearly) restricted models mentioned here to all elements of the measurement part, we obtain a model denoted as Model 1, in which the only restrictions come from the assumptions about the measurement part and from the assumption that S, G, T, and V do not directly influence the scores on the indicators.[7] If we take the restricted nature of the relations between the latent variables and the indicators for granted based on the analyses in Section 2, the test statistics for Model 1 in Table 5 provide information about the conditional independence of S, G, T, or V and the indicators. As such (and again), the test statistics for Model 1 tell us hardly anything meaningful. However, by defining models in which direct effects are added between variables S, G, T, or V and the indicators and comparing the test statistics obtained with those of Model 1, we find it possible to test the assumption of the absence of differential item functioning (Hagenaars, 1993). The test results (not reported here) did not show that including extra direct effects on the indicators was necessary. It is also important that the parameter estimates for the relations between the indicators and the latent variables in Model 1 are very similar to the estimates reported in Table 3, based on table $ABCDE$, and lead to the same interpretation of the latent variables. So, Model 1 forms a good starting point to test certain key features of the structural part. The stepwise testing procedure for model selection followed will be briefly explained later; discussion of the substantive results in terms of parameter estimates will take place after the final model has been chosen.

Following the logic of the causal order of the variables implied by Figure 2 and Equation (9), we find that the first subtables to be considered are the marginal tables *SG* and *SGT*. Because these subtables do not contain any latent variables, the relationships among *S*, *G*, and *T* can be investigated by simply defining the appropriate loglinear models for the observed tables *SG* and *SGT* (see also Table 4) without having to use the "complicated" EM approach described earlier. Application of the independence model $\{S, G\}$ to observed table *SG* made it clear that *S* and *G* are not statistically independent of each other ($L^2 = 16.855$, df $= 2$, $p = 0.00$). The preferred model for the effects of *S* and *G* on *T* using observed table *SGT* is the no-three-variable-interaction model $\{SG, ST, GT\} : L^2 = 7.87$, df $= 4$, $p = .10$; deleting the main effects of *S* or *G* on *T* yielded models that had to be rejected.

The same test results in terms of L^2 might have been obtained by following more closely the logic of DLM and the complicated EM approach. The starting point is Model 1 (Table 5), in which saturated models are defined for all structural elements in Equation (9). Model 2 is the same as Model 1, except that now for the first structural element π_{sg}^{SG} in Equation (9) Model $\{S, G\}$ is defined rather than saturated model $\{SG\}$. The test statistics are reported in Table 5, Model 2. Because the only difference between Models 1 and 2 is the independence of *S* and *G*, the conditional test statistic $L^2_{2/1} = 3{,}397.50 - 3{,}380.65 = 16.85$, $\text{df}_{2/1} = 8{,}249 - 8{,}247 = 2$, and $p = 0.000$ is the same as the unconditional L^2 obtained earlier. It is concluded (also from BIC) that the saturated model $\{SG\}$ should be applied to π_{sg}^{SG}. For determining the effects of *S* and *G* on *T*, Model 1 has been modified by defining model $\{SG, ST, GT\}$ for table *SGT* corresponding to the second structural element $\pi_{tsg}^{T|SG}$. The conditional test statistic for testing the validity of Model 3 against Model 1 yields $L^2_{3/1} = 3{,}388.52 - 3{,}380.65 = 7.87$, $\text{df}_{3/1} = 4$, and $p = .10$, the same test outcomes obtained earlier: Model 3 has to be preferred above Model 1 (also based on BIC) and becomes the new starting point. So far, the results agree with the hypotheses that led to the causal model in Figure 2.

The next effects to be investigated are the effects of *S*, *G*, and *T* on latent variable *Y*, System Involvement. Because of the presence of latent variable *Y*, no shortcuts using a simple observed table are possible. The new baseline Model 3 is now modified only by estimating the conditional probability $\pi_{ysgt}^{Y|SGT}$ in Equation (9) by means of model $\{SGT, STY, GY\}$ for table *SGTY*, in agreement with the initial hypotheses. The test outcomes for this modified Model 3 confirmed these initial hypotheses about the effects on *Y*: BIC $= -56{,}732.26$, $X^2 = 7{,}082.92$,

$L^2 = 3{,}400.50$, and df $= 8271$. However, the values for the three variable interaction parameters $\hat{\lambda}^{STY}_{sty}$ were very small and not significant. Model 4, in which the interaction effects of *S* and *T* on *Y* were left out, fitted the data equally well (see Table 5, Model 4). An inspection of the relevant parameters in Model 4 (not shown here) suggested a linear relationship between *S* and *Y* and between *T* and *Y*. In Model 5, the linear × linear association model for the direct effects of *T* on *Y* is defined, imposing restrictions on λ^{TY}_{ty} that are analogous to the restrictions in Equations (2) and (3). From inspection and computation of the (conditional) test statistics, the relationship *T–Y* may clearly be regarded as linear. Imposing linear restrictions on the relation between *S* and *Y* yielded borderline test results, and it was decided not to impose this restriction at this stage.

For the investigation of the effects on *Z*, Model 5 is used as the starting point. Saturated model {*SGTYZ*} for table {*SGTYZ*} is replaced by the main-effects-on-*Z*-only model {*SGTY, SZ, GZ, TZ, YZ*} (Model 6). Given the test outcomes, we find that Model 6 must be preferred above Model 5. Introducing additional three-variable effects involving *Z* did not yield models that fitted the data significantly better than Model 6. According to the original hypotheses, there should be no direct effects of *S* on *Z*. The effects *S–Z* in Model 6 were rather small, but because of the outcomes of the conditional tests, it was decided to keep them included. It was further concluded from the test outcomes that a model in which linear × linear association models were postulated for the relations *G–Z* and *T–Z* had to be preferred above Model 6 (see Table 5, Model 7). As before, the estimates of the direct effects of *Y* on *Z* (in Models 6 and 7) suggested a monotonically increasing direct relationship, but not a perfectly linear one. The best restricted model for the relationship *Y–Z* turned out to be a column association model in which the row variable *Y* is treated as an interval level variable and the column variable *Z* as nominal, similar to the restrictions in Equation (5) (Model 8; $L^2_{8/7} = 4.52$; $\text{df}_{8/7} = 2$, and $p = .10$).

This brings us to the effects on the final dependent variable *V*, using Model 8 as the starting point. For determining the effects on *V*, Voting, Model 8 was modified to include only the main effects on *V*, resulting in Model 9. According to the (conditional) test statistics, Model 9 did not fit the data significantly worse than Model 8 ($L_{8/9} = 231.56$, df $= 304$, $p = .999$; see also BIC). From the original hypotheses, it was expected that there would be no direct effects of *S*, *G*, and *T* on *V*. Models in which the effects of *G* or *T* on *V* were deleted, however, fitted the data significantly worse than Model 9. The effects of *S* on *V*, in contrast, were very small (as estimated in Model 9) and were not significant (compare the test statistics for Models 9 and 11). In Model 11 the effects of S on V have been deleted.

Whether including any higher-order interaction effects on V is necessary is not clear. Models 9 and 11 with no three- or higher-order effects on V do not fit the data worse than Model 8, which contains all possible higher-order effects on V. In contrast, a comparison of the main-effects-on-V-only Models 9 or 11 with Models 10 or 12 in which all three variable effects on V have been included yields ambiguous results: According to the BIC, the more parsimonious models must be preferred, but from the conditional L^2 tests it must be concluded that the higher-order interaction terms contribute significantly ($L^2_{9/10} = 112.00$ df $= 64$, $p = 0.00$, and $L^2_{11/12} = 94.56$, df $= 48$, $p = 0.00$). However, when estimated, these three-variable interaction effects $\hat{\lambda}^{GTV}_{gtv}$, $\hat{\lambda}^{GYV}_{gyv}$, and so on show an extremely capricious pattern. They are excessively (and unbelievably) large, but only a few are significant; finding any meaningful interpretation of the outcomes is impossible. Furthermore, if the three variable effects were added one at a time or deleted one at a time, the conditional L^2 tests yielded borderline results, that is, mostly p values between .005 and .070. In the end it was decided to leave out all three-variable interaction effects on V and to choose Model 11 as the new baseline model. Inspection of the parameter estimates in Model 11 led finally to Model 13, in which linear restrictions are imposed on the effects of G, T, and Y on the log odds of belonging to one category of V rather than another: Column association models analogous to the restrictions in Equations (4) or (5) were used with Voting as the nominal (column) variable and the G, T, and Y as interval level variables ($L^2_{13/11} = 8.32$, df $= 6$, $p = .22$).

Model 13 should not automatically be regarded as the definite model. Apart from the usual, but serious pitfalls of using statistical tests in a partly exploratory way, latent variable models have an additional drawback. Latent variables are technically "defined" by their relationships with the indicators and the other manifest, external variables in the model. If one introduces additional constraints on these relationships later in the causal model, earlier findings must be checked again. This is true even if the "final" model is valid in the population because of sampling fluctuations. In more technical terms, the estimated observed frequencies $\hat{f}$ change when additional restrictions are imposed that change the estimated probabilities $\hat{\pi}$ for the whole model (contrary to the normal observed frequencies f). Therefore, it was investigated whether imposing some additional restrictions or adding to or deleting from Model 13

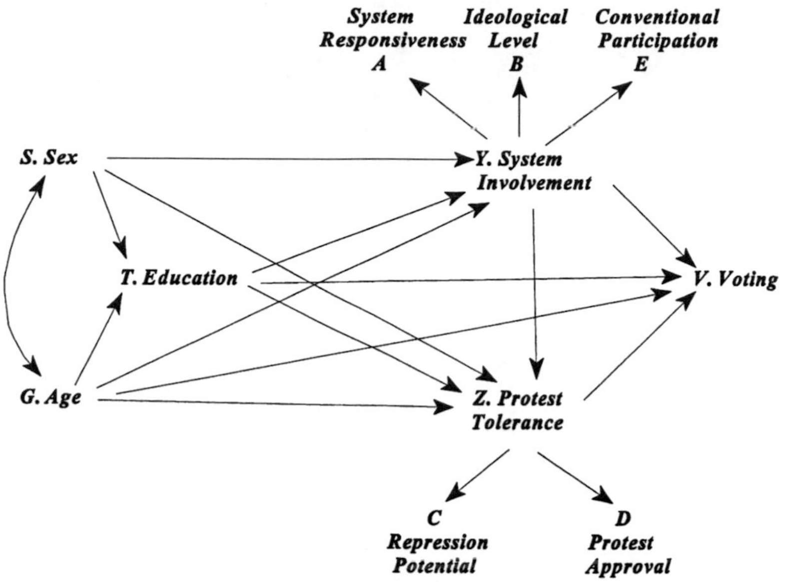

Figure 3. Causal model with latent variables (Model 14, Table 5).

particular effects would result in a "better" model. The (sub)models for tables *SG* and *SGT* will, of course, remain the same, as these pertain to observed variables only. With regard to the effects on *Y*, the only change compared with Model 5 concerns the relationship *S*–*Y*. From the parameter estimates $\hat{\lambda}_{sy}^{SY}$ in Model 5, it was concluded that the relationship between *S* and *Y* was more or less linear. Now the test outcomes confirmed that there is no significant departure from linearity. Furthermore, with regard to the effects on *Z* (Model 8), the linear × linear association model is now valid for the effects of *Y* on *Z*. For the effects on *V*, Model 13 remains the end model. All together, no serious changes need be made that would alter our main conclusions, the small exception perhaps being the clearly linear relationship between the two latent variables. Model 14 contains all last changes and represents the finally chosen model ($L^2_{14/13} = 2.93$, df $= 2$, $p = .23$), depicted in Figure 3.

E. Application: Interpreting Parameter Estimates

The parameter estimates found in Model 14 are reported in Tables 6 and 7. These estimates will be discussed mainly by giving "qualitative" interpretations of the magnitudes of the coefficients to make the discussion of the outcomes less burdensome. However, more (numerically) precise

Table 6. Parameter Estimates Model 14, Table 5; Measurement Part

		Probabilities			Loglinear Parameters			Standard Errors $\hat{\sigma}_\lambda$			
		Y			Y			Y			
		1	**2**	**3**	**1**	**2**	**3**	**1**	**2**	**3**	
			$\hat{\pi}_{ay}^{A\mid Y}$			$\hat{\lambda}_{ya}^{YA}$					
A	1	0.339	0.210	0.085	0.566	0	−0.566	0.100			Wald = 66.96
	2	0.519	0.393	0.194	0.369	0	−0.369	0.089			df = 2
	3	0.142	0.397	0.721	−0.936	0	0.936	0.114			$p = 0.000$
			$\hat{\pi}_{by}^{B\mid Y}$			$\hat{\lambda}_{yb}^{YB}$					
B	1	0.032	0.210	0.684	−1.048	0	1.048	0.136			
	2	0.968	0.790	0.316	1.048	0	−1.048				
			$\hat{\pi}_{ey}^{E\mid Y}$			$\hat{\lambda}_{ye}^{YE}$					
E	1	0.670	0.197	0.016	1.673	0	−1.673	0.209			
	2	0.287	0.449	0.189	0	0	0				
	3	0.043	0.355	0.796	−1.673	0	1.673				
			Z			Z			Z		
		1	2	3	1	2	3	1	2	3	
			$\hat{\pi}_{cz}^{C\mid Z}$			$\hat{\lambda}_{zc}^{ZC}$					
C	1	0.059	0.357	0.760	−1.392	0	1.392	0.126			
	2	0.262	0.391	0.207	0	0	0				
	3	0.679	0.252	0.033	1.392	0	−1.392				
			$\hat{\pi}_{dz}^{D\mid Z}$			$\hat{\lambda}_{zd}^{ZD}$					
D	1	0.640	0.057	0.000	3.092	0	−3.092	0.356			
	2	0.355	0.698	0.115	0	0	0				
	3	0.006	0.245	0.885	−3.092	0	3.092				

Table 7. Parameter Estimates Model 14, Table 5; Structural Part

Table Entries		Loglinear Parameters $\hat{\lambda}$			Standard Errors $\hat{\sigma}_{\lambda}$			
			G			*G*		
Π^{SG}_{sg}		1	2	3	1	2	3	
	*S*1	0.038	0.112	−0.150	0.038	0.038	0.038	Wald = 16.63
	2	−0.038	−0.112	0.150	0.038	0.038	0.038	df = 2
								$p = 0.00$
			T			*T*		
$\Pi^{T\|SG}_{tsg}$		1	2	3	1	2	3	
	*S*1	0.016	−0.128	0.112	0.040	0.040	0.038	Wald = 12.60
	2	−0.016	0.128	−0.112	0.040	0.040	0.038	df = 2
								$p = 0.00$
	*G*1	−0.530	0.195	0.336	0.060	0.056	0.053	Wald = 163.58
	2	−0.113	0.218	−0.105	0.056	0.055	0.055	df = 4
	3	0.643	−0.412	−0.231	0.054	0.063	0.058	$p = 0.00$
			Y			*Y*		
$\Pi^{Y\|SGT}_{ysgt}$		1	2	3	1	2	3	
	*S*1	−0.398	0	0.398	0.084			
	2	0.398	0	−0.398				
	*G*1	0.512	−0.023	−0.489	0.125	0.101	0.151	Wald = 19.62
	2	−0.342	−0.050	0.391	0.111	0.092	0.141	df = 4
	3	−0.170	0.072	0.098	0.126	0.110	0.170	$p = 0.00$
	*T*1	1.575	0	−1.575	0.191			
	2	0	0	0				
	3	−1.575	0	1.575				
			Z			*Z*		
$\Pi^{Z\|SGTY}_{zsgty}$		1	2	3	1	2	3	
	*S*1	−0.107	0.174	−0.067	0.081	0.062	0.094	Wald = 8.61
	2	0.107	−0.174	0.067				df = 2
								$p = 0.01$
	*G*1	−1.434	0	1.434	0.163			
	2	0	0	0				
	3	1.434	0	−1.434				
	*T*1	0.436	0	−0.436	0.142			
	2	0	0	0				
	3	−0.436	0	0.436				
	*Y*1	1.143	0	−1.143	0.310			
	2	0	0	0				
	3	−1.143	0	1.143				

(*continued*)

Table 7 (*continued*)

Table Entries		Loglinear Parameters $\hat{\lambda}$			Standard Errors $\hat{\sigma}_{\lambda}$			
		V			V			
$\Pi^{V\|SGTYZ}_{vsgtyz}$		1	2	3	1	2	3	
	G1	−0.051	−0.279	0.330	0.086	0.088	0.101	Wald = 13.04
	2	0	0	0				df = 2
	3	0.051	0.279	−0.330				$p = 0.00$
	T1	−0.252	0.232	0.020	0.093	0.093	0.127	Wald = 12.69
	2	0	0	0				df = 2
	3	0.252	−0.232	−0.020				$p = 0.00$
	Y1	−1.043	−0.649	1.692	0.239	0.197	0.318	Wald = 28.48
	2	0	0	0				df = 2
	3	1.043	0.649	−1.692				$p = 0.00$
	Z1	1.026	−0.513	−0.513	0.236	0.173	0.248	Wald = 31.78
	2	0.328	−0.386	0.058	0.134	106	0.134	df = 4
	3	−1.354	0.899	0.454	0.300	0.190	0.256	$p = 0.00$
Total Effects on V (see text)								
$\hat{\Pi}^{V\|SG}_{vsg}$	S1	0.067	0.056	−0.123				
	2	−0.067	−0.056	0.123				
	G1	−0.310	−0.016	0.326				
	2	0.106	−0.005	−0.101				
	3	0.204	0.020	−0.224				
$\hat{\Pi}^{V\|SGT}_{vsgt}$	T1	−0.384	−0.154	0.538				
	2	0.021	−0.029	0.008				
	3	0.363	0.183	−0.546				
$\hat{\Pi}^{V\|SGTY}_{vsgty}$	Y1	−0.702	−0.844	1.546				
	2	0.013	−0.030	0.018				
	3	0.689	0.874	−1.564				

interpretations may be given, analogous to the expositions of the outcomes in Table 3 for the relationships among Y and B and E presented earlier.

The relationships between the latent variables and the indicators have been extensively discussed before, and the same conclusions still apply. Note that the estimates in Table 6 show the same patterns as the ones found before in Table 3.

The effects in the structural part of the causal model have been reported in Table 7. Sex (S) and Age (G) are very weakly related to each other: The estimates $\hat{\lambda}^{SG}_{sg}$ are very small but statistically significant. Men tend to occupy to a somewhat larger extent the middle-age category (34–55 years), whereas women dominate a bit the oldest age category

(56–91 years), effects that as we saw earlier must be explained in terms of differential mortality rates or differential nonresponse for men and women.

The direct effects of Sex on Education (T) are very small (although statistically significant): Men have a somewhat bigger chance of attaining the highest educational level and a bit smaller probability of obtaining a middle level education than women. Age has rather strong effects on Education. Although the relationship between G and T is not exactly linear, there is a clear tendency that "the older one is, the lower the education," reflecting the general increase in educational level over successive generations.

System Involvement (Y) is mainly determined by Education: T has a very large linear influence on Y: the higher one's education, the higher is the system involvement. The direct effects of S and G on Y are much less strong, but not to be ignored. Men are more involved in the political system than women. Age and System Involvement have a curvilinear relationship: The youngest age group is the least involved in the political system, the middle age group the most, and the oldest people are more or less evenly spread over the three categories of Y. It would be interesting to speculate about the underlying reasons for this relationship between Age and System Involvement, both in terms of generational differences and life course changes.

In terms of background variables, Protest Tolerance (Z) is mainly an age or generational phenomenon: G has a very strong direct linear effect on Z, showing that the older one is, the less tolerant one is toward protests. Education is important, but much less than Age: The higher the education is, the more tolerant one is toward protests (*A* direct linear relationship). Sex has only a very small influence on Z: Men are a bit more inclined to take a middle position on Z, whereas women are a bit more outspoken, belonging to either the highest or the lowest category of Z. The final conclusion about the effects of Y on Z is that there exists a strong direct linear relationship in the sense that the more the respondents are involved in the political system, the more tolerant they are toward protests.

The main determinants of Voting Behavior (V) are the latent variables. As expected, System Involvement is very important in determining whether one votes (remembering that category $Z = 3$ mainly consists of nonvoters): The higher system involvement is, the more one is inclined to vote for the Republicans or the Democrats rather than not vote. With regard to party choice, there is the modest tendency (contrary to the original expectations of "no difference"): The more involved, the higher the odds of voting Republican rather than Democratic $[(\hat{\lambda}^{YP}_{11} - \hat{\lambda}^{YP}_{12}) = 0.394]$. Protest Tolerance is much more important for the party choice than for voting. The higher the protest tolerance is, the more one is inclined to vote Democratic rather than Republican, especially those with the highest tolerance level $[(\hat{\lambda}^{ZP}_{31} - \hat{\lambda}^{ZP}_{32}) = -2.253)$ compared with lower level of tolerance. Furthermore, there is the smaller tendency: The higher Protest Tolerance is, the less one is inclined to vote (for Republicans or Democrats) rather than not vote.

With regard to the influences of the background variables on voting, Sex has no direct influence at all, whereas Age and Education have small direct effects. The older one is, the more one is inclined to vote for one of the major parties rather than not vote; when the choice is between the two major parties, there is this tendency: the older one is, the larger the odds of voting Democratic rather than Republican. Education does not make much of a direct difference with regard to voting, but it is important for the party choice: The higher the educational level, the larger the odds of voting Republican rather than Democratic.

Besides these direct effects, indirect and total effects may be interesting. As DLM lacks a simple "calculus of path coefficients" comparable to the standard linear structural equation model (Fienberg, 1977, p. 91; Brier, 1978), the total (marginal) association between two variables (e.g., S and V) cannot be written as the simple sum of direct, indirect, and spurious (and unknown) effects. Several attempts have been made to somehow define indirect and total effects in nonlinear structural equation models. Winship and Mare (1983) use a Pearsonian approach to define a calculus of path coefficients for categorical variables in which it is assumed that the observed categorical variables are realizations of underlying continuous variables (Hagenaars, 1998; Aris, 2001). This is no solution for the Yulean approach used in DLM, in which categorical variables are treated as categorical and not as realizations of underlying continuous variables. Pearl (2000) proposes an interventionist strategy for finding causes and consequences. He investigates what happens with the (conditional) distributions of the dependent variable in general, linear, and nonlinear structural equation models when particular variables act as causes and are given, as it were from the outside, a particular value. He defines total and direct effects by indicating which conditional distributions of the dependent variables should be compared within categories of which variables given the interventionist strategy (Pearl, 2000, Section 5.4.2). Recently, he has further clarified the meaning of indirect effects (Pearl, 2001). Pearl's approach is very general and not confined to, for example, linear structural equation models. Nevertheless, some difficulties still arise when nonlinear parameterizations are used for the comparisons of the conditional distributions of the dependent variable, such as odds ratios instead of differences between percentages (or probabilities). These difficulties are similar to the DLM approach advocated by the author that will be exemplified later.

The parameters in the logit equations may be regarded as the direct effects. Total effects may be interpreted as the effects found after collapsing the full table with estimated expected frequencies $\hat{F}$ over the intervening, but not the antecedent variables (Hagenaars, 1993, pp. 49–50; Andress, Hagenaars, and Kühnel, 1997, p. 193). For example, the total effects of S on V controlling for the other exogenous variable G and the total effects of G on V controlling for S may be found by investigating the effects in table SGV with estimated frequencies $\hat{F}^{SGV}_{sgv} (= \hat{F}^{SGTVYZABCDE}_{sg+v+++++++})$ obtained from collapsing the estimated frequencies from Model 14. These total effects are also reported in Table 7. Although Sex did not have a direct influence on voting, there are small total effects (controlling for G), especially because S is related to Y, an important determinant of V: Men are more involved in the system than women, and so Sex is related to "voting preferences" following this differential involvement. The direct effects of Age were such that the older one is, the more one is inclined to vote rather than not vote. This is also true in terms of total effects, controlling for S. However, the direct effects also showed that when the choice is between the two major parties, there is the tendency for older people to prefer voting Democratic rather than Republican. With regard to the total effects, this conclusion is reversed: the older one is, the more Republican. This becomes understandable once one recognizes that the older one is, the less tolerant toward protests and the less tolerant, the more inclined to vote Republican. Similar differences between direct and total effects occur when one investigates the total influence of T and Y on V in tables $SGTV$, $SGTYV$ respectively (see Table 7). Through its strong relationship to System Involvement, the total effects of Education (T) are such that they mainly make a difference in voting or not, contrary to the direct effects that mainly concerned the difference in party choice. The indirect effects of System Involvement through Protest Tolerance make for total effects of Y that annihilate (or are opposite to) the direct effects with regard to party choice: The total effects are such that there is no difference in party preference for the higher and lower involved people (or a very slight growing preference for the Democrats with increasing involvement), whereas the direct effects showed a preference for the Republicans with increasing involvement.

Although this way of looking at total effects does not provide an exact calculus of direct, indirect, and total effects, it does provide a simple and clear interpretation for the total effects: How much do the odds of belonging to one category of the dependent variable rather than another change when the independent variable changes, controlling for all variables that cause (partly) spurious relations, but not for the intervening variables (for a much more extended and precise statement, see Pearl, 2000)? Nevertheless, the approach is not without difficulties or counterintuitive outcomes. For example, contrary to ordinary linear systems, even when there are no higher-order interactions in the complete model for the direct effects, such higher-order interactions may be present for the total effects using summed frequencies and marginal tables. For example, within table *SGV* the three-variable interaction terms $\hat{\lambda}_{sgv}^{SGV}$ are not exactly zero (although none are larger than 0.03). In the same way, all direct effects in the model may be linear (that is, logistic), and still the total effect can become nonlinear (nonlogistic). Also, and perhaps more fundamentally, the direct or total effects in terms of odds ratios have no simple and direct interpretation in terms of the individual causal effects as in Rubin's potential response model (Rubin, 1974; Steyer, Gabler, and Rucai, 1996; Aris, 2001). These are all direct consequences of the nonlinear nature of DLM and ultimately relate to simple things, such as the sum of a set of logistic curves is itself not logistic, or the sum of the logarithms of a set of numbers (e.g., frequencies) is not equal to the logarithm of the sum of these numbers. The logic and implications of nonlinear structural equation models may be different from the consequences of the linear ones.

All these causal conclusions depend, of course, on the validity of the finally selected model. The main purpose has been to show the logic at work of causal modeling and DLM, and not enough attention has been paid to the many pitfalls of a proper causal analysis, nor to the many problems surrounding model selection and to procedures to overcome these problems (see the references in Note 1 regarding causal analyses and, concerning model selection, among others, Aitkin, 1980; Bonett and Bentler, 1983; Santner and Duffy, 1989; Sen and Srivastava, 1990; Bollen and Long, 1993; Marcoulides and Schumacker, 1996; Fox, 1997).

An essential feature of the latent variable model discussed earlier is that it deals with random measurement error in the indicators in form of independent classification error (ICE) in agreement with the standard LCA assumption of local independence. Correction for ICE is important because random measurement errors in categorical variables generally cause biased estimation of univariate distributions, strengths of association, and the patterns of association. For example, it is possible to replace the latent variables by their strongest indicators, that is, latent variable Y by indicator E and latent variable Z by indicator D, and fit a model, denoted as Model 14′, to the observed table that is analogous to the final latent variable Model 14 in Table 5. It then turns out that, for example, the effects of D and E on Voting in Model 14′ are much smaller than the corresponding effects of Z and Y on V in Model 14: $\hat{\lambda}_{11}^{DV} = 0.245$ whereas $\hat{\lambda}_{11}^{ZV} = 1.026$ and $\hat{\lambda}_{11}^{EV} = -0.203$ whereas $\hat{\lambda}_{11}^{YV} = -1.043$. Furthermore, the direct effects of Age, G, and Education, T, are larger in Model 14′ than in Model 14; less of the influence of the background variables on Voting is mediated through the indicators (in Model 14′) than through the latent variables (in Model 14). For example, in Model 14′, $\hat{\lambda}_{11}^{GV} = -0.296$ and $\hat{\lambda}_{11}^{TV} = -0.389$, whereas in Model 14 the corresponding results were $\hat{\lambda}_{11}^{GV} = -0.051$ and $\hat{\lambda}_{11}^{TV} = -0.252$. In addition, some patterns of association are different in Model 14′ from Model 14, which has to do with different probabilities of misclassification for different categories and/or different sizes of the categories. For example, in latent variable Model 14, the higher protest tolerance is, the less are the odds of voting rather than nonvoting (see Table 7), but in Model 14′ there is no effect of D on the odds of voting or not.

The conclusion must be that random errors may do much more harm than "just" attenuate the correlations, and this is especially true in longitudinal research. In panel studies, ICE may cause very systematic looking patterns of change, whereas no true change at all takes place. Often encountered phenomena discussed and explained in theoretical terms, such as minorities and small groups – for example, nonvoters – change relatively much more than majorities and bigger groups – for example, voters, *or*, those who changed in the past will change more easily again in the future than those who were stable in the past – can be often explained completely in terms of purely random measurement error (Maccoby, 1956; Hagenaars, 1990, Section 4.4; Hagenaars, 1994).

To make things worse, measurement error is often not random, but systematic; systematic misclassifications by definition will bias the results. Fortunately, if one is willing to make certain assumptions about the nature of the systematic misclassifications, systematic measurement error can also be handled in a straightforward manner by applying directed loglinear models with latent variables. An explicit exposition of how one may take systematic misclassifications into account will be presented in the next section.

4. SYSTEMATIC, NONINDEPENDENT MISCLASSIFICATIONS

To exemplify the treatment of nonindependent classification errors (NICE), the Survey of Income and Program Participation (SIPP) panel will be used. However, unlike the example in the previous section, no data and analysis outcomes will be presented; they are provided by Bassi et al. (2000). The SIPP panel was started in 1984 by the U.S. Bureau of Census. Its main focus is on individual changes in labor market status (Not in the labor force, Unemployed, or Employed). To understand the kinds of response errors that appear to occur in the SIPP panel, one needs an elementary understanding of the SIPP interviewing scheme. The respondents of the SIPP panel are divided into four rotation groups. At the beginning of every month (e.g., May), the members of one rotation group R1 are interviewed and asked a number of questions about their labor status during the preceding four months (January through April), in which the respondent first remembers the month closest to the moment of interviewing (April), then the next closest (March), and so on (Bassi et al., 2000). Another rotation group R2 starts a month later (June) and has an analogous schedule, and so forth. Because of this rotation scheme, the information concerning the labor market status in one particular month is provided by four different rotation groups and gathered at the beginnings of four different consecutive months. Here, only one group (R1) will be considered. Because R1 is supposed to have been interviewed at the beginning of May, the information on labor market status gathered during the May interview refers to the labor market states in January, February, March, and April and will be denoted as *A*, *B*, *C*, and *D*, respectively.

The gross turnover in labor market states is often analyzed by means of a Markov model. The causal model corresponding with a first-order Markov chain is depicted in Figure 4(a). The corresponding "graphical" equation is

$$\pi_{abcd}^{ABCD} = \pi_a^A \pi_{ba}^{B|A} \pi_{cb}^{C|B} \pi_{dc}^{D|C}. \tag{10}$$

If for each of the marginal "tables" at the right side of Equation (10), the saturated loglinear model is postulated, the result is a first-order, nonstationary Markov chain. The Markov chain is a first-order chain because there are only direct effects from month $t-1$ to month t. Second-order Markov chains result from adding extra direct effects from month $t-2$ to month t; still higher-order chains are defined in an analogous way. The Markov chain will be stationary when (loglinear) models are valid in which the transition probabilities A–B, B–C, and C–D are restricted

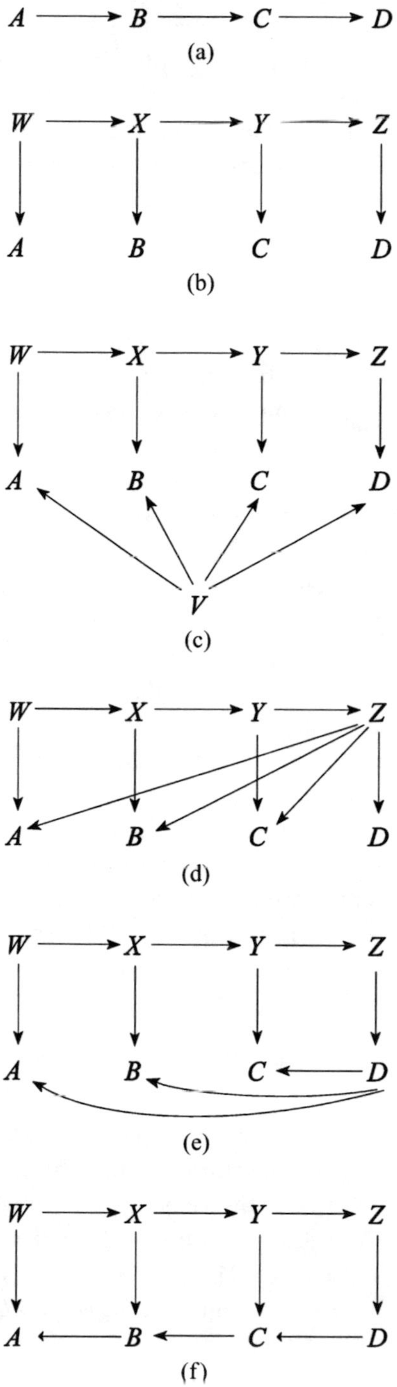

Figure 4. Models for SIPP panel data: first-order Markov model at the (a) manifest level and (b) latent level; latent Markov model with (c) unmeasured consistency trait, (d) consistency with true position at time of interview, (e) consistency with first given answer, and (f) consistency with last given answer.

to be equal to each other: $\pi_{ji}^{B|A} = \pi_{ji}^{C|B} = \pi_{ji}^{D|C}$ (Bishop, Fienberg, and Holland, 1975, Chapter 7; Vermunt, 1997a, pp. 43–45).

Using the data on labor market status from the SIPP panel, one can arrive at well-fitting causal loglinear models in the form of a Markov chain, if necessary by relaxing the first-order or stationarity assumptions. However, as explained by Bassi et al. (2000), the outcomes show some peculiar regularities. For example, the bivariate turnover tables for consecutive months determined for one rotation group within one particular interview (here, A–B, B–C, C–D) show too little gross change in labor status, less than in reality, that is, less than is expected from knowledge provided by other, more reliable sources. Furthermore, the amount of gross change between two consecutive months that are close to the moment of interviewing (C–D) is always larger than the amount of gross change between two consecutive months that are further away from the date the interview took place (A–B). These and other peculiarities lead to the conclusion that the observed labor status is not a completely valid and reliable measurement of the respondents' true labor status and that particular kinds of response errors must have occurred.

Latent Markov chains have been proposed to correct for response errors (Wiggins, 1955, 1973; Poulsen, 1982; Van de Pol, 1989; Van de Pol and Langeheine, 1990, 1997). The DLM approach makes this approach even more general and flexible.[8] A straightforward translation of the first-order (nonstationary) Markov model of Figure 4(a) into a latent Markov model is presented in Figure 4(b) and in Equation (11), in which W, X, Y, and Z denote the trichotomous latent labor market states in January, February, March, and April, respectively; the first part at the right side of Equation (11) represents the structural (Markovian) component of the latent variable model, and the second part represents the measurement model.

$$\pi_{wxyzabcd}^{WXYZABCD} = \left(\pi_{w}^{W}\pi_{xw}^{X|W}\pi_{yx}^{Y|X}\pi_{zy}^{Z|Y}\right)\left(\pi_{aw}^{A|W}\pi_{bx}^{B|X}\pi_{cy}^{C|Y}\pi_{dz}^{D|Z}\right). \quad (11)$$

When the latent Markov model in Equation (11) is applied to the SIPP data, using saturated loglinear models for each of the (conditional) probabilities at the right side of Equation (11) (and using extra equality restrictions to identify the model; see Bassi et al., 2000), the results are not satisfactory. The amount of gross change at the latent level turns out to be even smaller than at the manifest level, and none of the peculiarities referred to earlier disappeared at the latent level. Obviously, for this data, the latent class model in the form of the latent Markov model in Figure 4(b), in which the standard independent classification error (ICE) assumption has been made, deals with response error in the wrong way. Most probably the misclassifications are systematic and nonindependent. Given the SIPP interviewing scheme in which the information on the labor market status for the previous four months was collected during one particular interview session, it must be expected that the errors in these labor market reports are directly correlated with each other, as a result of conditioning effects, tendencies to appear consistent, memory effects, test–retest effects, and the like. Replacing the strict ICE assumption by a milder one and defining NICE or local dependence LCA models seems necessary (Hagenaars, 1988, 1990).

One way to account for possible direct associations among the response errors is the introduction of an extra latent variable V (response consistency) that is directly related to indicators A through D, but (at least here) uncorrelated with the proper latent variables W through Z, as in Equation (12), and as depicted in a more parsimonious form without three-variable or higher-order interactions in Figure 4(c):

$$\pi_{vwxyzabcd}^{VWXYZABCD} = \left(\pi_v^V \pi_w^W \pi_{xw}^{X|W} \pi_{yx}^{Y|X} \pi_{zy}^{Z|Y}\right) \times \left(\pi_{avw}^{A|VW} \pi_{bvx}^{B|VX} \pi_{cvy}^{C|VY} \pi_{dvz}^{D|VZ}\right). \tag{12}$$

Because an extra latent variable V has been introduced without additional observed indicators, the model of Equation (12) has to be examined very carefully for (extra) identifiability problems. One possible, but not necessarily sufficient, way of achieving identifiability is to define more restrictive loglinear models, for example, no-three-variable interaction models for the three-way tables in Equation (12) involving latent variable V.

Because V is treated here as a categorical latent variable, the number of categories of V must be determined. One usually starts with two categories and adds more and more categories until the model is no longer identified or a good fit with the data has been obtained. As there are seldom or never theoretical reasons for assuming that V has a certain number of categories, this is the only way to proceed. Actually, its atheoretical nature is the most serious objection one might raise against the introduction of an extra latent variable such as V: finding a compelling substantive interpretation of the latent variable is hard. Essentially, V is just a variable modeling unobserved heterogeneity (Heckman and Singer, 1982; DeSarbo and Wedel, 1993; Vermunt, 1996), taking care of that part of the association among the indicators A through E that is not explained by the other, proper latent variables without providing a direct clue for what V might stand. V just accounts for the "correlated error terms," but, from a substantive point of view, in an unknown way. Even when the model fits the data well, the results are usually not very informative because almost any interpretation can be given to V. Therefore, preference should be given to models that somehow incorporate the presumed nature of the systematic response errors in a more theoretical way.

One possible substantive reason for the presence of correlated error terms might be that during the May interview, the respondents of R1 have the tendency to make their answers about their labor status during the previous four months consistent with their true position at the time of the interview. Because latent variable Z in Figure 4 represents as closely as possible this true position, this hypothesis implies that Z has a direct influence not only on D, but also on A, B, and C. Equation (13) and (in more parsimonious form) Figure 4(d) represent this hypothesis:

$$\pi_{wxyzabcd}^{WXYZABCD} = \left(\pi_{w}^{W}\pi_{xw}^{X|W}\pi_{yx}^{Y|X}\pi_{zy}^{Z|Y}\right)\left(\pi_{awz}^{A|WZ}\pi_{bxz}^{B|XZ}\pi_{cyz}^{C|YZ}\pi_{dz}^{D|Z}\right). \tag{13}$$

Restricted (no-three-variable interaction) loglinear models may be defined for the conditional probabilities at the right-hand side of Equation (13) and are necessary to achieve identifiability.

However, this explanation of the dependent misclassifications in the SIPP panel may not be valid, for example, when respondents do not have the tendency to adapt their answers to their present true, latent position, but to the first given manifest answer. Because during the May interview the R1 respondents remember first their labor market positions in April represented by indicator D, it follows that D should have a direct effect on A, B, and C, as rendered in Equation (14) and (more parsimoniously) in Figure 4(e).

$$\pi_{wxyzabcd}^{WXYZABCD} = \left(\pi_{w}^{W}\pi_{xw}^{X|W}\pi_{yx}^{Y|X}\pi_{zy}^{Z|Y}\right)\left(\pi_{awd}^{A|WD}\pi_{bxd}^{B|XD}\pi_{cyd}^{C|YD}\pi_{dz}^{D|Z}\right). \tag{14}$$

Still another possibility is that a particular manifest response is not made consistent with the first given answer, but with the immediately previous one, leading to the model in Equation (15) and (more parsimoniously) in Figure 4(f).

$$\pi_{wxyzabcd}^{WXYZABCD} = \left(\pi_{w}^{W}\pi_{xw}^{X|W}\pi_{yx}^{Y|X}\pi_{zy}^{Z|Y}\right)\left(\pi_{awb}^{A|WB}\pi_{bxc}^{B|XC}\pi_{cyd}^{C|YD}\pi_{dz}^{D|Z}\right) \tag{15}$$

Which of the models in Equations (10)–(15) should be preferred depends on what the investigator deems plausible and on the fit of the model when empirically tested. And, of course, many variations of these models can be formulated (Bassi et al., 2000; Hubble and Judkins, 1987).

Once a particular (valid) measurement model for the (correlated) response errors has been obtained, the complete model provides the estimates for the parameters of the structural part, corrected for these response errors. The structural part itself may assume many forms, different from the one discussed earlier. For example, the assumptions of the latent Markov chain may be relaxed by defining higher-order Markov chains. The nature of the latent change may be restricted by excluding certain transitions between particular latent states, by assuming no latent change between particular points of time, or by assuming "linear" latent change and so on (Collins and Wugalter, 1992; Hagenaars, 1990, 1994; Jagodzinski, Kühnel, and Schmidt, 1987).

As will be clear from this example, the DLM approach offers a very flexible tool to model many kinds of latent changes and many types of response errors simultaneously. At the same time, and this cannot be emphasized strongly enough, flexibility changes from a virtue into a vice when the models are not theoretically based and are used only as an arbitrary data-fitting device. The introduction of correlated error terms, in one way or another, will almost always "save" a model that otherwise would have been rejected (Hagenaars, 1988). However, correlated error terms often violate the very basic measurement assumptions the analysis started with and should not be applied unless for a very good theoretical reason. Moreover, the problem of identifiability has been dealt with earlier in a very casual manner (Goodman, 1974b; DeLeeuw, Van der Heijden, and Verboon, 1990). However, this problem should not be underestimated. Without further restrictions, the parameters in Equations (11)–(15) are not identifiable. They can be made identifiable in many different ways; only some of these will make sense from a substantive, theoretical point of view.

5. CONCLUSIONS

Compared with Pearsonian approaches toward the analysis of categorical data in structural equation modeling in which a categorical variable is seen as a realization of a usually normally distributed, underlying latent variable (Aris, 2001; Schumacker and Marcoulides, 1998; Hagenaars, 1998), the Yulean approach incorporated in the DLM in which a categorical variable is treated as categorical has certain advantages. Higher-order interactions and nonlinear relationships, whether these involve only manifest or mixtures of latent and manifest variables, pose no special problems and no restrictive distributional assumptions that are alien to categorical data are required.

At the same time, its problematic aspects are also clear. From a causal point of view, the lack of a simple calculus of path coefficients makes the causal analysis using the DLM certainly not simpler. From a more general point of view, the most important problems are directly related to the major plague of categorical data analysis: small sample size and sparse tables. Testing a particular model, even carrying out conditional tests comparing several models, may become difficult with extremely sparse tables, as the asymptotics of the standard chi-square tests break down and the maximum-likelihood estimates cannot be expected to follow a normal distribution. Parametric bootstrapping methods may provide a way out of these problems (Agresti, 1992; Van der Heijden et al., 1997; Collins et al., 1993; Langeheine, Pannekock, and Van de Pol, 1996; Vermunt, 1999), although we still do not know enough of their behavior for the extremely sparse tables that often occur in applied research.

Observed sparse tables easily result in zero cells in the table of estimated expected frequencies, and certain parameters may not be defined. Furthermore, with zero estimated frequencies (and boundary parameter estimates), the appropriate number of degrees of freedom is not defined. Sampling zeros may become a kind of a posteriori structural zeros: the maximum-likelihood estimates obtained are terminal estimates or maximum-likelihood estimates based on the a priori assumption that the pertinent cell probabilities are indeed zeros in the populations. In a way, the degrees of freedom are not fixed by the model, but become dependent on the data and become random variables. Also, for very sparse tables, the "asymptotics" break down and no reliable estimates of the standard errors can be obtained. Adding small constants to the observed frequencies will, at least in regular loglinear models without latent variables, prevent the occurrence of estimated zero cells, but the parameter estimates may be very much dependent on the particular constants chosen. Perhaps a Bayesian approach may solve some of the problems mentioned (Rubin and Stern, 1994; Gelman et al., 1995; Schafer, 1997).

In a way, the introduction of latent variables contributes to the sparseness of the (complete) tables because extra variables and parameters are added without increasing the number of independent observations. To enhance the stability of latent variable models, having many indicators for each latent variable is desirable. At the same time, the many indicators contribute toward the sparseness of the observed table. More research is therefore needed into two-step procedures in which, first, the measurement model is analyzed separately and, then, once a satisfactory measurement model for the latent variable(s) is obtained, "observed" latent scores (factor scores) are computed, which, finally, are introduced into the causal, structural part of the causal model instead of the latent variables and their indicators. Following this procedure, it is no longer necessary to estimate simultaneously all relationships among all "causal" factors and indicators. Problematic in this two-step approach are the estimates of the standard errors of the parameter estimates in the causal model and the fact that such observed latent scores are usually biased and not (completely) identified. However, with regard to the latter problem, for several important situations the bias is known and can be corrected for (Hagenaars, 1990, Section 3.3; Bolck, Croon, and Hagenaars, 1998; Croon and Bolck, 1997).

Finally, one should be careful when interpreting models with latent variables as the meaning of the latent variables is not always clearly understood. Latent variables denote the respondents' "true" positions, corrected for response error by taking the probabilistic relationship between latent and observed variables (the indicators) into account. However, there may be two kinds of random error in the manifest data (confining the exposition here to purely random error, i.e., ICE models). On the one hand, there is a less-than-perfect relationship between a latent variable and its indicators because the respondent makes nonsystematic response errors and happens to misreport his or her true position. On the other hand, there is random error and a probabilistic relationship between the latent variable and its indicators because there is truly random behavior: the actual position taken and correctly observed contains a nonsystematic component.[9] Figure 4(b), which is about changes in (un)employment, can be used to illustrate this distinction. According to the model in Figure 4(b), there is probabilistic relationship between the true (un)employment status and the observed one. This "imperfection" is partly caused by the nonsystematic response errors the respondents make, and by nonsystematic recording errors made by interviewers, coders, and so on. However, another source of random error is the fact that some persons are "truly" (un)employed and are registered as such during most of the period, but at some point in time happen to lose their job for a short while or, on the contrary, happen to find employment for a while. Latent variable models as such cannot make the difference between someone who is usually employed but at one point in time happens to be erroneously recorded as unemployed and a person who is usually employed but at one point in time happens to be without a job. Whether this confusion is problematic depends on the purposes of the investigation. If one wants to measure the true underlying attitude of a person or, for example, a person's weight, in the sense of the value that is not affected by daily or even hourly random fluctuations (the operational true score; see Note 9), latent variable models are the right models to use. However, if one is interested in the real number of people who are actually unemployed at a certain moment of time (the platonic true score; see Note 9), regardless of whether they are "always" unemployed or typically employed but accidentally at this moment unemployed, latent variable models will generally not be appropriate because, in this particular example, they will underestimate the true number of unemployed people (being the smallest group) and underestimate the actual changes that are going on the labor market. At least part of the failure of latent variable models such as those in Figure 4(b) that generally underestimate the true changes on the labor market may be caused by the presence of random but real true change that is not detected by these models. If this is true, the proposed models that take systematic response error into account may still not provide the correct estimates of the true labor market changes, or at least not for the right reasons. Sometimes, latent variable models cannot be substitutes for the gold standard.

REFERENCES

Agresti, A. (1990). *Categorical Data Analysis.* New York: Wiley.

Agresti, A. (1992). "A survey of exact inference for contingency tables," *Statistical Science*, **7**, 131–77.

Aitkin, M. (1980). "A note on the selection of loglinear models," *Biometrics*, **36**, 173–8.

Andress, H.-J., Hagenaars, J. A., & Kühnel, S. (1997). *Analyse von Tabellen und kategorialen Daten.* Berlin: Springer.

Aris, E. (2001). "Statistical causal models for categorical data," Ph.D. dissertation, Tilburg, The Netherlands: Tilburg University.

Barnes, S. H., & Kaase, M., Eds. (1979). *Political Action: Mass Participation in Five Western Democracies*. Berlin: DeGruyter.

Bartholomew, D. J., & Knott, M. (1999). *Latent Variable Models and Factor Analysis*. London: Arnold.

Bassi, F., Hagenaars, J. A., Croon, M. A., & Vermunt, J. K. (2000). "Estimating true changes when categorical panel data are affected by uncorrelated and correlated errors," *Sociological Methods and Research*, **29**, 230–68.

Becker, M. P. (1994). "Analysis of cross-classifications of counts using models for marginal distributions: an application to trends in attitudes on legalized abortion." In P. V. Marsden (ed.), *Sociological Methodology 1994*. Washington, DC: American Sociological Association, pp. 229–66.

Becker, M. P., & Yang, Ilsoon. (1998). "Latent class marginal models for cross-classifications of counts." In A. E. Raftery (ed.), *Sociological Methodology 1998*. Washington, DC: American Sociological Association, pp. 293–326.

Bergsma, W. P. (1997). *Marginal Models for Categorical Data*. Tilburg, The Netherlands: Tilburg University Press.

Bishop, Y. M. M. (1969). "Full contingency tables, logits, and split contingency tables," *Biometrics*, **95**, 383–99.

Bishop, Y. M. M., Fienberg, S. E., & Holland, P. W. (1975). *Discrete Multivariate Analysis: Theory and Practice*. Cambridge: MIT Press.

Blalock, H. M. (1964). *Causal Inferences in Nonexperimental Research*. Chapel Hill, NC: University of North Carolina Press.

Bolck, A., Croon, M. A., & Hagenaars, J. A. (1998). *On the Use of Latent Class Scores in Causal Models for Categorical Data*. WORC Paper 98.7.005. WORC, Tilburg University, Tilburg, The Netherlands.

Bollen, K., & Long, J. S. (1993). *Testing Structural Equation Models*. Newbury Park: Sage.

Bonett, D. G., & Bentler, P. M. (1983). "Goodness-of-fit procedures for the evaluation and selection of loglinear models," *Psychological Bulletin*, **93**, 345–70.

Brier, S. S. (1978). "The utility of systems of simultaneous logistic response equations." In Karl. F. Schuessler (ed.), *Sociological Methodology 1979*, chap. 5. San Francisco: Jossey–Bassc, pp. XXX–XXX.

Camilli, G., & Shephard, L. A. (1994). *MMSS: Methods for Identifying Biased Test Items*. Thousand Oaks, CA: Sage.

Clogg, C. C. (1981a). "New developments in latent structure analysis." In D. J. Jackson & E. F. Borga (eds.), *Factor Analysis and Measurement in Sociological Research*. Beverly Hills: Sage, pp. 215–46.

Clogg, C. C. (1981b). "Latent structure models of mobility," *American Journal of Sociology*, **86**, 836–68.

Clogg, C. C. (1995). "Latent class models." In G. Arminger, C. C. Clogg, & M. E. Sobel (eds.), *Handbook of Statistical Modeling for the Social and Behavioral Sciences*. New York: Plenum, pp. 311–60.

Clogg, C. C., & Shihadeh, E. S. (1994). *Statistical Models for Ordinal Variables*. Thousand Oaks, CA: Sage.

Collins, L. M., & Wugalter, S. E. (1992). "Latent class models for stage-sequential dynamic latent variables," *Multivariate Behavioral Research*, **27**, 131–57.

Collins, L. M., Fidler, P. L., Wugalter, S. E., & Long, J. D. (1993). "Goodness-of-fit testing for latent class models," *Multivariate Behavioral Research*, **28**, 375–89.

Cox, D. R., & Wermuth, N. (1996). *Multivariate Dependencies: Models, Analysis and Interpretations*. London: Chapman & Hall.

Croon, M. A., & Bolck, A. (1997). *On the Use of Factor Scores in Structural Equation Models*. WORC Paper 97.10.012. WORC, Tilburg University. Tilburg, The Netherlands.

Croon, M. A., Bergsma, W., Hagenaars, J. A. (2000). "Analyzing change in categorical variables by generalized log-linear models," *Sociological Methods and Research*, **29**, 195–29.

De Leeuw, J., Van der Heijden, P. G., & Verboon, P. (1990). "A latent time–budget model," *Statistica Neerlandica*, **44**, 1–22.

DeSarbo, W. S., & Wedel, M. (1993). *A Review of Recent Developments in Latent Class Regression Models.* Research Memorandum 521, Groningen, University of Groningen.

Diggle, P. J., Liang, K.-Y., & Zeger, S. L. (1996). *Analysis of Longitudinal Data.* Oxford: Clarendon.

Fienberg, S. E. (1977). *The Analysis of Cross-Classified Data.* Cambridge: MIT Press.

Fox, J. (1997). *Applied Regression Analysis, Linear Models, and Related Methods.* Thousand Oaks, CA: Sage.

Gelman, A., Carlin, J. B., Stern, H. S., & Rubin, D. B. (1995). *Bayesian Data Analysis.* London: Chapman & Hall.

Glymour, C., & Cooper, G. F., Eds. (1999). *Computation, Causation and Discovery.* Menlo Parl, CA: AAAI Press/MIT Press.

Goodman, L. A. (1973). "The analysis of multidimensional contingency tables when some variables are posterior to others; a modified path analysis approach," *Biometrika*, **60**, 179–92.

Goodman, L. A. (1974a). "The analysis of systems of qualitative variables when some of the variables are unobservable. Part I. A modified latent structure approach," *American Journal of Sociology*, **79**, 1179–259.

Goodman, L. A. (1974b). "Exploratory latent structure analysis using both identifiable and unidentifiable models," *Biometrika*, **61**, 215–31.

Goodman, L. A. (1984). *The Analysis of Cross-Classified Data Having Ordered Categories.* Cambridge: Harvard University Press.

Goodman, L. A., & Hout, M. (1998). "Statistical methods and graphical displays for analyzing how the association between qualitative variables differs among countries, among groups, or over time: a modified regression-type approach (Comments and Rejoinder)." In A. E. Raftery (ed.), *Sociological Methodology 1998*, Vol. 28. Washington, DC: American Sociological Association, pp. 175–61.

Haberman, S. J. (1974). "Loglinear models for frequency tables derived by indirect observation. Maximum likelihood equations," *Annals of Statistics*, **2**, 911–24.

Haberman, S. J. 1979. *Analysis of Qualitative Data. Vol. 2. New Developments.* New York: Academic.

Hagenaars, J. A. (1988). "Latent structure models with direct effects between indicators: local dependence models," *Sociological Methods and Research*, **16**, 379–405.

Hagenaars, J. A. (1990). *Categorical Longitudinal Data; Loglinear Panel, Trend, and Cohort Analysis.* Newbury Park: Sage.

Hagenaars, J. A. (1993). *Loglinear Models with Latent Variables.* Sage University Papers. Newbury Park: Sage.

Hagenaars, J. A. (1994). "Latent variables in loglinear models of repeated observations." In A. Von Eye & C. C. Clogg (eds.), *Latent Variables Analysis: Applications for Developmental Research.* Thousands Oaks, CA: Sage, pp. 329–52.

Hagenaars, J. A. (1998). "Categorical causal modeling: latent class analysis and directed log-linear models with latent variables," *Sociological Methods and Research*, **26**, 436–86.

Heckman, J. J., & Singer, B. (1982). "Population heterogeneity in demographic models." In K. Land & A. Rogers (eds.), *Multidimensional Mathematical Demography*. New York: Academic Press, pp. 567–99.

Heinen, T. G. (1996). *Latent Class and Discrete Latent Trait Models: Similarities and Differences.* Thousand Oaks, CA: Sage.

Hubble, D. L., & Judkins, D. R. (1987). *Measuring the Bias in Gross Flows in the Presence of Autocorrelated Measurement Error.* SIPP Working Paper 8712.

Jagodzinski, W., Kühnel, S. M., & Schmidt, P. (1987). "Is there a Socratic Effect in nonexperimental panel studies," *Sociological Methods and Research*, **12**, 375–98.

Jennings, M. K., & van Deth, J. W. (1990). *Continuities in Political Action*. Berlin: DeGruyter.

Kiiveri, H., & Speed, T. P. (1982). "Structural analysis of multivariate data: a review." In S. Leinhardt (ed.), *Sociological Methodology 1982*. San Francisco: Jossey-Bass, pp. 209–89.

Koster, J. T. (1996). "Markov properties of nonrecursive causal models," *Annals of Statistics*, **24**, 2148–77.

Kuha, J., & Skinner, C. (1997). "Categorical data and misclassification." In L. Lyberg, P. Biemer, M. Collins, E. De Leeuw, C. Dippo, N. Schwarz, & D. Trewin (eds.), *Survey Measurement and Process Quality*. New York: Wiley, pp. 633–71.

Lang, J. B., & Agresti, A. (1994). "Simultaneous modeling joint and marginal distributions of multivariate categorical responses," *Journal of the American Statistical Association*, **89**, 625–32.

Langeheine, R., Pannekoek, J., & Van de Pol, F. (1996). "Bootstrapping goodness-of-fit measures in categorical data analysis," *Sociological Methods and Research*, **24**, 492–516.

Lauritzen, S. L. (1996). *Graphical Models.* Oxford: Clarendon.

Lazarsfeld, P. F. (1950). "The logical and mathematical foundation of latent structure analysis." In S. A. Stouffer, L. Guttman, E. A. Suchman, P. F. Lazarsfeld, S. A. Star, & J. A. Clausen (eds.), *Studies in Social Psychology in World War II: Vol. 4. Measurement and Prediction*. Princeton, NJ: Princeton University Press, pp. 362–472.

Lazarsfeld, P. F. (1955). "Interpretation of statistical relations as a research operation." In P. F. Lazarsfeld & M. Rosenberg (eds.), *The Language of Social Research*. New York: Free Press, pp. 115–25.

Lazarsfeld, P. F., & Henry, N. W. (1969). *Latent Structure Analysis*. Boston: Houghton Mifflin.

Liang, Kung-Yee, & Zeger, S. L. (1986). "Longitudinal analysis using generalized linear models," *Biometrika*, **73**, 13–22.

Lord, F. M., & Novick, M. R. (1968). *Statistical Theories of Mental Test Scores.* Reading, MA: Addison-Wesley.

Luijkx, R., & Hagenaars, J. A. (1990). *Social Mobility on the Latent Level: A Structural and a Measurement Model.* Working Paper No. 53. Tilburg, The Netherlands: Tilburg University.

Maccoby, E. E. (1956). "Pitfalls in the analysis of panel data; a research note on some technical aspects of voting," *American Journal of Sociology*, **61**, 359–62.

Manski, C. F. (1995). *Identification Problems in the Social Sciences*. Cambridge, MA: Harvard University Press.

Marcoulides, G.. A., & Schumacker, R. E., Eds. (1996). *Advanced Structural Equation Modeling: Issues and Techniques*. Mahwa, NJ: Erlbaum.

McCutcheon, A. L. (1993). "Latent logit models with polytomous effects variables." In A. Von Eye & C. C. Clogg (eds.), *Latent Variables Analysis; Applications for Developmental Research*. Thousands Oaks, CA: Sage, pp. 353–72.

McCutcheon, A. L. (1996). "Multiple group association models with latent variables: an analysis of secular trends in abortion attitudes, 1972–1988." In A. E. Raftery (ed.), *Sociological Methodology 1996*, vol. 26. Washington, DC: American Sociological Association, pp. 79–112.

McKim, V. R., & Turner, S. P., Eds. (1997). *Causality in Crisis? Statistical Methods and the Search for Causal Knowledge in the Social Sciences*. Notre Dame: University of Notre Dame Press.

Pearl, J. (2000). *Causality: Models Reasoning and Inference*. Cambridge: Cambridge University Press.

Pearl, J. (2001). *Direct and Indirect Effects*. Technical Report R-273, Cognitive Systems Laboratory, Computer Science Department, UCLA.

Poulsen, C. S. (1982). "Latent structure analysis with choice modeling applications," Ph. D. dissertation, The Århus School of Business Administration and Economics.

Raftery, A. E. (1993). "Bayesian model selection in structural equation models. In K. Bollen & J. S. Long (eds.), *Testing Structural Equation Models*. Newbury Park: Sage, pp. 163–80.

Raftery, A. E., Ed. (1998). "Causality in the social sciences, in honor of Herbert L. Costner," *Sociological Methods and Research*, **27**, 139–348.

Rosenbaum, P. R. (1995). *Observational Studies*. New York: Springer.

Rubin, D. B. (1974). "Estimating causal effects of treatments in randomized and nonrandomized studies," *Journal of Educational Psychology*, **60**, 688–701.

Rubin, D. B., & Stern, H. S. (1994). "Testing in latent class models using a posterior predictive check distribution," In A. Von Eye & C. C. Clogg (eds.), *Latent Variables Analysis: Applications for Developmental Research*. Thousands Oaks, CA: Sage, pp. 420–38.

Santner, T. J., & Duffy, D. E. (1989). *The Statistical Analysis of Discrete Data*. New York: Springer.

Schafer, J. L. (1997). *Analysis of Incomplete Multivariate Data*. London: Chapman & Hall.

Schumacker, R. E., & Marcoulides, G. A., Eds. (1998). *Interaction and Nonlinear Effects in Structural Equation Modeling*. Mahwah, NJ: Erlbaum.

Sen, A., & Srivastava, M. (1990). *Regression Analysis: Theory, Methods, and Applications*. New York: Springer.

Simon, H. A. (1957). "Spurious correlation: a causal interpretation." In H. Simon (ed.), *Models of Man*. New York: Wiley & Son, pp. 37–49.

Snijders, T. A. B., & Hagenaars, J. A., Eds. (2001). "Causality at work," *Sociological Methods and Research*, **30**, 3–114.

Sobel, M. E. (1994). "Causal inference in latent variable models." In A. Von Eye &

C. C. Clogg (eds.), *Latent Variables Analysis: Applications for Developmental Research*. Thousands Oaks, CA: Sage, pp. 3–35.

Steyer, R., Gabler, S., & Rucai, A. A. (1996). "Individual causal effects, average causal effects, and unconfoundedness in regression models." In F. Faulbaum & W. Bandilla (eds.), *Sofstat 1995, Advances in Statistical Software 5*. Stuttgart: Lucius and Lucius, pp. 203–310.

Suttcliffe, J. P. (1965a). "A probability model for errors of classification. I. General considerations," *Psychometrika*, **30**, 73–96.

Suttcliffe, J. P. (1965b). "A probability model for errors of classification. II. Particular cases," *Psychometrika*, **30**, 129–55.

Van de Pol, F. (1989). *Issues of Design and Analysis of Panels.* Amsterdam: Sociometric Research Foundation.

Van de Pol, F., & Langeheine, R. (1990). "Mixed Markov latent class models." In C. C. Clogg (ed.), *Sociological Methodology 1990*. Oxford: Blackwell, pp. 213–47.

Van de Pol, F., & Langeheine, R. (1997). "Separating change and measurement error in panel surveys with an application to labor market data." In L. Lyberg, P. Biemer, M. Collins, E. De Leeuw, C. Dippo, N. Schwarz, & D. Trewin (eds.), *Survey Measurement and Process Quality*. New York: Wiley, pp. 671–88.

Van de Vijver, F., & Leung, K. (1997). *Methods and Data Analysis for Cross-Cultural Research*. Thousand Oaks, CA: Sage.

Van der Heijden, P., Harm 't Hart, & Dessens, J. (1997). "A parametric bootstrap procedure to perform statistical tests in a LCA of anti-social behavior." In J. Rost & R. Langeheine (eds.), *Applications of Latent Trait and Latent Class Models in the Social Sciences*. New York: Waxmann, pp. 196–208.

Van der Linden, W. J., & Hambleton, R. K., Eds. (1997). *Handbook of Modern Item Response Theory*. New York: Springer.

Vermunt, J. K. (1996). *Loglinear Event History Analysis: A General Approach with Missing Data, Latent Variables, and Unobserved Heterogeneity.* Tilburg, The Netherlands: Tilburg University Press.

Vermunt, J. K. (1997a). *Loglinear Models for Event History Analysis.* Thousand Oaks, CA: Sage.

Vermunt, J. K. (1997b). *LEM: A General Program for the Analysis of Categorical Data*, Technical Report. Tilburg, The Netherlands: Methodology Department, Tilburg University.

Vermunt, J. K. (1999). "A general class of nonparametric models for ordinal categorical data." In M. E. Sobel & M. P. Becker (eds.), *Sociological Methodology 1999*. Washington, DC: The American Sociological Association, pp. 187–224.

Whittaker, J. (1990). *Graphical Models in Applied Multivariate Statistics.* Chichester: Wiley.

Wiggins, L. M. (1955). *Mathematical Models for the Analysis of Multiwave Panels*. Doctoral Dissertation Series No. 12.481, Ann Arbor, MI.

Wiggins, L. M. (1973). *Panel Analysis: Latent Probability Models for Attitude and Behavior Processes.* Amsterdam: Elsevier.

Winship, C., & Mare, R. D. (1983). "Structural equations and path analysis for discrete data," *American Journal of Sociology*, **89**, 54–110.

Yule, G. U., & Kendall, M. G. (1950). *Introduction to the Theory of Statistics*, 14th Ed. New York: Hofner.

NOTES

1. The term *causal* will be used here somewhat loosely to denote asymmetrical relationships among variables. The purpose of this chapter is to show how loglinear causal models may be built, not to claim a fundamental causal meaning for the relationships in the examples. The expositions of the causal modeling approach in this chapter owe a lot to some older (and sometimes forgotten) basic insights from traditional tables analysis (Yule and Kendall, 1950; Lazarsfeld, 1955) and from traditional regression methods for observational data (Simon, 1957; Blalock, 1964). Furthermore, Goodman's work has been decisive. He laid the foundations of causal loglinear modeling, introducing the loglinear modified path approach (Goodman, 1973) that has been extended to include latent variables (the modified LISREL model; Hagenaars, 1990, 1993). Also very important have been the developments in the field of (directed) graphical modeling, which have made clear the commonalities and differences between linear and general nonlinear (structural) equation models (Kiiveri and Speed, 1982; Whittaker, 1990; Lauritzen, 1996; Cox and Wermuth, 1996; Koster, 1996). Throughout the 1990s, there has been a growing interest in and investigation of the causal nature of these causal analyses; the names of Rubin, Glymour and his associates, and Pearl are singled out as probably the most influential contributors. Overviews of the contents of this work, of its potentialities for the analysis of nonexperimental, observational data, as well as the controversies (still) surrounding it, may be found in Manski (1995), Rosenbaum (1995), McKim and Turner (1997), Glymour and Cooper (1999), Pearl (2000), and in two special issues of *Sociological Methods and Research*, with guest editors Raftery (1998) and Snijders and Hagenaars (2001).
2. The maximum-likelihood estimates for this and all other models in this chapter have been obtained by means of the program LEM (Vermunt, 1997b). In Model 1, four (independent) estimates of the conditional response probabilities (not presented here) turned out to be zero. So, the (other) parameter estimates can only be regarded as the maximum-likelihood estimates under the condition that these four zero response probabilities are indeed (structural) zeros in the population. Although the values of these zero conditional response probabilities are in line with the substantive interpretations of the outcomes, these zero values were not expected a priori and just happened to occur in this particular sample. Therefore, it was decided not to add them as a priori restrictions to the number of degrees of freedom. Another problem connected to zero estimated probabilities is that they may give rise to the questioning of the identifiability of the model, in the sense that one may wonder whether the model would be identifiable without the zero estimates. Here, by use of simulated data, it was made certain that the model as such was identifiable: For simulated data without zero estimates, the variance–covariance matrix of the estimates was full rank.
3. In principle, one can define column, row, row plus column, and linear × linear association models with fixed scores for the interval level variable(s) (type 1 association models) or with scores to be estimated (type 2 association

models). Several models of type 1 are identical to a type 2 model, especially in two-way tables. In multiway tables, such equivalences also depend on the nature of the restrictions imposed on the scores over the subtables (Clogg and Shihadeh, 1994; Agresti, 1990, Chapter 8; Luijkx and Hagenaars, 1990; Vermunt, 1997a, 1997b). Finally, "in-between" models may be defined in which order restrictions are imposed in the form of (weak) inequality restrictions on the parameters α_a, and so on; for a recent overview of the many possibilities, see Vermunt (1999).

4. Actually, all subsequent analyses have been carried out by using both Model 5 and Model 7. Because no real differences were encountered between these two approaches, Model 7 remained the preferred model.
5. The effects of increasing degrees of system involvement on the odds of obtaining score z on Protest Tolerance rather than z' are determined by means of a logit model: $\Phi^{Z/Y}_{z/z'y} = \ln(\Pi^{Z/Y}_{zy}/\Pi^{Z/Y}_{z'y}) = \beta^{Z}_{z/z'} + \beta^{Z/Y}_{z/z'y}$. The logit model is essentially a reparametrization of the loglinear model and the estimates for β in the logit model can easily be obtained from the λ parameters of the loglinear model, which here is $\beta^{Z/Y}_{z/z'y} = \lambda^{YZ}_{yz} - \lambda^{YZ}_{yz'}$ (Bishop, 1969; Agresti, 1990, Section 6.3.3; Hagenaars, 1990, Section 2.7).
6. More complicated models can also be handled by means of marginal (or generalized loglinear) models in which (loglinear) models are defined simultaneously for the full cross-classification table and for marginal tables derived from the full table using either full maximum-likelihood (Lang and Agresti, 1994; Becker, 1994; Becker and Yang, 1998; Bergsma, 1997; Croon et al., 2000) or GEE methods (Liang and Zeger, 1986; Diggle et al., 1996). These recent developments will certainly make it easier in the near future to estimate still more complex causal models for answering still more complex substantive questions.
7. The pertinent Model 1 is denoted in Table 5 as loglinear model {*SGTYZV*, *AY*, *BY*, *CZ*, *DZ*, *EY*}. This loglinear model implies exactly the same restrictions on the data and yields exactly the same estimates for the joint probabilities in the complete table as application of the series of saturated submodels for the relations among *S*, *G*, *T*, *Y*, and *Z* together with the restricted models for the relations between the latent variables and the indicators, as discussed in the text. Sometimes it is possible to obtain the estimates for the joint probabilities in the full table directly by specifying one (loglinear) model rather than a sequence of submodels. This possibility has to do with the kinds of restrictions that are imposed on the submodels, with "collapsibility" of the loglinear model (Agresti, 1990, Section 5.4.2; Santner and Duffy, 1989, Section 4.6), and with the question whether the causal model satisfies the "Wermuth condition" and is a "moral graph" (Whittaker, 1990, Section 3.5). The notation used here in Table 5 is just used for practical reasons; in the text, all models will be discussed in terms of the sequential approach.
8. Both Poulsen and Van de Pol/Langeheine (see Chapter 11, this volume) defined not only Markov models at the latent level, but also mixed Markov models, dividing the respondents into different unobserved groups (latent classes) having different (Markov) parameters or into groups only some of which follow the Markov model, whereas for the others a different model is

postulated. These and other forms of unobserved heterogeneity can also be handled by the DLM, as illustrated later. In general, it may be interesting to note that almost all models proposed for misclassifications in categorical data, including those using external validation data (for a recent overview, see Kuha and Skinner, 1997), can be formulated in terms of categorical latent variable models and the DLM, making categorical misclassification models much more flexible than is usually recognized.

9. The distinction made here between random response error and random behavior is related to the distinction in platonic true score and operational true score, where the platonic true score model assumes that there exists a real and actual true score and the operational true score model defines the true score as the score a person obtains on average in a series of independently conducted experiments (see, e.g., Suttcliffe, 1965a, 1965b; Lord and Novick, 1968, Section 9; Hagenaars, 1990, Section 4.4.2; Sobel, 1994). Whether this distinction matters depends on the purposes of the analysis. Here, it is argued that sometimes it does matter.

TEN

Latent Class Models for Longitudinal Data

Linda M. Collins and Brian P. Flaherty

The work described in this chapter comes from a perspective rooted in psychological research,[1] where there is considerable interest in characteristics of the individual that cannot be directly measured, such as intelligence, abilities in various domains, attitudes, and complex private behaviors. The focus is on using manifest variables, usually subject self-reports or ratings by trained observers, to measure latent variables representing these unmeasured characteristics of individuals. In this chapter, we will discuss the use of latent class models to measure change over time in latent variables.

We will begin by drawing a conceptually important distinction between static and dynamic latent variables (Collins, 1996). Static latent variables are not expected to change over time, or else the change is not of interest in a particular study, as when group differences at a particular time are examined. In contrast, dynamic latent variables do change in systematic and important ways over time. Often the same variable may be thought of as static in one context and dynamic in another, depending on the objectives and interests of a particular research project. For example, drug use in eighth grade, cognitive ability in an elderly population, and gender role identification are all static latent variables. Change in drug use from seventh to ninth grade, age-related decline in cognitive ability, and gender role development throughout high school are all dynamic latent variables.

Growth in *continuous* dynamic variables proceeds in a strictly quantitative fashion over time. In other words, growth can be expressed in terms of an increase or decrease in the amount of something – ounces of alcohol consumed per week, reading ability, number of delinquent acts in a three-month period, elderly individuals' ability to care for themselves, and so on. Growth in continuous dynamic latent variables can be modeled by means of latent growth curve approaches (e.g., Willett and Sayer, 1994; McArdle and Hamagami, 1991). In contrast, *stage-sequential* dynamic latent variables involve a different kind of growth, expressed by movement between stages. This is often described as *qualitative growth*, although in many cases it is accompanied by or even made up primarily of quantitative growth. A classic example of a stage-sequential model of development is provided by Piaget, whose model of cognitive development specifies that children begin in the sensorimotor stage, then move to the preoperational stage, then enter the concrete operations stage, then move to the formal operations stage. In Piaget's model, children gain intellectual ability as they enter each new stage, so there is quantitative development. However, the critical feature of the model is the way each stage represents a qualitatively different way of organizing information.

Latent transition analysis (LTA; Graham et al., Hansen, 1991; Collins and Wugalter, 1992) is a latent class theory approach to measuring stage-sequential dynamic latent variables and estimating and testing models of stage-sequential development. (For an alternative formulation of latent class models for growth, emphasizing Markov chains, see, e.g., Langeheine, 1994; Langeheine, Stern, and van de Pol, 1994; Langeheine and van de Pol, Chapter 11, this volume). By means of LTA, a researcher can specify the number of stages in a model and the transitions between stages that are consistent with the model. Parameters are estimated that reflect the prevalence of stage memberships at the first time of measurement; the incidence of stage transitions; and the probability of particular item responses, conditional on stage membership. Models can be compared just as they are in other latent class applications.

1. STAGE-SEQUENTIAL MODELS AND LATENT CLASS THEORY

Consider the following hypothetical latent class situation. There are three categorical latent variables, representing mastery or nonmastery of three different skills, labeled P, Q, and R. Thus each individual is in the mastery or nonmastery latent class for skill P, the mastery or nonmastery latent class for skill Q, and the mastery or nonmastery latent class for skill R. Each of these skills is measured by different tasks or different instruments.

Now suppose there is a model specifying that the three skills are related in some way, so that they are part of an underlying skill acquisition process. For example, there might be an underlying process of acquisition, such that skill P is a prerequisite for skill Q, which in turn is a prerequisite for skill R. The set of an individual's latent class memberships at one particular time is referred to as a *latent status*. Only certain latent statuses out of all those possible are consistent with this model. They are nonmastery of all three skills; mastery of P and nonmastery of Q and R; mastery of P and Q and nonmastery of R; and mastery of all three skills. An example of a latent status that is not consistent with the model is mastery of Q and nonmastery of P and R because it contradicts the idea that P must be learned before Q. This is now a stage-sequential model of stage-sequential development, and the skill acquisition process is a stage-sequential dynamic latent variable. According to the model, individuals change latent status memberships over time in particular ways. For example, an individual who learns skill P moves from the latent status involving nonmastery of all three skills to the latent status involving mastery of P and nonmastery of the other two skills.

Let $a = 1, \ldots, A$ represent latent status membership at Time 1. The probability of membership in latent status a at Time 1 is denoted π_a^A. The array of these parameters represents the prevalence of latent status memberships at the first occasion of measurement. Because $\sum \pi_a^A = 1$, at most $A - 1$ of these parameters are estimated.

The probability of membership in latent status b at Time 2, conditional on membership in latent status a at Time 1, is $\pi_{ba}^{\bar{B}A}$. This represents the incidence of transitions between latent statuses from Time 1 to Time 2. These parameters are often called transition probabilities, and they are usually arranged in a matrix where the rows represent Time 1 latent status and the columns represent Time 2 latent status. Because $\sum_b \pi_{ba}^{\bar{B}A} = 1$, at most only $A(B - 1)$ of these parameters are independently estimated.

Suppose there are three manifest variables, for example, three tests, administered at each time. Item 1 has response categories $i = 1, 2, \ldots, I$, Item 2 has response categories $j = 1, 2, \ldots, J$, and Item 3 has response categories $k = 1, 2, \ldots, K$. The responses to the items at Time 2 will be denoted with the same letter with the addition of a prime. Then the probabilities of individual item responses at Time 1, conditional on membership in latent status a, are

$$\pi_{ia}^{\bar{I}A}, \quad \pi_{ja}^{\bar{J}A}, \quad \pi_{ka}^{\bar{K}A}.$$

The corresponding probabilities for Time 2 are

$$\pi_{i'b}^{\bar{I}B}, \quad \pi_{j'b}^{\bar{J}B}, \quad \pi_{k'b}^{\bar{K}B},$$

and so on; there would be corresponding probabilities for additional items. Because $\sum_i \pi_{ia}^{\bar{I}A} = 1$, at most $I - 1$ of these parameters are estimated for each combination of a latent status, item, and time.

It is possible to add a manifest or latent exogenous variable to the LTA model. This variable is also discrete, dividing the population into latent classes. For example, the model described earlier could involve an attitude variable, reflecting positive and negative attitudes, that is believed to have an effect on the rate at which individuals acquire the skills.

Let $x = 1, \ldots, X$ represent the exogenous variable. Then π_x^X represents the probability of membership in latent class x of the exogenous variable, and $\pi_{mx}^{\bar{M}X}$ represents the probability of response m to Item M, conditional on membership in latent class x of exogenous latent variable X.

As before, $a = 1, \ldots, A$ represents latent status membership at Time 1. The probability of membership in latent status a at Time 1, conditional on membership in latent class x of exogenous latent variable X, is $\pi_{ax}^{\bar{A}X}$. This represents the prevalence of latent status membership at the initial time of measurement, conditional on the exogenous variable. The probability of membership in latent status b at Time 2, conditional on membership in latent status a at Time 1 and latent class x of exogenous variable X, is $\pi_{bax}^{\bar{B}AX}$. This represents the incidence of transitions between latent statuses from Time 1 to Time 2, conditional on the exogenous variable. The probability of individual item responses at Time 1, conditional on membership in latent status a and latent class x of the exogenous variable, is $\pi_{iax}^{\bar{I}AX}$ and so on with additional probabilities for additional items and times. For models including an exogenous variable,

$$\sum_x \pi_x^X = 1, \quad \sum_i \pi_{iax}^{\bar{I}AX} = 1, \quad \sum_a \pi_{ax}^{\bar{A}X} = 1, \quad \sum_b \pi_{bax}^{\bar{B}AX} = 1.$$

Let $\pi_{mijki'j'k'}$ represent the probability of observing response m to a manifest variable measuring the static latent variable, responses i, j, and k at Time 1 to three manifest variables measuring the dynamic latent variable, and responses i', j', and k' at Time 2 to the same three manifest variables. Suppose there are $x = 1, \ldots, X$ latent classes, $a = 1, \ldots, A$ latent statuses at Time 1, and $b = 1, \ldots, B$ latent statuses at Time 2. Then

$$\pi_{mijki'j'k'} = \sum_{x=1}^{X} \sum_{a=1}^{A} \sum_{b=1}^{B} \pi_x^X \pi_{mx}^{\bar{M}X} \pi_{ax}^{\bar{A}X} \pi_{iax}^{\bar{I}AX} \pi_{jax}^{\bar{J}AX} \pi_{kax}^{\bar{K}AX} \times \pi_{bax}^{\bar{B}AX} \pi_{i'bx}^{\bar{I}BX} \pi_{j'bx}^{\bar{J}BX} \pi_{k'bx}^{\bar{K}BX}$$

2. MORE THAN TWO TIMES AND SECOND-ORDER MODELS

Additional times of measurement can easily be incorporated into LTA models, by adding parameters to express the incidence of transitions between Time 2 and Time 3, Time 3 and Time 4, and so on, and parameters to express the probability of particular item responses conditional on latent status membership at the additional times. When there are three or more times of measurement, it is possible for a second-order model to be specified. In second-order models, transitions between latent statuses are conditional not only on the immediately previous time, but on the time before that as well. In other words, in a second-order model, Time 3 latent status is conditional not only on Time 2, but on Time 1 as well. Suppose there are $c = 1, \ldots, C$ latent statuses at Time 3. Then

$$\begin{aligned}\pi_{mijki'j'k'i''j''k''} = \sum_{x=1}^{X} \sum_{a=1}^{A} \sum_{ab=1}^{AB} \sum_{c=1}^{C} & \pi_x^X \pi_{mx}^{\bar{M}X} \pi_{ax}^{\bar{A}X} \pi_{iax}^{\bar{I}AX} \pi_{jax}^{\bar{J}AX} \\ & \times \pi_{kax}^{\bar{K}AX} \pi_{bax}^{\bar{B}AX} \pi_{i'bx}^{\bar{I}BX} \pi_{j'bx}^{\bar{J}BX} \pi_{k'bx}^{\bar{K}BX} \\ & \times \pi_{cbax}^{\bar{C}BAX} \pi_{i''cx}^{\bar{I}CX} \pi_{j''cx}^{\bar{J}CX} \pi_{k''cx}^{\bar{K}CX},\end{aligned}$$

where $\pi_{cbax}^{\bar{C}BAX}$ is the probability of membership in latent status c at Time 3, conditional on latent status b at Time 2, latent status a at Time 1, and latent class x of static latent variable X.

A. Empirical Example

In order to demonstrate the use of LTA, we present an empirical example from the area of adolescent substance use. An important question in substance use research is the extent to which use of certain substances can be considered risk factors for the use of certain other substances. For example, the "gateway" theory of drug use (e.g., Kandel and Faust, 1975; Kandel, Kessler, and Margulics, 1978) suggests that alcohol and tobacco are gateways for the use of marijuana and more advanced substances (e.g., cocaine). In the analyses presented in this chapter, we test the gateway theory, confining our attention to the early part of the substance use onset process. We select, estimate, and test a prospective model of adolescent experimentation with alcohol, tobacco, and marijuana, and we examine the parameter estimates in order to determine whether experimentation with alcohol and tobacco is a risk factor for experimentation with marijuana.

3. METHODS

A. Subjects and Measures

These analyses are based upon data collected in the Adolescent Alcohol Prevention Trial (AAPT; Hansen and Graham, 1991; Graham, Marks, and Hansen, 1991). Participants were 5,242 junior high school students in the Los Angeles area who were surveyed annually between the seventh and ninth grades.

As part of the survey, each respondent was asked about tobacco, alcohol, and marijuana use. For purposes of the analyses reported here, each question was categorized to form a dichotomous item reflecting whether the respondent had ever tried the substance. The items were coded as 1 for never tried and 2 for have tried. Any level of use was considered trying for these data; that is, one puff of tobacco or marijuana or a sip of alcohol was considered trying. The subject's responses to each of these three questions over the three survey years constitute the data.

B. Estimation

A computer program called WinLTA (Collins et al., 1999) can be used for the analyses reported here. This program uses the expectation–maximization (EM) algorithm for estimation, which does not produce standard errors as a by-product of estimation. In order to have some way of assessing the stability of the parameter estimates, the sample was split randomly into two groups. All analyses were done separately on the two samples, and the results were compared. Missing data were handled by imputation prior to the use of LTA, by means of the EMCOV procedure (Graham, Hofer, and Piccinin, 1994).

C. Model 1

The first model tested involved all possible latent statuses in order to examine whether any of the latent statuses could be omitted. Given the three substances, there are eight possible latent statuses an individual can occupy, ranging from never trying a substance to trying all three substances. The eight possible latent statuses are as follows:

1. Never tried tobacco, alcohol, or marijuana.
2. Tried tobacco alone.
3. Tried alcohol alone.
4. Tried marijuana alone.

5. Tried tobacco and alcohol.
6. Tried tobacco and marijuana.
7. Tried alcohol and marijuana.
8. Tried all three substances.

We expected that there might be few, or even no, members in latent statuses 4 and 7, the marijuana alone and alcohol and marijuana statuses. This is because alcohol and tobacco are known risk factors for marijuana use, so it is unlikely that anyone would try marijuana without having tried these substances.

Parameter Restrictions

The WinLTA software allows the user the choice of estimating parameters freely, constraining sets of parameters to be equal to each other, or fixing parameters to be equal to a prespecified value. In the analyses reported here, some $\pi_{ba}^{\bar{B}A}$ parameters were fixed and some $\pi_{ia}^{\bar{I}A}$ parameters were constrained.

Some of the $\pi_{ba}^{\bar{B}A}$ parameters correspond to illogical transitions, for example, having tried alcohol at seventh grade and then not having tried it in eighth grade. The parameters corresponding to these illogical transitions were all fixed at zero. With these constraints in place, progression through the model is quite logical. No use is followed by the trying of one substance (probably tobacco or alcohol). Single substance experimentation is followed by experimentation with another substance (again, probably alcohol or tobacco). This in turn is followed by experimentation with all three substances.

Equality constraints were placed on the $\pi_{ia}^{\bar{I}A}$ parameters for two distinct purposes. First, these parameters were constrained to be equal across times. If these parameters are allowed to vary across times, the interpretation of the latent statuses can also vary, making interpretation of transitions difficult. Second, in order to ensure identification by reducing further the number of these parameters estimated, the constraints shown in Table 1 were imposed so that effectively only one parameter was estimated each for the alcohol, tobacco, and marijuana items.

Goodness of Fit

Goodness-of-fit in contingency table models is assessed by comparing the observed response proportions with the probabilities predicted by the model. Usually this is expressed in terms of either the likelihood ratio statistic L^2 or Pearson's X^2. For models involving very large, and

Table 1. Constraints on $\pi_{ia}^{\bar{I}A}$ Parameter Estimates in Model 1

Latent Status	Tried Tobacco	Tried Alcohol	Tried Marijuana
Parameters Corresponding to Probability of a "Yes" Response			
1. Never tried	2[a]	12	22
2. Tried tobacco	3	12	22
3. Tried alcohol	2	13	22
4. Tried marijuana	2	12	23
5. Tried tobacco & alcohol	3	13	22
6. Tried tobacco & marijuana	3	12	23
7. Tried alcohol & marijuana	2	13	23
8. Tried tobacco, alcohol, & marijuana	3	13	23
Parameters Corresponding to Probability of a "No" Response			
1. Never tried	3	13	23
2. Tried tobacco	2	13	23
3. Tried alcohol	3	12	23
4. Tried marijuana	3	13	22
5. Tried tobacco & alcohol	2	12	23
6. Tried tobacco & marijuana	2	13	22
7. Tried alcohol & marijuana	3	12	22
8. Tried tobacco, alcohol, & marijuana	2	12	22

[a] Parameters with the same number are constrained to be equal.

often sparse, contingency tables, there are problems with both of these goodness-of-fit statistics, because under these conditions they are not reliably distributed as a chi square (Read and Cressie, 1988). Collins et al. (1993) showed that under conditions typical in LTA models, the expectation of L^2 tends to be considerably lower than the associated degrees of freedom. They also showed that the X^2 has an unacceptably large variance. Following the convention of the present volume, we report both of these quantities. However, in our work we rely primarily on the more stable L^2.

Model Fit

The L^2 values for the two subsamples were, respectively, 488.1 and 480.7 (Pearson's X^2 were, respectively, 1,001.9 and 1,688.3) with 457 degrees of freedom. Although these L^2 values indicate a reasonably good fit, it seemed that the model could be simplified by reducing the number of latent statuses. Table 2 shows the seventh grade π_a^A parameters for the two samples. As described earlier, we had expected the marijuana

Table 2. π_a^A Parameter Estimates in Each Subsample

Latent Status	Subsample A	Subsample B
1. Never tried	0.35	0.36
2. Tried tobacco	0.03	0.03
3. Tried alcohol	0.28	0.28
4. Tried marijuana	0.00	0.00
5. Tried tobacco & alcohol	0.32	0.29
6. Tried tobacco & marijuana	0.00	0.00
7. Tried alcohol & marijuana	0.00	0.00
8. Tried tobacco, alcohol, & marijuana	0.03	0.04

alone and the alcohol and marijuana latent statuses to have expected frequencies near zero. Table 2 shows that the results were consistent with this expectation. In addition, the tobacco and marijuana latent status was nearly empty. It should be noted that in longitudinal data, it is possible for a latent status to be empty or nearly empty at an early observation, but for individuals to move into the latent status at a later observation. Such a latent status should not be removed. However, in the present data, these three latent statuses were empty across all three times.

D. Model 2

Because three of the latent statuses in Model 1 proved to be nearly empty, it appeared that a more parsimonious five latent status model would fit the data adequately. Accordingly, we fit Model 2, a five latent status model with the same pattern of constraints as described above. Model 2 had $L^2 = 502.5$ and 493.1 ($X^2 = 1{,}060.9$ and 2,636.6) in the two subsamples, with 486 degrees of freedom. It appears that removing the three nearly empty latent statuses results in a much more parsimonious model and does not worsen the fit very much.

E. Model 3

Although Model 2 fits fairly well, the $\pi_{ia}^{\bar{I}A}$ and other parameters are rather severely constrained. To investigate whether lifting some of the constraints on these parameters would improve the fit of the model, we changed the constraints to the pattern illustrated in Table 3. This pattern of constraints, introduced by Dayton and Macready (1976), is equivalent to allowing each item to be associated with two error parameters: the

Table 3. Constraints on $\pi_{ia}^{\bar{I}A}$ Parameter Estimates in Model 3

Latent Status	Tried Tobacco	Tried Alcohol	Tried Marijuana
Parameters Corresponding to Probability of a "Yes" Response			
1. Never tried	2[a]	12	22
2. Tried tobacco	3	12	22
3. Tried alcohol	2	13	22
4. Tried tobacco & alcohol	3	13	22
5. Tried tobacco, alcohol, & marijuana	3	13	23
Parameters Corresponding to Probability of a "No" Response			
1. Never tried	33	43	53
2. Tried tobacco	32	43	53
3. Tried alcohol	33	42	53
4. Tried tobacco & alcohol	32	42	53
5. Tried tobacco, alcohol, & marijuana	32	42	52

[a] Parameters with the same number are constrained to be equal.

probability of a "yes" response when according to the latent status it should have been a "no" (e.g., the probability of responding "yes" to the alcohol use item conditional on membership in Latent Status 1), and the probability of a "no" response when according to the latent status it should have been a "yes" (e.g., the probability of responding "no" to the alcohol use item conditional on membership in Latent Status 5). This model and Model 2 are nested. When compared with Model 2, Model 3 resulted in a loss of 3 degrees of freedom against a decrease in L^2 of 94.7 in Subsample A and 148.8 in Subsample B. (There was also a decrease in X^2 of 231.7 in Subsample A, but an *increase* in X^2 of 281.4 in Subsample B. Examination of the data showed that the increase in the Subsample B X^2 is attributable to one cell with a very small expectation and an observed frequency of one. The change in the model reduced the expectation in this cell, creating an enormous cell X^2 of 1,828.6. The corresponding cell L^2 is 15.02. This is one example of the instability of the X^2 statistic.) Based on the more stable L^2, it appears that estimating the additional three parameters is warranted.

Tables 4, 5, and 6 show the $\pi_{ia}^{\bar{I}A}$, π_a^A, and $\pi_{ba}^{\bar{B}A}$ parameters, respectively, for each of the two subsamples. As these three tables show, the overall pattern of results is identical across the two subsamples, and the parameter estimates themselves are highly similar. An examination of Table 4 shows how the $\pi_{ia}^{\bar{I}A}$ parameters provide the basis for the user

Table 4. $\pi_{ia}^{I A}$ **Parameter Estimates for Model 3**

Subsample	Probability of a "Yes" Response Tried Tobacco	Tried Alcohol	Tried Marijuana
A			
1. Never tried	0.05	0.17	0.01
2. Tried tobacco	0.94	0.17	0.01
3. Tried alcohol	0.05	0.96	0.01
4. Tried tobacco & alcohol	0.94	0.96	0.01
5. Tried tobacco, alcohol, & marijuana	0.94	0.96	0.81
B			
1. Never tried	0.05	0.15	0.01
2. Tried tobacco	0.94	0.15	0.01
3. Tried alcohol	0.05	0.96	0.01
4. Tried tobacco & alcohol	0.94	0.96	0.01
5. Tried tobacco, alcohol, & marijuana	0.94	0.96	0.77

to interpret the latent statuses. Items with $\pi_{ia}^{I A}$ parameters near zero or one are highly related to latent class a.

In this empirical example, the probability of a "yes" response to any item is low for those in Latent Status 1. This suggests that Latent Status 1 is characterized by individuals who have not tried any substance. For Latent Status 2, the probability of a "yes" response is high for Tried Tobacco and low for each of the other two items. This suggests that Latent Status 2 is

Table 5. π_{a}^{A} **Parameter Estimates for Model 3**

Subsample	Grade 7th	8th	9th
A			
1. Never tried	0.39	0.26	0.13
2. Tried tobacco	0.05	0.04	0.04
3. Tried alcohol	0.23	0.25	0.25
4. Tried tobacco & alcohol	0.29	0.37	0.45
5. Tried tobacco, alcohol, & marijuana	0.05	0.09	0.14
B			
1. Never tried	0.40	0.25	0.12
2. Tried tobacco	0.05	0.04	0.04
3. Tried alcohol	0.24	0.25	0.25
4. Tried tobacco & alcohol	0.26	0.37	0.45
5. Tried tobacco, alcohol, & marijuana	0.06	0.10	0.14

Table 6. $\pi_{ba}^{\bar{B}A}$ Parameter Estimates for Model 3

	Grade Latent Status				
Subsample	**1**	**2**	**3**	**4**	**5**
Subsample A: 7th Grade Latent Status			8th Grade		
1. Never tried	0.67	0.04	0.17	0.11	0.01
2. Tried tobacco	0[a]	0.47	0	0.51	0.01
3. Tried alcohol	0	0	0.81	0.17	0.02
4. Tried tobacco & alcohol	0	0	0	0.91	0.09
5. Tried tobacco, alcohol, & marijuana	0	0	0	0	1.00
Subsample B: 7th Grade Latent Status					
1. Never tried	0.64	0.04	0.19	0.13	0.01
2. Tried tobacco	0	0.45	0	0.48	0.07
3. Tried alcohol	0	0	0.74	0.24	0.02
4. Tried tobacco & alcohol	0	0	0	0.91	0.10
5. Tried tobacco, alcohol, & marijuana	0	0	0	0	1.00
Subsample A: 8th Grade Latent Status			9th Grade		
1. Never tried	0.51	0.06	0.24	0.18	0.01
2. Tried tobacco	0	0.54	0	0.45	0.02
3. Tried alcohol	0	0	0.74	0.24	0.02
4. Tried tobacco & alcohol	0	0	0	0.88	0.12
5. Tried tobacco, alcohol, & marijuana	0	0	0	0	1.00
Subsample B: 8th Grade Latent Status			9th Grade		
1. Never tried	0.49	0.08	0.26	0.16	0.01
2. Tried tobacco	0	0.45	0	0.52	0.02
3. Tried alcohol	0	0	0.73	0.25	0.02
4. Tried tobacco & alcohol	0	0	0	0.90	0.10
5. Tried tobacco, alcohol, & marijuana	0	0	0	0	1.00

[a] Cells containing an exact 0 represent parameters fixed at a value of zero.

characterized by individuals who have tried tobacco but have not tried alcohol or marijuana. The other latent statuses can be interpreted based on these parameters in the same way.

Table 5 contains the estimated probabilities of membership in each latent status at each time. Only the parameters corresponding to the seventh grade are independently estimated. The rest can be calculated from other parameter estimates and are shown here for comparison purposes. As is to be expected, the largest single latent status at seventh grade is Never Tried. However, the membership in this latent status declines over time, whereas membership in the two most advanced latent statuses increases between seventh and ninth grade. Table 6, which shows the $\pi_{ba}^{\bar{B}A}$ parameters, contains detailed information about how the movement among latent statuses takes place. Between seventh and eighth grade, those who have tried tobacco only are most likely to transition into a more advanced latent status, usually into the Tried Alcohol and Tobacco latent status. In all of the other latent statuses, the likelihood of not transitioning is greater than the likelihood of transitioning. Between eighth and ninth grade, the early part of the process has become somewhat faster, with those in the Never Tried latent status now about as likely to transition to a higher level of use as they are to remain in their current latent status. It is interesting to examine the role that tobacco and alcohol play in the progression to experimentation with marijuana. An individual who is in the Never Tried latent status has only about a 1% probability of trying marijuana within a year, whereas an individual who is in the Tried Alcohol and Tobacco latent status has a probability of trying marijuana ranging from 0.09 to 0.12.

4. DISCUSSION

A. Substantive Results

These results provide some interesting information on substance use onset between seventh and ninth grades. It appears that substance use experimentation begins with alcohol and tobacco. The substances may be tried in either order, but those who begin with tobacco are considerably more likely to transition to the next substance than are those who begin with alcohol. The results of this study are a reminder that both alcohol and tobacco are gateways to experimentation with marijuana. Although marijuana receives publicity as a gateway drug, our results indicate that alcohol and tobacco must also be called gateway drugs because few, if any, individuals experiment with marijuana without first having experimented with alcohol and tobacco. If a substantial number of individuals experimented with marijuana without having experimented with the other substances, there would have been a need for a marijuana only latent status in our model. However, when Model 1, which included this latent status, was estimated, it was clear that the latent status was empty.

As Collins (in press) points out, merely noting that individuals first try alcohol and tobacco and then try marijuana is not very meaningful. After all, most of these individuals tried broccoli before they tried alcohol or tobacco, but no one would advance the idea that broccoli is a risk factor for adolescent substance use. Rather, it is important to investigate whether an individual who has tried a substance is at increased risk for trying another substance. (For a more detailed discussion of these ideas, see

Collins, in press.) Although we do not have the data to test this, we would speculate that an individual who has tried broccoli is no more likely to experiment with alcohol or tobacco than an individual who has not tried broccoli. In contrast, our results show that an individual who has tried both alcohol and tobacco is 4.5 to 6 times more likely to experiment with marijuana than an individual who has tried alcohol only. This suggests that as adolescents experiment with substances, the risk that they will experiment with more substances increases. We do not have the data or design to determine whether this is causal. However, these results are consistent with the idea that early intervention programs aimed at legal substances, such as the AAPT (Hansen and Graham, 1991; Graham et al., 1991), are likely to be successful in reducing the onset of experimentation and use of illegal substances.

B. Latent Transition Analysis

A multiway contingency table is the raw material for all latent class analyses. In latent transition analyses involving multiple times and items, these contingency tables can become very large. Unless the sample size is correspondingly large, this multiway contingency table will be sparse. For example, four dichotomous items measured at three times results in a contingency table of 4,096 cells, requiring a sample size of more than, 4,000 merely to achieve an average cell expectation of 1. This leads to a question about whether parameter estimation can be carried out successfully in the kinds of data likely to be available to most researchers. In a series of simulation studies, Collins et al. (1996) and Collins and Tracy (1997) have investigated this issue. They found that parameter estimation using the EM algorithm is surprisingly robust, even in situations in which the average cell expectation is very small.

Another important question is whether adding a variable to an LTA model, for example a redundant indicator of a latent variable (assuming it fits well with the other variables), to an LTA analysis is a benefit because it provides better estimation of the latent variable, or a cost because it increases the size, and, given a fixed N, the sparseness of the contingency table. Collins and Wugalter (1992) found that all else being equal, the benefits of adding indicators to latent transition analyses outweigh the costs. However, this may not be true when the total number of items becomes unmanageably large, say, above 40.

Another problem, which may be less tractable, is model selection in LTA. (The problems with goodness of fit that arise in LTA are unlikely to be a concern for most latent class models that use smaller contingency tables.) In a search for a better goodness-of-fit statistic, Collins et al. (1993) tried substituting a variety of members of the power-divergence family (Read and Cressie, 1988), but none of them worked better than L^2. It may be that the difference L^2 is more likely to be distributed as a chi square (Agresti and Yang, 1987; Haberman, 1977), but this has been debated for latent class models (Holt and Macready, 1989). When there is a clearly delineated choice of alternatives among several models, we recommend cross-validation. With this procedure, the user divides the sample into two subsamples. Estimation is done separately on each of the samples, but the L^2 is computed by taking the parameter estimates from one subsample and assessing their goodness-of-fit in the other subsample (Cudeck and Browne, 1983; Collins et al., 1994). In this way, the degrees of freedom and the chi-square distribution become less relevant. In selecting a model, the user looks for a low cross-validation L^2 and a parsimonious model.

5. CONCLUSIONS

In this chapter, we have described latent transition analysis, a latent class procedure for estimating and testing stage-sequential models of development. With the use of this procedure, it is possible to estimate the prevalence of stage memberships and the incidence of stage transitions. It is also possible to compare these quantities across groups, which may be either manifest or latent. In this chapter, we illustrated the application of LTA to the investigation of adolescent substance use. LTA has some notable limitations. First, it is limited to relatively small models; very large numbers of variables or several times can create an unmanageably large contingency table. Second, model selection can be a problem, as the L^2 goodness-of-fit statistic may not be reliably distributed as a chi square for many LTA applications. Despite these limitations, LTA can provide the user with an interesting look at the dynamic latent structure underlying longitudinal categorical data.

REFERENCES

Agresti, A., & Yang, M. C. (1987). "An empirical investigation of some effects of sparseness in contingency tables." *Computational Statistics and Data Analysis*, **5**, 9–21.

Collins, L. M. (1996). "Measurement of change in research on aging: old and new issues from an individual growth perspective." In J. E. Birren & K. W. Schaie (eds.), *Handbook of the Psychology of Aging*, 4th ed. San Diego, CA: Academic Press, pp. XXX–XXX.

Collins, L.M. (In press). "Using latent transition analysis to examine the gateway hypothesis." In D. Kandel & M. Chase (eds.), *Stages and Pathways of Involvement in Drug Use: Examining the Gateway Hypothesis*. Cambridge: Cambridge University Press, pp. XXX–XXX.

Collins, L. M., Fidler, P. L., & Wugalter, S. E. (1996). "Some practical issues related to estimation of latent class and latent transition parameters." In A. von Eye & C. C. Clogg (eds.), *Analysis of Categorical Variables in Developmental Research*. San Diego: Academic Press, pp. XXX–XXX.

Collins, L. M., Fidler, P. L., Wugalter, S. E., & Long, J. D. (1993). "Goodness-of-fit testing for latent class models." *Multivariate Behavioral Research*, **28**, 375–89.

Collins, L. M., Graham, J. W., Long, J. D., & Hansen, W. B. (1994). "Crossvalidation of latent class models of early substance use onset." *Multivariate Behavioral Research*, **29**(2), 165–83.

Collins, L. M., & Tracy, A. J. (1997). "Estimation in complex latent transition models with extreme data sparseness." *Kwantitatieve Methoden*, **18**, 57–71.

Collins, L. M., & Wugalter, S. E. (1992). "Latent class models for stage-sequential dynamic latent variables." *Multivariate Behavioral Research*, **27**, 131–57.

Collins, L. M., Flaherty, B. P., Hyatt, S. L., & Schafer, J. L. (1999). *WinLTA User's Guide (Version 2.0)*. University Park, PA: The Pennsylvania State University.

Cudeck, R. A., & Browne, M. W. (1983). "Crossvalidation of covariance structures." *Multivariate Behavioral Research*, **18**, 147–67.

Dayton, C. M., & Macready, G. B. (1976). "A probabilistic model for validation of behavioral hierarchies." *Psychometrika*, **41**, 189–204.

Graham, J. W., Collins, L. M., Wugalter, S. E., Chung, N, K., & Hansen, W. B. (1991). "Modeling transitions in latent stage-sequential processes: a substance use prevention example." *Journal of Consulting and Clinical Psychology*, **59**, 48–57.

Graham, J. W., Hofer, S. M., & Piccinin, A. M. (1994). "Analysis with missing data in drug prevention research." In L. M. Collins & L. A. Seitz (eds.), *Advances in Data Analysis for Prevention Intervention Research* (NIDA Research Monograph No. 142). Washington, DC: National Institute on Drug Abuse.

Graham, J. W., Marks, G., & Hansen, W. B. (1991). "Social influence processes affecting adolescent substance use." *Journal of Applied Psychology*, **76**(2), 291–98.

Haberman, S. J. (1977). "Loglinear models and frequency tables with small expected cell counts." *Annals of Statistics*, **5**, 1148–69.

Hansen, W. B., & Graham, J. W. (1991). "Preventing alcohol, marijuana, and cigarette use among adolescents: peer pressure resistance training versus establishing conservative norms." *Preventive Medicine*, **20**, 414–30.

Holt, J. A., & Macready, G. B. (1989). "A simuation study of the difference chi-square statistics for comparing latent class models under violation of regularity conditions." *Applied Psychological Measurement*, **13**, 221–31.

Kandel, D., & Faust, R. (1975). "Sequences and stages in patterns of adolescent drug use." *Archives of General Psychiatry*, **32**, 923–32.

Kandel, D., Kessler, R., & Margulies, R. (1978). "Antecedents of adolescent initiation into stages of drug use: a developmental analysis." In D. B. Kandel (ed.),

Longitudinal Research on Drug Use: Empirical Findings and Methodological Issues. Washington, DC: Hemisphere–Wiley, pp. 73–99.
Langeheine, R. (1994). "Latent variable Markov models." In A. von Eye & C. C. Clogg (eds.), *Latent Variables Analysis: Applications for Developmental Research*. Thousand Oaks, CA: Sage, pp. 373–95.
Langeheine, R., Stern, E., & van de Pol, F. (1994). "State mastery learning: dynamic models for longitudinal data." *Applied Psychological Measurement*, **18**, 277–91.
McArdle, J. J., & Hamagami, F. (1991). "Modeling incomplete longitudinal and cross-sectional data using latent growth structural models." In L. M. Collins & J. L. Horn (eds.), *Best Methods for the Analysis of Change*. Washington, DC: American Psychological Association, pp. 276–304.
Read, T. R. C., & Cressie, N. A. C. (1988). *Goodness-of-Fit Statistics for Discrete Multivariate Data*. New York: Springer-Verlag.
Schafer, J. L. (1997). *Analysis of Incomplete Multivariate Data*. London: Chapman & Hall.
Willett, J. B., & Sayer, A. G. (1994). "Using covariance structure analysis to detect correlates and predictors of individual change over time." *Psychological Bulletin*, **116**(2), 363–81.

NOTE

1. Linda M. Collins and Brian P. Flaherty are with The Methodology Center and the Department of Human Development and Family Studies, The Pennsylvania State University. The authors thank John W. Graham for making the AAPT data available to them. This research was supported by Grant 1 P50 DA10075 from the National Institute on Drug Abuse.

ELEVEN

Latent Markov Chains

Rolf Langeheine and Frank van de Pol

1. INTRODUCTION

This chapter deals with longitudinal categorical data. Such data may be of two types, in general. Type I refers to the situation in which a few subjects are measured repeatedly at many occasions. The models and methods relevant in this case are those of time series analysis. Type II refers to the reverse situation in which many subjects are measured repeatedly at a few (typically three to five) occasions, resulting in so-called panel data. If such data are categorical, one approach of analysis is by means of Markov chain models. These are the kind of models this chapter focuses on.

The goal then is to make statements about what happens from one point in time to the next, for example, whether consumers stay with a given brand (are brand loyal) or switch to some other brand. Apart from consumer behavior, Markov chain models have been applied in many other settings, such as attitude change; income, geographic, and industrial labor mobility; interpersonal relationships; voting behavior; animal behavior; learning; cognitive development; meteorology; and epidemiology.

It seems to be unanimous that the term *Markov chain* refers to discrete variables (i.e., categorical variables, items, or, more generally, indicators with response categories such as Yes/No; Agree/Disagree; Pass/Fail; Employed/Unemployed/Not in the labor force; Democrat/Republican/Don't Know) measured repeatedly over time with the same sample of subjects (panel data) and that the dynamics across time are modeled by assuming a discrete time process in order to make statements about change, stability, or both.

The views expressed in this paper are those of the authors and do not necessarily reflect the policies of Statistics Netherlands.

Whereas the first models considered refer to repeated measurements of a single indicator from *N* subjects at *T* points in time, recent developments mentioned in the final section show that the methodology can be extended to cope with the situation of multiple indicators that are considered as manifest indicators of some latent categorical variable (construct). That is, both the manifest and the latent variables considered in this chapter are categorical (discrete). However, models assuming ordered categorical latent variables may be conceived of for multiple indicator models.

In addition to the assumption of a discrete space, Markov chain models make the assumption of a process operating in discrete time. That is, Markov chain models make statements about transitions from one point in time to the next. They do not describe the process of change between occasions in the way that continuous time models do. Pros and cons of discrete and continuous time representations of dynamic models are dealt with by Singer and Spilerman (1974, 1976), Beck (1975), Bartholomew (1981), Kohfeld and Salert (1982), Kalbfleisch and Lawless (1985), Plewis (1985), and Hagenaars (1990), among others.

As the title of this chapter indicates, the focus will be on latent Markov chains. Figure 1 gives a hierarchy of models that all operate on the latent level in one way or the other, with the exception of the simple Markov model at the bottom of Figure 1. In the next section, we will start off with the most general model (latent mixed Markov for several groups) because this offers a reference point for all following models. We will then proceed with the simplest of all submodels and work bottom-up through the more comprehensive models, thus following the historical development of these models and the efforts made by various researchers to overcome the shortcomings at each stage of the hierarchy.

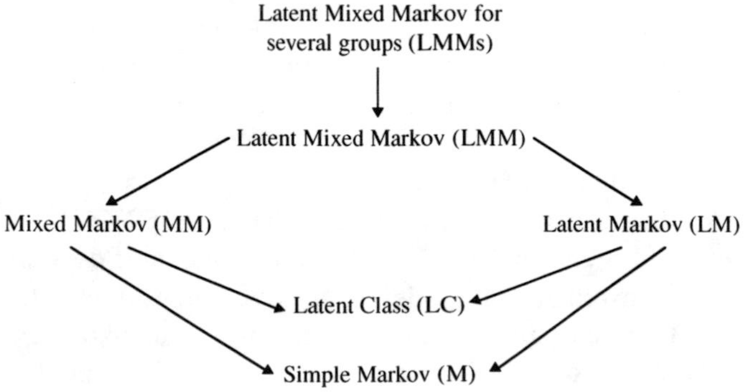

Figure 1. A hierarchy of Markov chain models (adapted from Langeheine, 1994).

Next we will introduce the models briefly. The simple Markov model assumes that the probability to be in a specific state only depends on one's state at the previous time point. The process ignores what happened before that. However, in nearly all instances it will turn out that this restrictive assumption is an oversimplification of the dynamics across time; that is, the model almost never turns out to be in agreement with a given set of data. There may be several reasons for this. One reason is that the model assumes population homogeneity; in other words, the dynamics across time tapped by the model are assumed to hold for all subjects. This assumption is relaxed by extending the simple Markov model to a mixture of several chains (mixed Markov), thus allowing each chain to follow its own dynamics. Another drawback of the simple Markov model is that it assumes the data to be free of measurement error, which renders this model unrealistic especially for social science researchers. This flaw may be coped with by extending the simple Markov model to a so-called latent Markov model that allows one to correct for measurement error. At this stage it is interesting to note that the classic latent class model can be shown to be a special case of both the mixed and the latent Markov models.

Just as the simple Markov model, the latent Markov model is a single chain model, thus sharing the same assumption with the former with respect to population homogeneity. The latent mixed Markov model therefore extends the latent Markov model to a mixture of several latent chains, thus allowing for unobserved population heterogeneity on the latent level. As we will see later, this model reduces to the mixed Markov if measurement is assumed to be perfect. On top, finally, we have the latent mixed Markov model for several groups, which extends the latent mixed Markov model to cope with the situation of having data from several groups defined by additional external discrete variables (such as Male/Female; Low/High socioeconomic status; Experimental/Control group). The feature of this model is that it allows for a simultaneous analysis of all groups in which the researcher is free, of course, to choose from all of the models considered so far.

Researchers who want to apply these models must have access to software to do so. A fairly general program package (PANMARK: PANel analysis using MARKov chains; Van de Pol, Langeheine, and de Jong, 1991) allows one to estimate all models considered in this chapter. Although PANMARK was originally developed to estimate Markov models, it also allows one to estimate a wide variety of latent class models. Another general program well suited to fit the class of models considered in this chapter is LEM (Vermunt, 1993).

In the following sections, the models of Figure 1 will be treated in detail and illustrated with examples where appropriate. The final section will address specific problems that are related to most of these models and also present some new results on how to cope with these problems.

2. THE LATENT MIXED MARKOV MODEL FOR SEVERAL GROUPS

Here we present the most general longitudinal latent class model known at the moment, the latent mixed Markov model for several groups of Van de Pol and Langeheine (1990), and we show how more simple (more restrictive) models are related to this general model. That is, we follow a top-down strategy. The reason simply is that this unifies and simplifies notation throughout. For those who prefer a bottom-up strategy, we recommend a cursory reading of this section only, and really starting with the most simple model, the classical one-chain Markov model, in the next section.

Consider the case in which some categorical variable having J categories is repeatedly measured at $T = 3$ occasions. This gives a J^3 table. Such a table may not only be available for a single group of subjects but for several, say H, groups that are defined by some additional external categorical variable such as gender, region, or experimental versus control group, in which the assumption is that the H groups so defined differ in their dynamics across time in the characteristic of interest. The latent mixed Markov model for several populations or groups is then given by

$$P_{hijk} = \gamma_h \sum_{s=1}^{S} \sum_{a=1}^{A} \sum_{b=1}^{B} \sum_{c=1}^{C} \pi_{s|h} \delta^1_{a|sh} \rho^1_{i|ash} \tau^{2\,1}_{b|ash} \rho^2_{j|bsh} \tau^{3\,2}_{c|bsh} \rho^3_{k|csh}. \quad (1)$$

The quantities in Equation (1) are that P_{hijk} is the model expected proportion in the population in cell (h, i, j, k), where h denotes group h and i, j, and k refer to the categories that some subject may belong to in the variable of interest at time points 1, 2, and 3 (superscripts denote time points). The proportion of subpopulation h is denoted γ_h. It is assumed that each subject belongs to one of the H subpopulations and that membership in subpopulation h remains unchanged at all T occasions. All other parameters are considered conditional on subpopulation h. Each member of subpopulation h belongs to one of one or several (S) Markov chains. We write $\pi_{s|h}$ for the proportion of subjects who belong to chain $s (s = 1, \ldots, S)$, given membership in subpopulation h. At time point 1 a member of subpopulation h and class s can belong to one of A latent classes, with $\delta^1_{a|sh}$ denoting the proportion of class $a (a = 1, \ldots, A)$. The vector δ^1_{sh} thus gives the initial (marginal) latent distribution for members in chain s and subpopulation h. The A latent classes are characterized by conditional response probabilities, where $\rho^1_{i|ash}$ is the probability that some subject belongs to category i of the manifest variable at time point 1, given membership in class a, chain s, and group h.

Transition is denoted by $\tau^{2\,1}_{b|ash}$, the probability of belonging to class $b(b = 1, \ldots, B)$ at time point 2, given membership in class a at time point 1, chain s, and group h. These τ, which are called *transition probabilities*, play an important role in Markov chain theory because they allow one to quantify the proportion of subjects who stay with their class ($b = a$) and the proportion of those who switch to another class across time ($b \neq a$).

As at occasion 1, the B latent classes at time point 2 are characterized by conditional response probabilities, $\rho^2_{j|bsh}$. The process is further modeled with transition probabilities for a change from latent class b at occasion 2 to latent class c at occasion 3 ($\tau^{3\,2}_{c|bsh}$) as well as conditional response probabilities at time point 3 ($\rho^3_{k|csh}$). Extension to more occasions is straightforward. (This is why we will give equations for $T = 3$ in general only.) Normally, the four-way population table with proportion P_{hijk}, for specific h, i, j, and k, is not available and a sample estimate is used, with proportion P_{hijk}. From this sample table, the parameters γ, π, δ, ρ, and τ may be estimated. The general latent mixed Markov model for several groups and special cases thereof can be estimated by using the PANMARK program. Note that in order to be uniquely defined, some identifiability conditions must be met for the parameters. A necessary, although not sufficient, condition is that all sets of parameters sum to unity, given any combination of variables that is conditioned on, for example, $\sum_h \gamma_h = \sum_s \pi_{s|h} = \sum_a \delta^1_{a|sh} = \sum_i \rho^1_{i|ash} = \sum_b \tau^{2\,1}_{b|ash} = 1$.

Additional conditions of identifiability will be referred to later.

We next give a cursory overview on how the more simple models presented in the hierarchy of Figure 1 derive from the general model. If there is one group only ($H = 1$), we obtain the latent mixed Markov model, a model that assumes S latent Markov chains. This model reduces to the latent Markov model if a single latent Markov chain ($S = 1$) is postulated. The mixed Markov model, in contrast, is obtained from the latent mixed Markov model by assuming a mixture of S manifest Markov chains, that is, Markov chains that are obtained when the data are assumed to be free of measurement error. Put differently, the mixed Markov model is equal to a latent mixed Markov model in which all latent distributions are assumed to be measured without error. The mixed Markov model thus is equal to a latent mixed Markov model in which all matrices of response probabilities (to be denoted by **R** in the sequel) are equal to the identity matrix, I. The classic latent class model turns out to be a special case of both the latent and the mixed Markov model. It is derived from the latent Markov model by assuming no latent change at all, that is, by equating all transition matrices **T** with the identity matrix, **I**. It is obtained from the mixed Markov model by assuming that transition probabilities to a state at time point t do not depend on the state at the previous time point. That is, instead of assuming dependence between consecutive points in time, local independence (the defining property of classic latent class model) is assumed to hold. Finally, the simple Markov chain model turns out to be a special case of both the mixed Markov and the latent Markov model. It is equal to a mixed Markov model with a single chain only ($S=1$) or equal to a latent Markov model without provision for measurement error (cf. the derivation of the mixed Markov model from the latent mixed Markov model).

3. THE SIMPLE MARKOV CHAIN

The data in Table 1 come from the German Socio-Economic Panel (SOEP), a household panel started in 1984 (cf. Hanefeld, 1984), and they refer to the question of how satisfied respondents are, on the whole, with their lives. Although respondents were originally given an 11-category ordinal scale ranging from fully unsatisfied to fully satisfied, we have dichotomized these data by collapsing categories 0–7 and categories 8–10, which corresponds to about a median split of the marginal distribution at each point in time. Category 1 in Table 1 thus refers to "unsatisfied" and category 2 to "satisfied." In order to have a handy table for demonstrative purposes, we consider data only from five waves (waves 3–7) out of the eight-wave data set originally available to us. Our sample of 5,147 individuals is subject to the following restrictions: The head of household is a West German adult citizen and gave an interview at all eight waves with complete responses. As we see from Table 1, 891 out of $N=5{,}147$ respondents report being unsatisfied with their lives at all five waves (response pattern 1 1 1 1 1). This is an observed proportion of .173, whereas the observed proportion of those indicating satisfaction with their lives at all occasions (response pattern 2 2 2 2 2) is 1,066/5,147= .207. The total manifest proportion of people staying with their original category across time thus amounts to .380. This leaves a considerable observed proportion (.620) of total change when one does not take into account whether someone switched from one category to another only once or on several occasions.

Table 1. General Life Satisfaction

t					Frequency			Obs. Frequency	
1	2	3	4	5	Observed	Model Expected	Chi-Square Contrib.	Males	Females
1	1	1	1	1	891	553.838	205.255	428	463
1	1	1	1	2	176	231.401	13.264	86	90
1	1	1	2	1	119	92.849	7.365	66	53
1	1	1	2	2	106	235.235	71.000	45	61
1	1	2	1	1	111	92.849	3.548	46	65
1	1	2	1	2	60	38.794	11.592	23	37
1	1	2	2	1	52	94.388	19.035	22	30
1	1	2	2	2	92	239.132	90.526	46	46
1	2	1	1	1	120	92.849	7.939	56	64
1	2	1	1	2	64	38.794	16.378	22	42
1	2	1	2	1	51	15.566	80.662	23	28
1	2	1	2	2	67	39.436	19.265	31	36
1	2	2	1	1	54	94.388	17.281	29	25
1	2	2	1	2	50	39.436	2.830	22	28
1	2	2	2	1	49	95.951	22.974	20	29
1	2	2	2	2	176	243.093	18.518	75	101
2	1	1	1	1	237	288.855	9.309	109	128
2	1	1	1	2	107	120.688	1.552	53	54
2	1	1	2	1	68	48.426	7.912	27	41
2	1	1	2	2	107	122.687	2.006	56	51
2	1	2	1	1	80	48.426	20.587	37	43
2	1	2	1	2	75	20.233	148.245	41	34
2	1	2	2	1	51	49.228	0.064	18	33
2	1	2	2	2	200	124.719	45.439	98	102
2	2	1	1	1	136	293.640	84.629	73	63
2	2	1	1	2	95	122.687	6.248	53	42
2	2	1	2	1	64	49.228	4.433	30	34
2	2	1	2	2	187	124.719	31.101	84	103
2	2	2	1	1	99	298.504	133.338	49	50
2	2	2	1	2	165	124.719	13.009	65	100
2	2	2	2	1	172	303.449	56.942	81	91
2	2	2	2	2	1,066	768.792	114.898	512	554

Notes: The 2^5 tables originate from the dichotomized data from the SOEP. $t = 1, \ldots, 5$ refer to waves 3–7 of the SOEP. Model expected frequencies and chi-square contributions refer to the simple Markov model with time homogeneous transition probabilities. The two columns on the right side split the total observed frequency into males and females.

Researchers who are interested in making statements about the dynamics across time captured by a single indicator often start fitting a very simple model, the Markov chain model, which, for $T = 3$ points in time, is given by

$$P_{ijk} = \delta_i^1 \tau_{j|i}^{2\,1} \tau_{k|j}^{3\,2}, \tag{2}$$

where all parameters are manifest quantities: δ^1 is the observed initial marginal distribution at occasion 1 and $\tau_{j|i}^{2\,1}$ and $\tau_{k|j}^{3\,2}$ are observed transition probabilities for a transition from time point 1 to time point 2 and from time point 2 to time point 3, respectively. Model (2) is derived from Model (1) by dropping several sets of parameters from the latent mixed Markov model for several populations: The γ because $H=1$; the π because $S=1$; and the ρ because the chain is assumed to be a manifest one. As a consequence, the latent classes, that is, a, b, and c in Equation (1), coincide with the manifest categories, i, j, and k in Equation (1).

In order to elucidate the philosophy inherent in this model, which is – in principle – common to all Markov chain models considered in this chapter, we will refer in some detail to the transition counts from occasion 1 to occasion 2 as well as the estimated parameters that are given in Table 2. As we see from Table 2, 2,238 out of 5,147 respondents are

Table 2. General Life Satisfaction

		Trans. Counts $t+1$				**Trans. Prob.** τ, $t+1$	
t	**Cat.**	**1**	**2**	**Marginals**	**Init. Distrib.** δ^1	**1**	**2**
1	1	1,607	631	2,238	0.435	0.718	0.282
	2	925	1,984	2,909	0.565	0.318	0.682
				$N=5{,}147$			
2	1	1,811	721	2,532		0.715	0.285
	2	784	1,831	2,615		0.300	0.700
3	1	1,826	769	2,595		0.704	0.296
	2	694	1,858	2,552		0.272	0.728
4	1	1,728	792	2,520		0.686	0.314
	2	626	2,001	2,627		0.238	0.762

Notes: Transition counts are from t to $t+1$ and estimated parameter values in the simple Markov chain model with time heterogeneous transition probabilities.

found in category 1 at $t = 1$, whereas the rest (2,909 respondents) belong to category 2 (cf. marginals). The process thus starts with an initial distribution $\delta^1 = (0.435, 0.565)$. What happens from $t = 1$ to $t = 2$? Looking at the first row of the transition counts from $t = 1$ to $t = 2$ tells us that 1,607 subjects out of 2,238, those who reported to be unsatisfied at $t = 1$, stay with this category at $t = 2$. This is a proportion of .718, whereas the rest, 631 out of 2,238 (i.e., 28.2%) switch to category 2. Doing the same computations for the row 2 transition counts shows that 68.2% of those saying that they were satisfied with their lives at occasion 1 do so as well at time point 2, whereas 31.8% change to category 1. We thus see that transition probabilities are conditional probabilities:

$$\tau_{j|i} = n_{j|i}^{t+1,t} / n_i^t, \tag{3}$$

with $n_{j|i}^{t+1,t}$ as the number of cases in category j at time point $t+1$, given membership of category i at time t. That is, $\tau_{j|i}^{2\,1}$ is the probability to answer j at $t = 2$, given membership in category i at $t = 1$. As a consequence, transition probabilities have to sum to 1.0 rowwise.

If we were interested in the marginal distribution at $t = 2$, δ^2, we could obtain it from the respective observed marginals or by postmultiplying the initial distribution vector, δ^1, by the matrix of transition probabilities from $t = 1$ to $t = 2$, $\mathbf{T}^{2\,1}$:

$$\delta^2 = \delta^1\,\mathbf{T}^{21}$$

$$(0.492, 0.508) = (0.435, 0.565)\begin{bmatrix} 0.718 & 0.282 \\ 0.318 & 0.682 \end{bmatrix}. \tag{4}$$

Table 2 also gives the rest of the transition probabilities from $t = 2$ to $t = 3$, $t = 3$ to $t = 4$, and $t = 4$ to $t = 5$. The estimated parameter values given in Table 2 thus refer to a model that allows for time heterogeneous transition probabilities. That is, the model is consistent with the formulation given in Equation (2), whereas the classic definition of the model restricts transition probabilities to be stationary (homogeneous or constant in time) so that superscripts may be dropped.

$$\tau_{j|i}^{2\,1} = \tau_{j|i}^{3\,2} = \cdots = \tau_{j|i}. \tag{5}$$

In fact, inspection of the four matrices of transition probabilities in Table 2 shows that they look rather similar, so that it appears meaningful to fit model (2) with restriction (5). This gives

$$\mathbf{T} = \begin{bmatrix} 0.705 & 0.295 \\ 0.283 & 0.717 \end{bmatrix}$$

where the τ are obtained as the sum of all corresponding transition counts, divided by the respective marginals:

$$\tau_{j|i} = \sum_{t=1}^{T-1} n_{j|i}^{t+1,t} \bigg/ \sum_{t=1}^{T-1} n_i^t, \qquad (6)$$

with summation over all $T - 1$ matrices of one-step transition counts (cf., e.g., Anderson, 1954; Anderson and Goodman, 1957; Kullback, Kupperman, and Ku, 1962). For example, $\tau_{1|1} = (1{,}607 + 1{,}811 + 1{,}826 + 1{,}728)/(2{,}238 + 2{,}532 + 2{,}595 + 2{,}520) = 0.705$.

The classic Markov chain model with time homogeneous transitions thus turns out to be a very parsimonious model. Irrespective of the number of time points under consideration, the number of nonredundant estimated parameters is the number of categories minus one, $J - 1$ δ plus $J(J - 1)\tau$ only, that is, three in our example.

It should be emphasized, however, that the model makes a number of restrictive assumptions.

(a) As the path diagram in Figure 2 shows, the model makes the assumption of a first-order process. That is, the state someone is in at time point t depends on the state he or she was in at the most recent time point $t - 1$ only but not on earlier points in time, for example, $t - 2$. This is also called *a process without memory*. Higher-order models will be referred to in the final section.
(b) The classic version of the model assumes transition probabilities to be time homogeneous. The adequacy of this assumption may be evaluated by comparing the fit of this model with the one of a model allowing for time heterogeneous transitions.
(c) The model assumes population homogeneity, that is, all subjects are characterized by the same one set of parameters describing the process across time. As a consequence, unobserved heterogeneity is not an

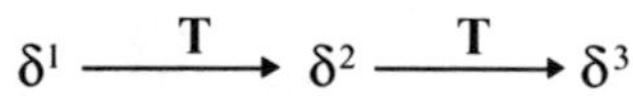

Figure 2. A path diagram for the classic Markov chain model. (Note: The δ^t are the marginal distributions at time point t. **T**, without superscripts, is the matrix of time homogeneous transition probabilities.)

issue. Models that allow for unobserved heterogeneity are considered in the next section.

(d) Although measurement error is a crucial issue in nearly all instances in social science studies, the model considers the data to be free of measurement error. In Section 5, we will therefore consider the latent Markov model, a model that allows one to correct for measurement error.

As a result of these restrictive assumptions, which have been noted by various authors (e.g., Anderson, 1954; Blumen, Cogan, and McCarthy, 1955; Goodman, 1962; Coleman, 1964; Wiggins, 1973; Singer and Spilerman, 1976; Bartholomew, 1981; Logan, 1981; Plewis, 1981; Poulsen, 1982, 1990; Bye and Schechter, 1986; Van de Pol and de Leeuw, 1986; Langeheine, 1988), it is not surprising that the classic Markov model almost never fits a given set of data. The fit of any model considered in this chapter may be evaluated by using various inferential statistics and descriptive fit measures developed for contingency table analysis. We use the likelihood ratio chi square (L^2), Bayesian information criterion (or BIC; see Schwarz, 1978) and the index of dissimilarity Δ (cf. Clogg, 1981; Shockey, 1988).

Table 3 gives L^2, X^2 (the Pearson chi square), BIC, and Δ for a number of models fitted to the life satisfaction data. Note that all of these models, which are referred to in this and the next few sections, are first-order models. As would be expected, the classic Markov model (denoted by M in Table 3) fares badly ($L^2 = 1{,}266.0$ with df $= 28$) with roughly 22% misspecification according to Δ. Detailed information about misfit is obvious from Table 1, which also contains model-expected frequencies as well as chi-square contributions of each cell to the Pearson X^2. Allowance for time heterogeneous transition probabilities results in a significant decrease of the deviance ($1{,}266.0 - 1{,}209.3 = 56.7$; df $= 28 - 22 = 6$) and a slightly better BIC. However, this model is clearly inadequate as well.

Note that given a model with time homogeneous transition probabilities fits a given set of data, this would enable us to make a number of predictions with respect to the future process of the system (cf., e.g., Anderson, 1954; Goodman, 1962). First

$$\mathbf{T}^{t,1} = \mathbf{T}^{(t-1)}; \tag{7}$$

that is, the matrix of transition probabilities from $t = 1$ to some other point t is given by the $(t - 1)$th power of $\mathbf{T}$. As we will see later, transitions from the first to the last occasion are of special interest because they give a quick response about what happens during the time span considered.

Table 3. General Life Satisfaction

Model	Time Hom. Trans. Prob.	L^2	X^2	df	BIC	Δ
M	yes	1,266.0	1,287.1	28	1,026.7	0.219
	no	1,209.3	1,239.5	22	1,021.3	0.219
MS	yes	323.6	322.9	26	101.4	0.087
	no	270.8	272.3	20	99.9	0.084
MM	yes	146.2	148.9	24	−58.9	0.069
	no	94.9	97.5	12	−7.7	0.056
MMS	yes	62.3	61.9	22	−125.7	0.035
	no	6.6	6.6	10	−44.7	0.030
IPIPS	yes	130.8	135.8	24	−74.3	0.048
	no	77.2	80.4	18	−76.6	0.033
MMM	yes	47.1	46.6	20	−123.8	0.030
	no	1.7	1.7	2	−15.4	0.004
LM	yes	235.9	243.8	26	13.7	0.082
	no	130.2	131.0	20	−40.7	0.070
LC		414.1	426.6	28	174.8	0.116
LMRr	yes	112.8	116.9	21	−66.7	0.049
LMM	yes	46.0	45.4	20	−124.9	0.031
LMS	yes	68.7	67.7	22	−119.3	0.038
pLMS	yes	119.2	122.4	24	−85.9	0.051
	no	11.0	10.9	18	−142.8	0.013

Notes: Fit statistics are for various models fitted to the data of Table 1. M = simple Markov; MS Mover–Stayer; MM = two-chain mixed Markov; MMS = 2 Markov and 1 Stayer chains; IPIPS = 2 independence and 1 Stayer segments; MMM = three-chain mixed Markov; LM = latent Markov; LC = latent class; LMRr = latent Markov plus random response; LMM = latent mixed Markov; LMS = latent Mover–Stayer; pLMS = partially latent Mover–Stayer.

Second, the marginal distribution at some time point t is given by

$$\delta^t = \delta^1 \mathbf{T}^{t,1} \tag{8}$$

with $\mathbf{T}^{t,1}$ as defined in Equation (7). If we would compute a series of marginal distributions for our example, we would note that there is no further change in δ from $t = 6$ to $t = 7$. That is, the system is in equilibrium at $t = 6$ or has reached what some people call a *steady state*.

4. THE MIXED MARKOV MODEL

As mentioned previously, the simple Markov model makes the rather unrealistic assumption that the dynamics across time mirrored by this model are valid for all individuals. One may easily imagine, however, that several subgroups exist, each of which is characterized by a specific process across time. The goal then would be to identify such groups that are characterized by intragroup homogeneity and intergroup differences. Historically, there have been three attacks to the heterogeneity problem.

Several authors (e.g., Anderson, 1954; Goodman, 1962) have proposed splitting the sample according to one or more discrete exogenous variables and to fit the simple Markov model to each of these strata. The results may then be compared either informally or by a formal test on (the equality of) the transition matrices. Although such a stratification may be advisable on theoretical grounds (cf. the section on models for several groups), the stratifying variable may nevertheless turn out not to be the relevant variable accounting for heterogeneity.

Whereas stratification by a manifest variable may reveal observed heterogeneity, the other two attacks approach the problem by postulating unobserved heterogeneity. The basic idea here is that there is an unobserved, latent discrete variable that splits the total group of subjects into a (preferably small) number of subgroups to be identified by maximizing intragroup homogeneity and intergroup heterogeneity. In fact, this is exactly what latent class models (Lazarsfeld, 1950; see also Lazarsfeld and Henry, 1968; Goodman, 1974a, 1974b) do, and we will see later that latent class models are one of the constituting elements of the mixed (as well as the latent) Markov model.

To our knowledge, the first attack on the problem of unobserved heterogeneity in categorical panel data is from Blumen et al. (1955). These authors presented the Mover–Stayer model, which is defined by two Markov chains: Movers follow an ordinary Markov chain, whereas Stayers stay in their initial category with probability 1 across time (i.e., their transition matrix is equal to the identity matrix, **I**).The goal then is to unmix the observed frequency distribution into two subsets by using some algorithm that allows one to estimate the proportions of the two subsets as well as the parameters within these subsets. Although parameter estimation was a problem that the Mover–Stayer model was plagued with over many years (cf. Langeheine and Van de Pol, 1990), Poulsen (1982) conceives of this model as a special case of the mixed Markov model that attacks the problem of unobserved heterogeneity in a general way (assuming a mixture of a finite number of Markov chains and using efficient maximum likelihood estimation methods).

There is no doubt that several authors (e.g., Anderson, 1954; Blumen et al., 1955; Spilerman, 1972; Singer and Spilerman, 1974) have imagined a general mixed Markov model. However, an exact formulation of the model was first presented by Poulsen (1982). He also mentions several authors who attempted to incorporate unobserved heterogeneity into a Markov model, and he notes that all of these attempts are limited in one way or the other. His mixed Markov model is given by

$$P_{ijk} = \sum_{s=1}^{S} \pi_s \delta_{i|s}^{1} \tau_{j|is}^{2\,1} \tau_{k|js}^{3\,2}, \tag{9}$$

where – in comparison with the simple Markov model (2) – we now assume S chains. We therefore have one additional set of parameters, the π, which are the chain proportions, with all other sets of parameters considered conditional on chain membership. This model is derived from our most general model (1) in a way similar to that of the simple Markov model (2), except that we now have S chains. Because these chains are assumed to be free of measurement error, they may be called *manifest chains.* It should be noted, however, that the mixed Markov model operates on the latent level with a latent S-categorical variable that unmixes the observed frequency distribution into S chains.

Model (9) is the version allowing for time heterogeneous transition probabilities. Of course, stationarity of transition probabilities may be assumed, as has been done for the simple Markov model. In the present context this means

$$\tau_{j|is}^{2\,1} = \tau_{j|is}^{3\,2} = \cdots = \tau_{j|is}. \tag{10}$$

Note that the general mixed Markov model contains a number of special cases well known in the literature as submodels:

(a) If $S=1$, the simple Markov model is obtained.
(b) If $S=2$ and approriate restrictions are imposed on the transition probabilities and the initial probabilities models such as the Mover–Stayer (Blumen et al., 1955), strict Black and White (Converse, 1964, 1970) or biased coin Black and White (Converse, 1974) are obtained.
(c) The mixed Markov model contains the classic latent class model of Lazarsfeld (1950). If conditioning on a previous point in time is dropped from model (9), we obtain

$$P_{ijk} = \sum_{s=1}^{S} \pi_s \delta_{i|s}^{1} \tau_{j|s}^{2} \tau_{k|s}^{3}, \tag{11a}$$

or by using a more consistent notation,

$$P_{ijk} = \sum_{s=1}^{S} \pi_s \delta_{i|s}^{1} \tau_{j|s}^{2} \tau_{k|s}^{3}. \tag{11b}$$

That is, instead of a model assuming local dependence between consecutive points in time (or a first-order Markov process) within chain s as in model (9), we have local independence (or a zero-order process) within classes in model (11b). Note that, formally, model (11b) refers to the situation of a single item measured repeadly at three points in time. However, time points may be substituted by any set of categorical indicators.

(d) Of course, concepts such as movers, stayers, independence, and random response may be combined in a variety of ways resulting, for example, in two-chain models like Mover–Stayer, Mover–Independence, Independence–Stayer, Mover–Random Response, and the like (Van de Pol and Langeheine, 1990). Flexibility in defining models is even higher if three or more chains can be assumed.

Now that we have seen that the mixed Markov model may be considered as a latent class model with (a specific kind of) local dependence, it is not surprising that the EM algorithm (Dempster et al., 1977), which is a widely used algorithm for parameter estimation in latent class analysis, is also made use of in Markov chain analysis. Details are given in Poulsen (1982) and Van de Pol and Langeheine (1989).

We next consider results of some mixed Markov models fitted to the life satisfaction data of Table 1. Fit statistics are given in Table 3. First note that the classical Mover–Stayer model (denoted by MS in Table 3) results in a considerable improvement in fit ($L^2 = 323.6$, df $= 26$, time homogeneous transition probabilities) over the simple Markov model at the cost of estimating a few additional parameters only (two in the present case). Because this is sort of a general observation, Davies and Chrouchley's (1986) recommendation that the Mover–Stayer model "requiescat in pace" may not be well taken. However, in absolute terms, the fit of the model is not satisfactory, neither with nor without time homogeneous transition probabilities ($L^2 = 270.8$, df $= 20$). This is also true for the two-chain mixed Markov model with (model MM in Table 3, $L^2 = 146.2$, df $= 24$) or without ($L^2 = 94.9$, df $= 12$) stationary transition probabilities. We have nevertheless included parameter estimates for both the Mover–Stayer and the two-chain mixed Markov model in Table 4. The Mover–Stayer model allocates 70.7% of the subjects to the mover chain with a slight majority of 58% being satisfied with life at $t = 1$. The stationary transition probabilities would lead one to conclude that there is considerable change across time in both directions. That is, at each point in time, 41.4% of the subjects switch from "unsatisfied" to "satisfied" and

Table 4. General Life Satisfaction

Model	Chain, Cat.	Chain Prop. π	Init. Distrb. at $t = 1, \delta$	Transition Prob. τ From t to $t+1$ Cat. 1	2	From $t = 1$ to $t = 5$ Cat. 1	2
MS	1	0.707					
	1 = unsat.		0.420	0.586	0.414	0.494	0.506
	2 = satisf.		0.580	0.403	0.597	0.493	0.507
	2	0.293					
	1		0.471	1	0	1	0
	2		0.529	0	1	0	1
MM	1	0.464					
	1		0.725	0.827	0.173	0.805	0.195
	2		0.275	0.714	0.286	0.805	0.195
	2	0.536					
	1		0.184	0.312	0.688	0.211	0.789
	2		0.816	0.184	0.816	0.211	0.789

Notes: Estimated parameter values are for the Mover–Stayer model and the two-chain mixed Markov model. There are stationary transition probabilities. Parameter values of 1 and 0 are fixed by definition.

40.3% go in the opposite direction. As the transition probabilities from $t = 1$ to $t = 5$ show, the rows of this matrix are practically equal. That is, the state someone is in at $t = 5$ does not depend on where he or she was in at $t = 1$. In other words, the process has reached a steady state at about $t = 5$ with a marginal distribution $\delta^5 = (0.494, 0.506)$. A comparison of δ^1 and δ^5 would thus lead us to conclude that there is a slight amount of *net change* $(0.494 - 0.420 = 0.074)$ toward being unsatisfied with life in general. The rest of the subjects (29.3%) are found in the stayer segment, with an approximately equal split into subjects being permanently satisfied or unsatisfied with their lives.

If we had to decide between the Mover–Stayer model and the two-chain mixed Markov model as rival candidates, all statistics in Table 3 would clearly favor the MM model as the "better" one. A comparison of the estimated parameters of the latter model with those of the Mover–Stayer model reveals that statements about change differ considerably under these two conceptualizations. Under the MM model there is a split of the sample into two chains of about equal size. At time point 1, these chains are by far more clearly characterized now. The vast majority in chain 1 are people saying that they are unsatisfied (.725 in proportion), whereas the opposite is true for chain 2, which contains 81.6% of the subjects reporting to be satisfied with their life in general. As the transition probabilities from $t = 1$ to $t = 5$ reveal, this picture is relatively unchanged at the last occasion. In both cases a steady state has been reached at least at $t = 5$ with a small net change toward being unsatisfied with life.

Net change, of course, obscures the proportion of *total change*. Whereas total change goes with the mover chain of the MS model only, there are two mover chains in the MM model. Chain 1 is characterized by a high repeat probability of category 1 (0.827) and a high probability of switching from category 2 to category 1 (0.714) at successive time points. The opposite is true for chain 2 with a high repeat probability for category 2 (0.816) and a high probability of switching to category 2 (0.688). It may thus appear surprising that net change in the long run is toward "unsatisfied" for both chains. This becomes obvious, however, when one computes total proportions of change from one point in time to the next. From $t = 1$ to $t = 2$, for example, these proportions are – within chains –

chain 1,	from cat. 1 to cat 2:	$0.725 \times 0.173 = 0.125$,
chain 1,	from cat. 2 to cat 1:	$0.275 \times 0.714 = 0.196$,
chain 2,	from cat. 1 to cat 2:	$0.184 \times 0.688 = 0.127$,
chain 2,	from cat. 2 to cat 1:	$0.816 \times 0.184 = 0.150$.

The message thus is not to look at the transition probabilities exclusively. A small proportion ($\tau_{1|22} = 0.184$) of a large proportion ($\delta^1_{2|2} = 0.816$) may outweigh a large proportion ($\tau_{2|12} = 0.688$) of a small proportion ($\delta^1_{1|2} = 0.184$). Because two-chain mixed Markov models do not fare well in explaining the data in Table 1, we next consider models that assume even more unobserved heterogeneity by specifying three chains.

Model MMS in Tables 3 and 5 assumes two mover chains and one stayer segment. Because the rows within most of the mover matrices of transition probabilities of the MMS model appeared to be rather similar, we next fitted the more restrictive IPIPS model postulating two independence (zero-order) instead of two first-order Markov chains. Model MMM, finally, assumes three mover chains.

If one compares the fit statistics given in Table 3, it is obvious that three-chain models do a much better job than the respective two-chain counterparts. On purely statistical grounds, none of these models assuming time homogeneous transition probabilities fits the data well (MMS, $L^2 = 62.3$, df $= 22$; IPIPS, $L^2 = 130.8$, df $= 24$; MMM, $L^2 = 47.1$, df $= 20$), but both the MMS and the MMM model do if one allows for nonstationary

Table 5. General Life Satisfaction

			Transition Prob. τ			
			From t to $t+1$ Cat.		From $t=1$ to $t=5$ Cat.	
Chain Cat.	**Chain Prop.** π	**Init. Distrib. at** $t=1, \delta$	**1**	**2**	**1**	**2**
1	0.418					
1 = unsat.		0.602	0.757	0.243	0.659	0.341
2 = satisf.		0.398	0.656	0.344	0.659	0.341
2	0.419					
1		0.205	0.288	0.712	0.202	0.798
2		0.795	0.212	0.788	0.202	0.798
3	0.163					
1		0.595	1	0	1	0
2		0.405	0	1	0	1

Notes: Estimated parameter values are for the Mover–Mover–Stayer model. There are stationary transition probabilities. Parameter values of 1 and 0 are fixed by definition.

transition probabilities (MMS, $L^2 = 6.6, \text{df} = 10$; MMM, $L^2 = 1.7, \text{df} = 2$). Either version of the IPIPS model appears to be too restrictive for these data. Reliance on inferential statistics would thus lead us to accept the MMS model with nonstationary transition probabilities. This model is nested in the more general MMM model, but the L^2 difference $(6.6 - 1.7)$ is not statistically significant $(\text{df} = 10 - 2)$. Note, however, that BIC favors both MMS and MMM with stationary transition probabilities, which are by far more parsimonious in terms of estimated parameters. All three-chain models may be considered OK according to the index of dissimilarity. The message thus is that reliance on a single statistic may be inconclusive.

As an aside, note that the MMM model with nonstationary τ turned out to be weakly identified only with one parameter being estimated with a large standard error of 0.58 despite the fact that two estimated parameters converged to a boundary value of zero. (When counting degrees of freedom, we do not count boundary values as many authors do. The reason is that we consider boundary values as estimated parameters. For details, see the PANMARK manual.)

For completeness, we give estimated parameter values for the MMS model in Table 5. The structure of the two mover chains is rather similar to the one of the two-chain MM model (cf. Table 4). We therefore refrain from commenting on Table 5.

Before this chapter is closed, some general comments on the mixed Markov model appear to be in order. The distinct feature of this model is that it allows one to model unobserved heterogeneity. This is realized by extending the simple Markov model into the latent class framework, that is, by assuming a mixture of a finite number of Markov chains. Latent classes and Markov chains are thus the basic constituents of mixed Markov models. Although items (or time points) are assumed to be (conditionally) independent within classes of a latent class model, mixed Markov models allow for dependence between consecutive points in time within chains. Mixed Markov models may therefore also be called a special kind of local dependence latent class models.

There is no doubt that the mixed Markov model is a fairly general model. It contains a variety of other models well known in the literature as special cases (e.g., Mover–Stayer, Black and White, classic latent class models). These special cases are obtained by putting constraints on the parameters of the general model. As other programs for latent class analysis do, the PANMARK program allows one to fix parameters at predetermined values or to equate (sets of) parameters. In combination, this is what makes mixed Markov models really flexible to cope with situations in which specific assumptions about the form of a stochastic process are derived from theoretical considerations.

Mixed Markov models are not without problems, however. Although a more technical issue, identifiability must be mentioned in this context. In general, identifiability will depend on the number and nature of chains specified. For strong models – that is, models that request the data to conform to a simple structure with very few parameters to be estimated as in the Mover–Stayer model – identifiability will be no issue. Weak models – that is, models with many free parameters to be estimated – may turn out not to be identified for a given data set despite the fact that (many) degrees of freedom are left. In the latter case one will also observe that convergence of the iterative algorithm is comparatively slow with the risk of running into a second-best solution without getting to the global maximum of the likelihood function. It is advisable therefore to use multiple sets of starting values for most kinds of mixed Markov models.

A more critical issue is that mixed Markov models implicitly assume the data to be free of measurement error, an assumption that is not justified in most cases in the social sciences. Finally, mixed Markov models assume chain membership to be constant throughout the period of analysis. Poulsen (1982) calls this a simplifying assumption regarding the underlying stochastic process. That is, mixed Markov models do not allow one to evaluate the hypothesis of "real" change (i.e., switching from one chain to another through time). As we will see in Chapter 12, both measurement error and latent change are complementary elements of the latent Markov model.

5. THE LATENT MARKOV MODEL

As we have seen in Chapter 10, mixed Markov models combine certain features of the latent class model and the simple Markov chain. The latent Markov model, which is from a dissertation of Wiggins (1955), published later in book form (Wiggins, 1973), does so as well, albeit in a different manner. The gist of Wiggins' model is to consider manifest responses at each point in time as fallible indicators of some unobservable latent state, with statements about change now being made on the latent level. For three points in time, Wiggins' model is given by

$$P_{ijk} = \sum_{a=1}^{A} \sum_{b=1}^{B} \sum_{c=1}^{C} \delta_a^1 \rho_{i|a}^1 \tau_{b|a}^{2\,1} \rho_{j|b}^2 \tau_{c|b}^{3\,2} \rho_{k|c}^3. \tag{12}$$

According to this model, the process thus starts at time point 1 with a latent distribution, $\boldsymbol{\delta}^1$, having A latent states (or classes) to which subjects manifest responses are probabilistically related by means of response probabilities $\rho_{i|a}^1$. As a consequence, change from time point 1 to time point 2 is now modeled on the latent level by means of $\tau_{b|a}^{2\,1}$. The latent distribution at $t = 2$, $\boldsymbol{\delta}^2 = \boldsymbol{\delta}^1 \mathbf{T}^{2\,1}$ [see Equation (4)] is again characterized by response probabilities, $\rho_{j|b}^2$, and so on.

A schematic representation of this model is given in the upper panel of Figure 3, where $\mathbf{u}^t$ refers to the manifest marginal distribution at occasion

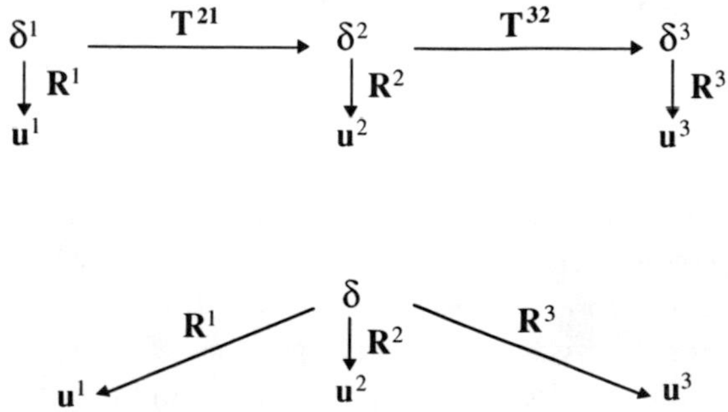

Figure 3. A path diagram for the latent Markov model (upper panel) and the latent class model (lower panel).

t, which is related to the latent distribution δ^t by response probabilities (matrices $\mathbf{R}^t$). Because, at least in most cases, it will be assumed that the number of latent states equals the number of manifest categories, some $\mathbf{R}$, for example, at $t = 1$, looks like

	$i = 1$	$i = 2$
$a = 1$	$\rho^1_{1\|1}$	$\rho^1_{2\|1}$
$a = 2$	$\rho^1_{1\|2}$	$\rho^1_{2\|2}$

in case of two categories/states. $\mathbf{R}^t = \mathbf{I}$ would mean that the latent variable is perfectly measured by the manifest indicator. This is what the manifest Markov model does assume at all points in time. (We could thus drop the $\mathbf{R}^t$ from the path diagram in the upper panel of Figure 3 and would obtain the path diagram of Figure 2 because the $\mathbf{u}^t$ coincide with the δ^t.) Departure of $\mathbf{R}$ from $\mathbf{I}$ thus points to the amount of measurement error. Because measurement error is ubiquitous in the social sciences and may have disastrous effects, especially in the case of repeated categorical data (Beck, 1975; Schwartz, 1985; Bye and Schechter, 1986), the latent Markov model appears to be a favorite model in the analysis of categorical panel data because of its capacity to correct manifest data for measurement error. The notion of measurement error as well as the path diagram in Figure 3 also make it evident that the latent Markov model is a categorical variable analog of linear structural equation models for continuous data. The measurement part, the $\mathbf{R}$, relate the latent variables to the manifest ones, whereas the structural part is given by the δ and $\mathbf{T}$.

A closer inspection of the model as given in Equation (12) will reveal that it is not identified unless some additional restrictions are imposed on the parameters. Wiggins therefore assumed that both the response probabilities and the transition probabilities are time homogeneous, that is

$$\mathbf{R}^1 = \mathbf{R}^2 = \mathbf{R}^3 = \mathbf{R}, \tag{13}$$

$$\mathbf{T}^{2\,1} = \mathbf{T}^{3\,2} = \mathbf{T}. \tag{14}$$

Model (12) with restrictions (13) and (14) thus is the latent analog of the simple manifest Markov model. According to this model, the whole population changes according to a single Markov chain, but measurement error causes the observed cross-table to diverge from what would be observed if reliability were perfect. Note that restriction (14) is not a necessary restriction because of identifiability, whereas restriction (13) is in case of three points in time only. For four or more occasions, one need only assume time homogeneous response probabilities for the first and last time points. However, this assumption is nevertheless a reasonable one, saying that the measurement model does not change across time. A model allowing for change in both subjects and indicators may be difficult to interprete, given that it is identified (for five waves, e.g., a model allowing for both time heterogeneous response and transition probabilities is not identified).

Although some authors (e.g., Beck, 1975; Duncan, 1975) were right in criticizing Wiggins because of a lack of discussion concerning model identifiability, the application of this model was prevented because of an even more serious problem: Wiggins made use of one of the older, unefficient methods of parameter estimation with the risk of obtaining estimated probabilities outside the 0–1 range. In fact, it took nearly 30 years until problems of parameter estimation were solved for the latent Markov model. Poulsen (1982) as well as Van de Pol and de Leeuw (1986) showed how to use the EM algorithm of Dempster et al. (1977) in order to estimate the parameters of the latent Markov model; Bye and Schechter (1986) used another algorithm that forced the parameters within the 0–1 range by use of a transformation.

We have already mentioned that the latent Markov (LM) model contains the simple Markov model as a special case, assuming LM without measurement error (i.e., $\mathbf{R} = \mathbf{I}$). As Figure 1 shows, the classic latent class model turns out to be a special case of the latent Markov model as well. This is easily seen from Figure 3. If all $\mathbf{T}$ in the upper panel of Figure 3 are replaced by the identity matrix, $\mathbf{I}$, the $\boldsymbol{\delta}^t$ of the latent Markov model coincide in a single $\boldsymbol{\delta}$, that is, a single latent distribution that is unreliably measured by the same item at three occasions (as in the lower panel of Figure 3) or, more generally, by three indicators. This is exactly what a latent class model is defined like. The latent class model thus may be considered being LM without latent change ($\mathbf{T} = \mathbf{I}$), giving

$$P_{ijk} = \sum_{a=1}^{A} \delta_a \rho^1_{i|a} \rho^2_{j|a} \rho^3_{k|a} \tag{15}$$

instead of Equation (12). Likewise, the latent Markov model may be called latent class model with latent change.

Fitting the latent Markov model (assuming time homogeneous response probabilities) to the life satisfaction data gives $L^2 = 235.9$, df $=$ 26 in case of stationary transition probabilities and $L^2 = 130.2$, df $= 20$ for nonstationary transition probabilities (cf. Table 3). According to all

Table 6. General Life Satisfaction

			Resp. Prob. ρ, Cat.		Transition Prob. τ			
					From t to $t+1$ Class		From $t=1$ to $t=5$ Class	
Model	Class	Class Prop. at $t=1$, δ	1 (Unsat.)	2 (Satisf.)	1	2	1	2
M	1	0.435	1	0	0.705	0.295	0.506	0.494
	2	0.565	0	1	0.283	0.717	0.474	0.526
LM	1	0.445	0.841	0.159	0.944	0.056	0.814	0.186
	2	0.555	0.153	0.847	0.064	0.936	0.215	0.785

Notes: Estimated parameter values are for the simple Markov model and the latent Markov model. There are time homogeneous response probabilities and transition probabilities. Parameter values of 1 and 0 are fixed by definition.

statistics (L^2, BIC, and Δ), both of these models do not describe the data adequately. Note, nevertheless, that there is a dramatic improvement in fit over the simple manifest Markov model at the cost of two additional parameters (two ρ) only to be estimated in the LM model.

For the sake of comparison, estimated parameter values of both the simple and the latent Markov model are given in Table 6. Note that model M in Table 6 is defined in terms of a latent Markov model assuming, however, perfect measurement of both classes (fixing response probabilities at one and zero, respectively). A comparison of the two matrices of response probabilities reveals that measurement accuracy is obviously overestimated by the simple Markov model. Whereas both manifest categories 1 and 2 are deterministically related to classes 1 and 2, respectively, in model M, parameter estimates (ρ) of the LM model show that both classes are well defined, although with an error rate of approximately 15%. A comparison of the transition matrices from both models clearly demonstrates the consequences of neglecting measurement error: change is overestimated by the simple Markov model. Although the simple Markov model would say that nearly 50% of those initially in either class 1 or 2 are found in the opposite class at $t = 5$, these figures reduce to approximately 20% according to the LM model.

Suppose for a moment that the latent Markov model did result in an adequate fit. Does this imply that we can argue in favor of latent change? The answer is no, as long as we have not fitted the rival model postulating no latent change, that is, LM with all $\mathbf{T} = \mathbf{I}$, which – as we know – is a classical latent class (LC) model. For completeness, we have included this two-class LC model in Table 3. Note that this is a model postulating no change at all: there is no latent change by definition and response probabilities are assumed to be constant across time as in the LM. Imposing these restrictions gives $L^2 = 414.1$, df $= 28$ for the LC model, thus pointing in the direction of latent change as modeled by the LM model.

Although it is common to equate the number of latent states in a latent Markov model with the number of categories of the manifest variable, this need not necessarily be the case. For two-categorical data, Langeheine and Van de Pol (1994) consider a three-class model called *latent Markov plus random response* (LMRr). In addition to the two classes of the LM model, the LMRr has a third class containing subjects who give random response. That is, response probabilities of this class are fixed at 0.5 (equiprobability). Although this model is not identified when allowing for time heterogeneous transition probabilities, it leads to a considerable improvement in fit over the respective LM model with time homogeneous transition probabilities at the cost of five additional parameters (one δ, four τ): $L^2 = 112.8$, df $= 21$ (see Table 3). Although the fit of this model is not satisfactory in absolute terms, we give parameter estimates in Table 7 for comparative purposes. The most striking point obviously is that approximately 41% of the respondents give random response at $t = 1$ (or say that they are as satisfied with their life as they are unsatisfied). Both classes 1 and 2 are measured more reliably now (cf. the ρ of both models), with the consequence that there is less latent change between classes 1 and 2 (cf. the respective τ). Most change is due to switching into or out of the random response class with $\boldsymbol{\delta}^5 = (0.308, 0.378, 0.314)$ showing that the overall percentage of random respondents decreases over time.

In passing, we want to stress the pros and cons of the latent Markov model. The distinctive feature of this model is that it allows one to correct manifest responses for measurement error. As a consequence, we no longer make statements about change on the manifest, but on the latent level. This change may be called *true change* because it implies switching from one latent class to a different one, whereas subjects class membership is constant throughout the period of investigation by definition in mixed Markov models. However, before arguing in favor of latent change, we should confront this model with the rival model postulating no change. If measurement error plays a significant role in a given set of data, a comparison of the results obtained from fitting the simple and the latent Markov model will reveal a considerable part of change as per the simple Markov model to be simply spurious change (or change caused by unreliability).

Table 7. General Life Satisfaction

		Resp. Prob. ρ, Cat.		Transition Prob. τ					
				From t to $t+1$ Class			From $t=1$ to $t=5$ Class		
Class	Class Prop. at $t=1$, δ	1 (Unsat.)	2 (Satisf.)	1	2	3	1	2	3
1	0.241	0.942	0.058	0.926	0.020	0.054	0.762	0.078	0.160
2	0.353	0.092	0.908	0.012	0.957	0.031	0.054	0.846	0.100
3	0.406	0.5	0.5	0.088	0.046	0.866	0.259	0.149	0.592

Notes: Estimated parameter values are for the latent Markov model with random response. Parameter values of 0.5 are fixed by definition.

In sum, the latent Markov model thus appears to have attractive properties. Note, however, that whereas population heterogeneity enters into the model by means of the response probabilities defining the classes, latent transitions are assumed to hold for all individuals. That is, the model is a one-chain model with a single chain of latent transition probabilities. In the next section we will therefore show how the single-chain latent Markov model may be extended to a multiple-chain model.

6. THE LATENT MIXED MARKOV MODEL

The restrictive property of the single-chain latent Markov model may be easily overcome by extending this model to a mixture of S latent chains:

$$P_{ijk} = \sum_{s=1}^{S} \sum_{a=1}^{A} \sum_{b=1}^{B} \sum_{c=1}^{C} \pi_s \delta^1_{a|s} \rho^1_{i|as} \tau^{2\,1}_{b|as} \rho^2_{j|bs} \tau^{3\,2}_{c|bs} \rho^3_{k|cs}. \tag{16}$$

As compared with the latent Markov model (12), we have one additional set of parameters here, the π, that give the proportions of the S latent chains. All other sets of parameters are considered conditional on chain membership, consequently. This is the latent mixed Markov model of Langeheine and Van de Pol (1990), which contains both the latent Markov model ($S = 1$) and the mixed Markov model (9) as special cases. The mixed Markov model is obtained from model (16) by assuming no measurement error. That is, all $\mathbf{R}^t = \mathbf{I}$, so that indices a, b, and c coincide with indices i, j, and k, making summation over a in Equation (16) superfluous.

For a data set pertaining to five waves only, latent mixed Markov models assuming two or more chains are faced with identification problems very quickly. Both the model assuming two latent chains (denoted by LMM in Table 3) and the model assuming one latent mover chain and one latent stayer chain (denoted LMS in Table 3) – that is, measurement is assumed to be imperfect in both chains in both of these models – are formally identified with transition probabilities specified to be constant in time. However, convergence is very slow in both cases despite the fact that only 11 parameters (LMM) and 9 parameters (LMS) are to be estimated. Both of these models may therefore be termed weak ones, which is also reflected by relatively large standard errors (approximately 0.20) of a number of estimated parameters. Although improvement in fit over the single-chain latent Markov model is considerable (as is improvement in fit of the LMM over the MM and of the LMS over the MS), we will therefore not consider the LMM and the LMS in detail. Note that both of these models are not identified when transition probabilities are assumed to be time heterogeneous.

This is not to say that two-chain latent Markov models are totally irrelevant for five-wave data. A viable candidate is the partially latent Mover–Stayer model (pLMS) of Langeheine and Van de Pol (1990). This model is defined by one latent mover chain (i.e., with measurement error) and one manifest stayer chain (i.e., without measurement error) – hence the term *partially latent*. With transition probabilities assumed to be stationary, this model ends up with $L^2 = 119.2$, df $= 24$ (cf. Table 3). This is a considerable improvement in fit in comparison to the single-chain latent Markov (LM) model as well as the two-chain manifest Mover–Stayer (MS) model at the cost of two additional parameters each only. Because the fit of this model still appears unsatisfactory, we allowed for nonstationary transition probabilities. This model not only shows an excellent fit ($L^2 = 11.0$, df $= 18$), but also has the lowest BIC of all models listed in Table 3 and would require only 1.3% reclassification according to the index of dissimilarity Δ.

Table 8 contains estimated parameter values. Stayers (chain 2) amount to 16.6%, with the majority of these (60.4%, class 1) saying that they are permanently unsatisfied with life in general. Out of the movers (chain 1), the majority (62.5%, class 2) are characterized by a high probability of being satisfied; those being unsatisfied are found in class 1. Latent transition probabilities from one point in time to the next reveal that the probabilities of switching from class 1 to class 2 increase over time, whereas the opposite is true for switching from class 2 to class 1. We might thus expect more and more people to become satisfied over the years. However, ignoring information about the latent distribution at each point in time will obscure the picture. Out of those 37.5% in class 1 at $t = 1$, only 1.7% switch to class 2 at $t = 2$, whereas 18.9% out of the 62.5% in class 2 at $t = 1$ switch to class 1 at $t = 2$. The latent marginal distributions δ^t of chain 1

	Class	
t	**1**	**2**
1	0.375	0.625
2	0.487	0.513
3	0.512	0.488
4	0.484	0.516
5	0.421	0.579

Table 8. General Life Satisfaction

Chain	Chain Prop. π	Class	Class Prop. at $t = 1, \delta$	Resp. Prob. ρ, Cat. 1 (Unsat.)	Resp. Prob. ρ, Cat. 2 (Satisf.)	t	Transition Prob. τ: $t + 1$ Class 1	Transition Prob. τ: $t + 1$ Class 2	Transition Prob. τ: From $t = 1$ to $t = 5$ Class 1	Transition Prob. τ: From $t = 1$ to $t = 5$ Class 2
1	0.834	1	0.375	0.779	0.221		0.983	0.017		
						1				
		2	0.625	0.174	0.826		0.189	0.811		
							0.938	0.062		
						2				
							0.108	0.892		
							0.896	0.104		
						3				
							0.052	0.948		
							0.870	0.130	0.724	0.276
						4				
							0.0	1.0	0.240	0.760
2	0.166	1	0.604	1	0		1	0	1	0
		2	0.396	0	1		0	1	0	1

Notes: Estimated parameter values are for the partially latent Mover–Stayer model, with nonstationary transition probabilities. Parameter values of 1 and 0 are fixed by definition (0.0 and 1.0 are boundary estimates).

show that the proportion of those being unsatisfied increases from $t = 1$ to $t = 3$ and decreases at $t = 4$ and 5 with a slight net change of roughly 5% in favor of class 1 (reporting being unsatisfied) over the whole time span (cf. also transitions from $t = 1$ to $t = 5$ in Table 8). Obviously, this is the reason that makes incorporation of nonstationary transition probabilities into the model necessary.

Although a variety of other models may be thought of (see Langeheine and Van de Pol, 1994, for several examples), we will not go into further detail here. However, we would like to comment on two observations that are obvious from the models listed in Table 3. First, irrespective of the model considered, the version allowing for time heterogeneous transition probabilities does a better job (according to L^2 and Δ, although not always according to BIC). Second, out of these, three (MMS, MMM, and pLMS) result in an acceptable fit on statistical grounds alone, with the pLMS model being the most parsimonious one.

7. LATENT MIXED MARKOV MODELS FOR SEVERAL GROUPS

Both the mixed Markov model and the latent mixed Markov model allow one to pinpoint unobserved heterogeneity. That is, the sample is split into several subsamples according to a latent (unobserved) variable quantifying the chain proportions (the π). Although an optimal split is aimed at (i.e., a split that minimizes the differences between observed and expected frequencies, given a specific model), nothing is known about the substantive meaning of the latent variable. The only thing we can conclude is that there are subsamples differing in their dynamics across time. Very often, however, additional external categorical variables such as gender, age, or education are available that enable an a priori split of the total sample into subgroups. If subgroups defined by such variables are expected to differ in their dynamics on the target variable, it is advisable to perform a simultaneous analysis for all subgroups instead of separate analyses for each subgroup or even a single analysis of the total sample pooled across subgroups.

To enable this type of analysis, that is, to incorporate observed heterogeneity, the latent mixed Markov model (16) is extended by a parameter γ_h referring to the proportion of each of the H subgroups. Moreover, the rest of the parameters of model (16) are considered conditional on subgroup h. Doing so, we reach the top of the hierarchy, that is, model (1) presented earlier.

Note that this model allows for both observed heterogeneity (H groups) and unobserved heterogeneity (S chains within groups). In addition, the kind of models that may be thought of explodes in comparison with the one-group case, depending on which sets of parameters (π, δ, ρ, and τ) are assumed to be equal across subgroups. On one hand, each subgroup may be left with its own parameters. This would be identical to fitting a different model to each subgroup separately. On the other hand, all sets of parameters may be requested to be equal across all subgroups, thus assuming complete homogeneity. A large number of models may be defined that fall in between these two extremes by allowing for partial homogeneity, that is, by requesting all (or some) subgroups to be equal in certain sets of parameters only. Finally, particular within-group restrictions may be specified. This kind of analysis is not restricted to a single external variable because two or more such variables may be combined into one composite variable.

Note that the multiple-group case has been addressed earlier. Both Anderson and Goodman (1957) and Kullback et al. (1962) consider the case of H groups assumed to be identical in their dynamics according to the simple Markov model. However, as we have shown, this is the most simple (and unrealistic) special case of the general model of Van de Pol and Langeheine (1990). The latter is built in accordance with the Clogg and Goodman (1984) simultaneous latent class analysis.

As an example, consider the data given in Table 1, where the two columns at the right contain observed frequencies for males and females. Fit statistics for the simultaneous analysis of these groups for a few models are given in Table 9. These models are among those considered earlier (cf. Table 3) for the total group pooled across males and females: LM = one-chain latent Markov model, LMM = two-chain latent Markov model, and pLMS = partially latent Mover–Stayer model. The following results are obvious from Table 9:

1. Allowance for time heterogeneous transition probabilities results in a considerably better fit, both according to the L^2 statistic and the BIC index. This is in line with the results reported earlier (cf. Table 3).
2. Out of the many possible versions of a specific model that may be thought of by constraining the two groups to be equal in sets of parameters, we have given the two extremes only. The first row pertains to the situation in which each group is left with its own parameters, whereas the second row gives the results for all sets of parameters

Table 9. General Life Satisfaction

Model	Time Hom. Trans. Prob.	Equal. Across Groups π	δ	ρ	τ	L^2	X^2	df	BIC	Δ
LM	yes	*	−	−	−	266.2	277.1	52	−178.2	0.085
		*	+	+	+	270.7	280.9	57	−216.4	0.087
	no	*	−	−	−	157.7	160.0	40	−184.1	0.071
		*	+	+	+	165.0	167.2	51	−270.9	0.074
LMM	yes	−	−	−	−	95.5	96.3	40	−246.3	0.047
		+	+	+	+	107.6	109.0	51	−328.3	0.051
pLMS	yes	−	−	−	−	146.5	150.0	48	−263.7	0.054
		+	+	+	+	154.0	150.0	55	−316.0	0.061
	no	−	−	−	−	36.5	36.3	36	−271.2	0.022
		+	+	+	+	45.7	45.1	49	−373.1	0.030

Notes: Fit statistics for some models are fitted to the males and females data of Table 1. See text for the definition of these models. Across-group equality constraints for sets of parameters: *, not in the model; −, no; +, yes.

constrained to be equal across groups. Irrespective of the model considered and irrespective of whether transition probabilities are assumed to be stationary or not, the L^2 difference between these two versions is small and statistically not significant.

3. The partially latent Mover–Stayer model with time heterogeneous transition probabilities does the best job according to all statistics. With all sets of parameters constrained to be equal across groups, estimated parameters are equal to those given in Table 8.

The result thus is that pooling across males and females does not obscure heterogeneity. Note, however, that this need not be the case generally. A well-fitting model for a pooled data set does not imply that the same model will hold in a simultaneous analysis of subgroups. See Langeheine and Van de Pol (1994) for an example. Neither this does imply, of course, that the same result will hold for other external variables. Theoretical reflections about observed heterogeneity should therefore guide the analysis. In view of the multitude of potential models in a simultaneous analysis, theory may also help restrict the number of models considered. The point is that sample sizes for subgroups will be smaller. As a consequence, parameter estimation will be less accurate (i.e., estimated parameters will be associated with larger standard errors). Some across-group restrictions are therefore advisable. For the case that theory is lacking, a procedure for selecting appropriate across-group equality constraints has been proposed in Van de Pol and Langeheine (1990).

8. EXTENSIONS, PROBLEMS, AND SOME SOLUTIONS

So far, models for a only single variable or indicator measured repeatedly over time have been presented. Extension to multiple indicators that are considered as manifest indicators of some latent categorical variable (construct) are straightforward, however. Instead of a single indicator as in the upper panel of Figure 3, we would have several indicators at each point in time. Put differently: At each time point we would have a latent class model, as in the lower panel of Figure 3. For the formalization and some applications of multiple-indicator Markov models, see Langeheine (1991, 1994), Graham et al. (1991), Collins and Wugalter (1992), Langeheine and Van de Pol (1993, 1994), Collins et al. (1994), Langeheine et al. (1994), and Macready and Dayton (1994). Apart from being more realistic as concerns the data situation, multiple-indicator models have at least two additional advantages. First, conditional response probabilities need not be constrained to be time homogeneous for model identification. Note, however, that an interpretation of results may be difficult if change is allowed for both indicators or items (i.e., time heterogeneous ρ) and subjects (i.e., time heterogeneous τ) because the meaning of the latent variable may change. Second, multiple indicators allow for a more flexible definition of the measurement model (i.e., the ρ defining a latent variable). Apart from purely categorical latent variables, ordered categorical ones may be conceived. A problem that one may be faced with very quickly with multiple indicators, however, is that a cross-table may become very large, thus causing sparseness problems (see later).

One way to explain individual differences in the probability of being in a particular state at a particular time point is by extending a model by observed heterogeneity and performing a multiple-group analysis (see previous section). This is equivalent to using time-constant exogenous covariates. However, one of the strong points of longitudinal data is the availability of information on time-varying covariates. Vermunt et al. (1999) have therefore presented a latent Markov model in which the latent states are regressed on both time-constant and time-varying covariates by means of a system of logit models, thus extending Goodman's (1973) causal loglinear model for observed states to the latent level. On one hand, this approach may turn out more parsimonious in terms of parameters to be estimated as compared with the multiple-group model (1). On the other hand it offers great flexibility in conceiving specific models. Note that this kind of model cannot be estimated by using the PANMARK program. However, the LEM program (Vermunt, 1993) can do so. Again, sparseness may be a problem with a given data set because the number of variables involved will quickly span a large table.

All of the models considered previously are so-called first-order models. That is, the (latent) state someone is in at some time point t does depend only on the most recent time point $t-1$, but not on time points that are more remote. Higher-order models may be formulated easily. Consider, for example, the extension of model (2) by second-order transitions

$$P_{ijk} = \delta_i^1 \tau_{j|i}^{2\,1} \tau_{k|ji}^{3\,2\,1}, \tag{17}$$

saying that the state someone occupies at $t=3$ depends on both $t=2$ and $t=1$. Higher-order models have found considerable interest both in the statistical and applied literature, confined to the manifest case, however. Note that any type of higher-order Markov model of the general model (1) may be fitted by using current software. With PANMARK, this can be done by using a trick; such models can be specified right away by using LEM. However, a note of caution is in order. Higher-order models have a rather bad image in the literature. As Gregory et al. (1992, p. 1146) put it, "just because a higher order model gives a better fit ... does not necessarily mean that the real system has explicitly higher-order dependence." The point is that higher-order models add a considerable number of parameters to first-order models, thus enabling a better fit. For three points in time, for example, model (17) is saturated, that is, fit is perfect. Simply adding parameters, however, need not imply modeling the real system. From information from recovery studies about how different models fare in fitting known true structures, Langeheine and Van de Pol (2000) advocate using higher-order models only if some substantive theory suggests doing so.

Current high-speed personal computers and modern software allow one to analyze increasingly large tables. Such tables may have thousands or millions of cells. Even with extremely large sample sizes, such tables will almost inevitably contain a certain amount of cells with very small numbers of observed frequencies and with zero counts. Although such sparseness usually does not cause problems in estimating a model, evaluation of model fit by means of chi-square-based statistics will be invalidated because the true distribution of these statistics is badly approximated by the theoretical X^2 distribution. In the past, several attempts were made to overcome this problem. Some authors suggest the pooling of cells having an expected frequency below a certain value. Others fully abandon statistical tests and rely on descriptive fit indices such as BIC instead. However, all of these strategies are not without problems. A promising solution of the problem is simulating the reference distribution(s) of the fit measure(s) used, for the model–data combination at hand, using bootstrap methods. Langeheine et al. (1996) therefore presented the so-called nonnaive bootstrap for contingency table analysis, which is available as an option in the most recent version of the PANMARK program. As their results show, the asymptotic X^2 distribution is not at all valid for sparse data. Although this procedure may appear to be costly at first sight (for a given model–data combination, the model must not only be estimated for the original sample, but also for each of the numerous bootstrap samples), these authors show how computation time may be reduced drastically, thus making the bootstrap a practicable alternative. As a by-product, bootstrapping allows one to forget about potentially very complicated situations requiring adjustment of the number of degrees of freedom caused by fitted zero parameters and fitted zero expected frequencies.

Fitting a model and interpreting the obtained results is one thing researchers may do. Yet even this may not be without pitfalls. The iterative algorithms used for parameter estimation may end up with a local instead of a global maximum of the likelihood function. This pitfall may be avoided by using multiple sets of starting values. Nevertheless, this does not guarantee that a model is identified. Because identification problems are a crucial issue in mixture models, and especially in mixtures of (latent) Markov chains, we have referred to this issue at several places earlier. Fortunately, the PANMARK program offers an option to check for identifiability.

When describing the life satisfaction data used for demonstrative purposes in this chapter, we mentioned that we retained only subjects with complete responses at all waves, despite the fact that information about nonresponse is available. This is what researchers normally do to simplify matters. However, ignoring nonresponse (which may be caused by an interview being rejected, temporary dropout, or panel attrition in the SOEP data) may invalidate one's results. Ignoring nonresponse is equivalent to assuming that subjects without response on some occasion(s) are a random sample of all subjects. The proper attack of the problem would be to check whether nonresponse is ignorable. A general approach to attack this problem is given by Vermunt (1996).

REFERENCES

Anderson, T. W. (1954). "Probability models for analyizing time changes in attitudes." In P. F. Lazarsfeld (ed.), *Mathematical Thinking in the Social Sciences*. New York: The Free Press, pp. 17–66.

Anderson, T. W., & Goodman, L. A. (1957). "Statistical inference about Markov chains," *Annals of Mathematical Statistics*, **28**, 89–110.

Bartholomew, D. J. (1981). *Mathematical Methods in Social Science*. Chichester: Wiley.

Beck, P. A. (1975). "Models for analyzing panel data: a comparative review," *Political Methodology* **2**, 357–80.

Blumen, I. M., Kogan, M., & McCarthy, P. J. (1955). *The Industrial Mobility of Labor as a Probability Process*. Ithaca: Cornell University Press.

Bye, B. V., & Schechter, E. S. (1986). "A latent Markov model approach to the estimation of response error in multiwave panel data," *Journal of the American Statistical Association*, **81**, 375–80.

Clogg, C. C. (1981). "Latent structure models of mobility," *American Journal of Sociology*, **86**, 836–68.

Clogg, C. C., & Goodman, L. A. (1984). "Latent structure analysis of a set of multidimensional contingency tables," *Journal of the American Statistical Association*, **79**, 762–71.

Coleman, J. S. (1964). *Models of Change and Response Uncertainty*. Englewood Cliffs, NJ: Prentice-Hall.

Collins, L. M., & Wugalter, S. E. (1992). "Latent class models for stage-sequential dynamic latent variables," *Multivariate Behavioral Research*, **27**, 131–57.

Collins, L. M., Graham, J. W., Long, J. D., & Hansen, W. B. (1994). "Crossvalidation of latent class models of early substance use onset," *Multivariate Behavioral Research*, **29**, 165–83.

Converse, P. E. (1964). "The nature of belief systems in mass publics." In D. E. Apter (ed.), *Ideology and Discontent*. New York: The Free Press, pp. 206–61.

Converse, P. E. (1970). "Attitudes and non-attitudes: continuation of a dialogue." In R. Tufte (ed.), *The Quantitative Analysis of Social Problems*. Reading, MA: Addison-Wesley, pp. 168–89.

Converse, P. E. (1974). "Comment: the status of nonattitudes," *American Political Science Review*, **68**, 650–60.

Davies, R. B., & Crouchley, R. (1986). "The Mover–Stayer model: requiescat in pace," *Sociological Methods & Research*, **14**, 356–80.

Dempster, A. P., Laird, N. M., & Rubin, D. B. (1977). "Maximum likelihood from incomplete data via the EM algorithm," *Journal of the Royal Statistical Society, Series B*, **39**, 1–38.

Duncan, O. D. (1975). "Review of 'Panel analysis: latent probability models for attitudes and behavior processes,'" *Journal of the American Statistical Association*, **70**, 959–60.

Goodman, L. A. (1962). "Statistical methods for analyzing processes of change," *American Journal of Sociology*, **68**, 57–78.

Goodman, L. A. (1973). "The analysis of multidimensional contingency tables when some variables are posterior to others: a modified path analysis approach," *Biometrika*, **60**, 179–92.

Goodman, L. A. (1974a). "Exploratory latent structure analysis using both identifiable and unidentifiable models," *Biometrika*, **61**, 215–31.

Goodman, L. A. (1974b). "The analysis of systems of qualitative variables when some of the variables are unobservable. Part I – a modified latent structure approach," *American Journal of Sociology*, **79**, 1179–1259.

Graham, J. W., Collins, L. M., Wugalter, S. E., Chung, N. K., & Hansen, W. B. (1991). "Modeling transitions in latent stage-sequential processes: a substance use prevention example," *Journal of Consulting and Clinical Psychology*, **59**, 48–57.

Gregory, J. M., Wigley, T. M. L., & Jones, P. D. (1992). "Determining and interpreting the order of a two-state Markov chain: application to models of daily precipitation," *Water Resources Research*, **28**, 1443–6.

Hagenaars, J. A. (1990). *Categorical Longitudinal Data: Loglinear Panel, Trend, and Cohort Analysis*. Newbury Park: Sage.

Hanefeld, U. (1984). "The German socio-economic panel." In American Statistical Association. (ed.), *1984 Proceedings of the Social Statistics Section*, Washington: American Statistical Association, pp. 117–24.

Kalbfleisch , J. D., & Lawless, J. F. (1985). "The analysis of panel data under a Markov assumption," *Journal of the American Statistical Association*, **80**, 863–71.

Kohfeld, C. W., & Salert, B. (1982). "Discrete and continuous representations of dynamic models," *Political Methodology*, **8**, 1–32.

Kullback, S., Kupperman, M., & Ku, H. K. (1962). "Tests for contingency tables and Markov chains," *Technometrics*, **4**, 573–608.

Langeheine, R. (1988). "Manifest and latent Markov chain models for categorical panel data," *Journal of Educational Statistics*, **13**, 299–312.

Langeheine, R. (1991). "Latente Markov-Modelle zur Evaluation von Stufentheorien der Entwicklung," *Empirische Pädagogik*, **5**, 169–89.

Langeheine, R. (1994). "Latent variables Markov models." In A. von Eye & C. C. Clogg (eds.), *Latent Variables Analysis. Applications for Developmental Research*. Thousand Oaks, CA: Sage, pp. 373–95.

Langeheine, R., Pannekoek, J., & Van de Pol, F. (1996). "Bootstrapping goodness-of-fit measures in categorical data analysis," *Sociological Methods & Research*, **24**, 492–516.

Langeheine, R., Stern, E., & Van de Pol, F. (1994). "State mastery learning: dynamic models for longitudinal data," *Applied Psychological Measurement*, **18**, 277–91.

Langeheine, R., & Van de Pol, F. (1990). "A unifying framework for Markov modeling in discrete space and discrete time," *Sociological Methods & Research*, **18**, 416–41.

Langeheine, R., & Van de Pol, F. (1993). "Multiple indicator Markov models." In R. Steyer, K. F. Wender, & K. F. Widaman (eds.), *Psychometric Methodology. Proceedings of the 7th European Meeting of the Psychometric Society in Trier*. Stuttgart: Fischer, pp. 248–52.

Langeheine, R., & Van de Pol, F. (1994). "Discrete-time mixed Markov latent class models." In A. Dale & R. Davies (eds.), *Analyzing Social and Political Change: A Casebook of Methods*. London: Sage, pp. 170–97.

Langeheine, R., & Van de Pol, F. (2000). "Fitting higher order Markov chains," *Methods of Psychological Research Online*, **5**, 32–55.

Lazarsfeld, P. F. (1950). "The logical and mathematical foundation of latent structure analysis." In S. A. Stouffer, L. Guttman, E. A. Suchman, P. F. Lazarsfeld, S. A. Star, & J. A. Clausen (eds.), *Measurement and Prediction*. Princeton, NJ: Princeton University Press, pp. 362–412.

Lazarsfeld, P. F., & Henry, N. W. (1968). *Latent Structure Analysis*. Boston: Houghton Mifflin.

Logan, J. A. (1981). "A structural model of the higher-order Markov process incorporating reversion effects," *Journal of Mathematical Sociology*, **8**, 75–89.

Macready, G. B., & Dayton, C. M. (1994). "Latent class models for longitudinal assessment of trait acquisition." In A. von Eye & C. C. Clogg (eds.), *Latent Variables Analysis. Applications for Developmental Research*. Thousand Oaks, CA: Sage, pp. 245–73.

Plewis, I. (1981). "Using longitudinal data to model teachers' ratings of classroom behavior as a dynamic process," *Journal of Educational Statistics*, **6**, 237–55.

Plewis, I. (1985). *Analysing Change: Measurement and Explanation Using Longitudinal Data*. Chichester: Wiley.

Poulsen, C. S. (1982). *Latent Structure Analysis with Choice Modelling Applications* (Ph.D. Dissertation, University of Pennsylvania). Aarhus: Aarhus School of Business Administration and Economics.

Poulsen, C. S. (1990). "Mixed Markov and latent Markov modelling applied to brand choice behaviour," *International Journal of Research in Marketing*, **7**, 5–19.

Schwartz, J. E. (1985). "The neglected problem of measurement error in categorical data," *Sociological Methods & Research*, **13**, 435–66.

Schwarz, G. (1978). "Estimating the dimension of a model," *Annals of Statistics*, **6**, 461–4.

Shockey, J. W. (1988). "Latent class analysis: an introduction to discrete data models with unobserved variables." In J. S. Long (ed.), *Common Problems/Proper Solutions: Avoiding Error in Quantitative Research*. Beverly Hills: Sage, pp. 288–315.

Singer, B., & Spilerman, S. (1974). "Social mobility models for heterogeneous populations." In H. L. Costner (ed.), *Sociological Methodology 1973–1974*. San Francisco: Jossey–Bass, pp. 356–401.

Singer, B., & Spilerman, S. (1976). "Some methodological issues in the analysis of longitudinal surveys," *Annals of Economic and Social Measurement*, **5**, 447–74.

Spilerman, S. (1972). "The analysis of mobility processes by the introduction of independent variables into a Markov chain," *American Sociological Review*, **37**, 277–94.

Van de Pol, F., & de Leeuw, J. (1986). "A latent Markov model to correct for measurement error," *Sociological Methods & Research*, **15**, 118–41.

Van de Pol, F., & Langeheine, R. (1989). "Mixed Markov models, Mover–Stayer models and the EM algorithm. With an application to labor market data from the Netherlands Socio-Economic Panel." In R. Coppi and S. Bolasco (eds.), *Multiway Data Analysis*. Amsterdam: North-Holland, pp. 485–95.

Van de Pol, F., & Langeheine, R. (1990). "Mixed Markov latent class models." In C. C. Clogg (ed.), *Sociological Methodology 1990*. Oxford: Blackwell, pp. 213–47.

Van de Pol, F., Langeheine, R., & de Jong, W. (1991). *PANMARK User*

Manual: PANel Analysis Using MARKov Chains. Voorburg: Netherlands Central Bureau of Statistics.

Vermunt, J. K. (1993). "LEM: loglinear and event history analysis with missing data using the EM algorithm." WORC Paper 93.09.015/7, Tilburg University, The Netherlands.

Vermunt, J. K. (1996). "Causal loglinear modeling with latent variables and missing data." In U. Engel & J. Reinecke (eds.), *Analysis of Change: Advanced Techniques in Panel Data Analysis*. Berlin: de Gruyter, pp. 35–60.

Vermunt, J. K., Langeheine, R., & Böckenholt, U. (1999). "Discrete-time discrete-state latent Markov models with time-constant and time-varying covariates," *Journal of Educational and Behavioral Statistics*, **24**, 179–207.

Wiggins, L. M. (1955). *Mathematical Models for the Analysis of Multi-Wave Panels* (Ph.D. Dissertation, Columbia University). Ann Arbor: University Microfilms.

Wiggins, L. M. (1973). *Panel Analysis: Latent Probability Models for Attitudes and Behavior Processes*. Amsterdam: Elsevier.

UNOBSERVED HETEROGENEITY AND NONRESPONSE

TWELVE

A Latent Class Approach to Measuring the Fit of a Statistical Model

Tamás Rudas

1. INTRODUCTION

Traditional approaches to measuring the fit of a statistical model are based on comparing the actual data to the expectation of the observations under the assumption that the model is true. The resulting statistics can be given a strict *test of fit* or a *measure of fit* interpretation. Carrying out a test of fit is possible only if the distribution of the statistic is known. A proper interpretation of a measure of fit also requires knowledge regarding its distribution. These procedures are error prone for two reasons. First, when the model is not true, a comparison of the data to what could only be expected if it was, is of very little meaning. Second, the actual distribution of the statistic may be very different from the reference distribution if some of the underlying assumptions are violated. A more detailed account of these problems in the context of models for contingency tables is given in Section 2; in this case, serious additional problems may arise because of small or large samples.

In this chapter, an alternative approach to measuring the fit of a statistical model will be described, which is not affected by most of the problems referred to earlier. We will abandon the idea that a simple model may describe the entire population. Consequently, we will not try to assess whether the data provide evidence against this assumption in terms of being too unlikely, or in terms of showing large deviations from what would be expected under the model. It is assumed that there are two latent classes in the population. In one of them, the model of interest holds true whereas the other one is completely unrestricted. The sizes of these latent classes are not known to us; rather, we are interested in finding out what possible maximum size the first latent class may have, that is, what is the largest fraction of the population where the model of interest may be true. The larger this fraction, the better the model fits the underlying population. The size of this fraction is the *mixture index of fit*, and it is described, together with several of its properties, in Section 3. Advantages of the mixture index of fit include that its definition does not rely on assumptions that may not be true, and its estimated values do not depend on the sample size in the way chi-square-related quantities do.

In Section 4, analyses based on the mixture index of fit of the citation practices of two operations research journals (Fienberg, 1980a, 1980b) and of the death penalty data of Radelet (1981) will be presented. These sets of data have been analyzed by other authors. With the use of our approach, new insight will be gained regarding the fit of simple models.

Finally, Section 5 shows that the ideas underlying the mixture index of fit can be applied to problems other than contingency table analysis.

2. DIFFICULTIES WITH CHI-SQUARED TESTS OF FIT

The standard method of testing the fit of a statistical model for a contingency table is to compute the table of estimated, or expected, frequencies, and to compare it with the table of observed frequencies. The expression *expected frequencies* is somewhat misleading here. One does not expect to observe this table, even if the model is true. Rather, this table represents the distribution that, if the model were true, would have made the observations more likely than any other distribution. Then, one is interested in knowing whether, even under these "most favorable" conditions, the actual observations belong to the group of the most unlikely observation. If the answer is yes, the hypothesis is rejected. To implement this procedure, one must know the distribution of the statistic that is used to compare the observed and estimated frequencies. The statistic that is used most frequently for this comparison is the Pearson chi-squared statistic.

The distribution of the chi-squared statistic depends on several factors, including the size of the table, the sample size, and the actual true probabilities. Therefore, using the true distribution of the statistic in the previously mentioned procedure is not feasible, or at least used to be not feasible when sufficient computational capacity was not available to every scientist. Fortunately, as the sample size increases, the true distribution of the statistic is getting closer and closer to a distribution that depends only on one quantity, which can be easily computed from certain properties of the model and the size of the table (namely, the number of degrees of freedom). This *asymptotic* distribution is a chi-squared distribution on the relevant number of degrees of freedom. The critical values of these distributions are tabulated and routinely used in testing.

How well the true distribution of the chi-squared statistic is approximated by its asymptotic distribution depends on exactly the same factors that influence the distribution itself. There have been various "practical" rules suggested to decide whether this approximation is good enough. Obviously, the simpler these rules are, the more likely it is that they are wrong. When the critical value is taken, instead of the true distribution of the test statistic, from its asymptotic distribution, the true level of the decision will be different from the nominal level chosen.

Some of the criteria suggest that the smallest expected frequency, in order to be able to use the critical value from the asymptotic distribution instead of the actual one, should be at least 5 (Fisher, 1941, p. 82); or that it should be at least 10 (Cramer, 1946, p. 420). These criteria are not very practical because the question of whether they are satisfied is answered only after the data collection has been finished. Other criteria have been suggested in terms of the ratio of the sample size to the number of cells of the table, including that this should be at least 4 or 5 (Fienberg, 1979), or that at least 3 (Rudas, 1986). For a review of some of the other "rules" suggested in the literature, see Rudas (1984).

The Pearson chi-squared statistic is not the only one used. The likelihood ratio statistic and some others may also be applied to compare the observed and estimated frequencies. For a unified treatment of these goodness-of-fit statistics, see Read and Cressie (1988). A great deal of research has been done to find out which of these statistics converges faster to its limiting distribution, but the findings have not been conclusive.

When a test of fit must be performed, and the size of the sample may be too small for the application of the critical values of the asymptotic distribution, computer simulations may be used to approximate the true distribution of the test statistic.

There are several measures of fit that are derived from the Pearson chi-squared statistic. Their correct interpretation poses problems similar to those described earlier.

When the sample size is "large," the application of the asymptotic critical values is not problematic, but one runs into other kinds of difficulties. These come from the fact that the Pearson chi-squared (and many related) statistics are "proportional" to the sample size. This means that if two observed distributions are the same, one with sample size N_1, and the other one with sample size N_2, then the ratio of the relevant Pearson chi-squared statistic values is N_1/N_2. In other words, after the sample size is multiplied by 2, say, the value of the statistic will be multiplied by 2 as well. Practically, with large samples, which are desirable otherwise, one must reject the simple model of interest most of the time.

This does not mean that something is wrong with chi-squared tests; rather, it means that they do something different from what many of the users expect them to do. The test, correctly, detects relatively weak effects, in terms of deviations from the model, when applied to large samples. The procedure is a test for statistical significance. What most of the users of this procedure are interested in, however, is not statistical, but subject-matter significance.

Testing for subject matter significance requires a precise specification of how large effects are considered important from the point of view of the scientific problem at hand, and the statistical testing procedure be adjusted accordingly. This usually leads to nonstandard statistical problems.

One advantage of the approach presented here is that it makes the handling of subject-matter significance possible. Other advantages include that it is derived from a framework that is always valid, it has a straightforward interpretation, and it does not depend on the sample size in the usual sense.

3. THE MIXTURE INDEX OF FIT

The mixture index of fit is defined within a specific latent class context, which will be outlined first. One important property of this framework is that it is always valid; therefore, it is not a restriction of reality, but rather a specific way to look at *any* population. There is a model of interest H, and our goal is to measure the ability of this model to account for the population underlying the data.

It is assumed that there are two latent classes in the population. In one of them, the model of interest holds true; that is, the distribution describing the first latent class belongs to model H. There are no assumptions regarding the other latent class. The sizes of the two latent classes, for the time being, are kept unspecified. The relative size of the first latent class (where H holds true) is denoted by $1 - \pi$, and of the second (unrestricted) latent class by π, where $0 \leq \pi \leq 1$. Here, $1 - \pi$ and π are the respective probabilities of the two latent classes. It follows that an observation comes from a distribution that belongs to H (i.e., from the first latent class) with probability $1 - \pi$, and with probability π it comes from another distribution (i.e., from the second latent class). In other words, the distribution P that describes the population has the following mixture representation:

$$P = (1 - \pi)Q + \pi R, \tag{1}$$

where $Q \in H$ is the distribution in the first latent class and R is the distribution in the second latent class. The only restriction in Equation (1) is that the distribution Q belongs to model H. How serious this restriction is depends on the magnitude of $1 - \pi$, that is, the relative size of this latent class. The smaller $1 - \pi$, the less restrictive the assumption in Equation (1). In fact, if $1 - \pi = 0$, which is its smallest value, Equation (1) simply says that the size of the fraction where model H is valid is zero, and the entire population is described by an unrestricted distribution R. That is, if $1 - \pi = 0$, then Equation (1) does not restrict the distribution P at all. When $1 - \pi = 1$, then the assumption in Equation (1) is that model H describes the entire population. This is the usual null hypothesis when model H is tested.

As $1 - \pi$ moves from 0 to 1, the assumption in Equation (1) moves from being not restrictive at all to assuming that model H describes the entire population. Whether or not Equation (1) is true for any specified value of $1 - \pi$ depends on several factors. But in every case, whatever the population, the model of interest and the true distribution are Equation (1) is true for at least one value of $1 - \pi$ (when it is equal to zero). If there are several values of $1 - \pi$ for which Equation (1) holds, then there is a largest one from among these.

As a way to illustrate this, suppose that the model of interest is independence and the distribution in the population is described by the 2×2 cross-classification given in Table 1.

Then, as it is easy to see, there may be two latent classes in the population, with respective probabilities of 0.357 ($= 1 - \pi$) and 0.643 ($= \pi$) and with the distributions (probabilities multiplied by 1,000) given in Table 2.

In Table 2, the first distribution is independent. The marginal probability in the upper left-hand side cell is $280 \times 0.357/1{,}000 = 0.1$, and in the upper right-hand side cell is $(120 \times 0.357 + 157.14 \times 0.643)/1000 = 0.2$.

Table 1. Hypothetical Distribution of a Population

0.1	0.2
0.3	0.4

Table 2. Two Components of a Mixture Yielding the Distribution in Table 1

280.00	120.00	0.00	157.14
420.00	180.00	150.00	64.29

Note: Probabilities are multiplied by 1,000. The first component is independent. See text for mixing weights.

Or, there may be two latent classes with respective probabilities of 0.555 and 0.445, with the distributions (multiplied by 1,000) given in Table 3.

Or, there may be two latent classes with respective probabilities of 0.714 and 0.286, and with the distributions (multiplied by 1,000) given in Table 4.

All the three latent class structures lead to the same distribution for the entire population. The distributions in the first latent classes are such that the two variables are independent. The hypothesis of independence of the two variables describes a fraction of 0.357, or a fraction of 0.555, or a fraction of 0.714 of the population. Can this hypothesis describe the *entire* population? The answer is no (as long as the distribution is the one given in Table 1); in fact, no fraction of the population greater than 0.933 can be described by independence (see the related comments in Section 4).

In other words, the maximal value of $1-\pi$ is 0.933 for the distribution in Table 1 and the hypothesis of independence. The largest possible fraction of the population where independence of the two variables may be true is $1-\pi^*=0.933$, or the smallest possible fraction that cannot be described by independence is $\pi^*=0.067$. With a different model, or with a different distribution, π^* would take on different values. Therefore, $\pi^* = \pi^*(P, H)$ is a measure of misfit, and $1-\pi^*=1-\pi^*(P, H)$ is a measure of model fit.

The quantity $\pi^* = \pi^*(P, H)$ depends both on the true distribution that is valid in the population (P) and on the model, or hypothesis of

Table 3. Two Components of Another Mixture Yielding the Distribution in Table 1

20.00	80.00	199.78	349.66
180.00	720.00	449.66	0.90

Note: Probabilities are multiplied by 1,000. The first component is independent. See text for mixing weights.

Table 4. Two Components of Yet Another Mixture Yielding the Distribution in Table 1

60.00	140.00	199.86	349.79
240.00	560.00	449.79	0.56

Note: Probabilities are multiplied by 1,000. The first component is independent. See text for mixing weights.

interest (H). This quantity measures, in the sense outlined above, the ability of hypothesis H to account for a population in which the true distribution is P. The π^* index of fit was suggested in Rudas et al. (1994). It is based on a framework that is always valid and has a straightforward interpretation; it has many other appealing properties that will be considered later.

The mixture index of fit is defined as a *population parameter*, that is, one that can only be computed if the true distribution in the population is known. This is not the case in practical applications, when the true distribution is not known; rather, it has to be estimated from a sample. When the observed frequencies in the contingency table are divided by the total sample size, one obtains a distribution in the contingency table that is an estimate of the underlying distribution. In fact, this is the maximum-likelihood estimate of the true distribution, if there are no assumptions restricting P. This estimate is usually denoted by $\hat{P}$. Note that model H is not assumed to hold true when $\hat{P}$ is computed. Now, the maximum-likelihood estimate of $\pi^*(P, H)$ is $\pi^*(\hat{P}, H)$; that is, the maximum-likelihood estimate of the π^* of the true distribution is the π^* of the maximum-likelihood estimate of the true distribution.

Therefore, the problem of maximum likelihood estimation of π^* reduces to the problem of calculation of the index for a given (estimated) distribution. For this problem, and for *any* model H, Rudas et al. (1994) suggested an algorithm that involved repeated application of the expectation–maximization (EM) algorithm (Dempster, Laird, and Rubin, 1977). For the case when H is a loglinear model for the contingency table, Xi and Lindsay (1997) suggested a more efficient algorithm. The Appendix describes other computational possibilities.

Because the estimate is a function of the observed distribution (and not of the observed frequencies), the estimated values do not depend on the sample size. For two samples with different sizes but that yield the same distribution, the estimated values of π^* will be identical (see the numerical example in Section 4). This is in sharp contrast with the chi-squared values, which in this case would be proportional to the sample sizes. In what way does sample size effect statistical inference regarding π^* if the estimates do not depend on it? One expects a larger sample to provide more information than a smaller one, and in the present case this means that the confidence interval for the true value π^* will be shorter for the larger sample than for the smaller sample. A method of computing confidence intervals for the mixture index of fit was described in Rudas, Clogg, and Lindsay (1994), and examples will be presented in the next section.

A further appealing property of the mixture index of fit is related to the problem of nested models. Clogg, Rudas, and Xi (1995) considered various simple models for social mobility tables, and their respective ability to account for well-known sets of mobility data. The models investigated were independence, quasi-independence, and quasi-uniform association. These are nested in the sense that for two-way tables of a fixed size, all the independent distributions are contained among the quasi-independent ones and these are contained among the ones where quasi-uniform association holds true. Therefore, the chi-squared statistics measuring deviation from the expectation under the model will give decreasing values when the model is independence, quasi-independence, or quasi-uniform association, respectively. The statistical inference is usually not based on the values of the chi-squared statistics solely, but the number of degrees of freedom is also taken into consideration. In the case of three nested models, the numbers of degrees of freedom are decreasing as well. Therefore, rejection of the larger (less restrictive) model does not imply the result of the test of fit of the smaller (more restrictive) model, and when the smaller model is not rejected, this does not imply the result of the test of fit of the larger model. Many users find this fact counterintuitive.

The mixture index of fit is monotone in the sense that if $H_1 \subseteq H_2$, then $\pi^*(P, H_1) \geq \pi^*(P, H_2)$, and, consequently, the same inequality holds for the estimated values, no matter what the observations are. In other words, the smaller model cannot appear to be able to account for a larger fraction of the population than the larger model, when the mixture index of fit is used to measure the ability of a model to describe the population. Clogg et al. (1995) found that some of the well-known social mobility tables could be described fairly well by the simple models referred to earlier, according to the mixture index of fit, and they reported a natural assessment of the gain in fit when one moves from a smaller (more restrictive) to a larger (less restrictive) model.

The monotonicity property of the mixture index of fit was utilized by Rudas and Zwick (1997), in comparing two competing nested hypotheses for educational testing data: the hypotheses of no differential item functioning and of uniform differential item functioning. In educational testing, differential item functioning is said to be present if individuals having the same ability, but belonging to different groups in the population, have different chances of responding to a question correctly. The residual analysis based on the mixture index of fit, outlined later, facilitated the identification of those parts of the population, where differential item functioning may occur.

The mixture index of fit leads to a decomposition of the observed distribution (or of the observed frequencies) in the following form:

$$\hat{P} = [1 - \hat{\pi}^*(P, H)]\hat{Q}^* + \hat{\pi}^*(P, H)\hat{R}^*, \tag{2}$$

where $\hat{Q}^*$ is the maximum-likelihood estimate of the distribution in that part of the population where model H holds true, if it describes a fraction as large as possible, and $\hat{R}^*$ is the maximum-likelihood estimate of the distribution in the smallest possible fraction of the population where model H does not hold true. In the decomposition in Equation (2), $\hat{R}^*$ can be considered as a residual distribution, and $\hat{\pi}^*$ is the estimate of the fraction where it is valid. This approach to residuals is fundamentally different from the usual approach, in which the deviations between observed and expected frequencies (or distributions) are considered. In Section 4, the interpretation of the residual distribution $\hat{R}^*$ and related analyses will be discussed.

The new definition of residuals, based on the mixture index of fit, is always valid in the sense that it does not rely on the assumption that the model of interest describes the (entire) population, as in the usual approach. When, on one hand, the latter assumption does not hold true, the estimates and the residuals have very little meaning. When, on the other hand, one concludes that hypothesis H does not have to be rejected, the residuals are deemed to be attributable to random variation, and their analysis is of limited importance. Furthermore, the residuals defined here have a straightforward interpretation (as the distribution in the smallest possible fraction of the population where H is not true) and can be used to estimate the fraction of the observations in every cell that have come from the part of the population where H is not true. The importance or weight of the residuals is measured by $\hat{\pi}^*$, and the measure of misfit of H is identical to the importance or weight of the residuals that appear to be present after as much as possible from the observations were attributed to H. The new definition of residuals, outlined above, was considered in detail in Clogg et al. (1995); related graphical diagnostic methods for the model of interest were described by Clogg, Rudas, and Matthews (1998).

In a certain aspect, the idea underlying the mixture index of fit is related to the idea underlying the traditional approach, when testing is based on maximum-likelihood estimation: in both cases, the best chance is given to the model of interest to account for the data, and the relevant hypothesis is discarded if it does not perform well even under these most favorable conditions. With maximum-likelihood estimation, parameters are estimated so that the data will have the highest possible likelihood under the model, and when the mixture index of fit is applied, the parameters are estimated so that the model can be valid in the biggest possible fraction of the population. These ideas lead naturally to a minimum distance estimation procedure based on the mixture index of fit (see Rudas, 1998b).

4. APPLICATIONS

Before presenting applications of the mixture index of fit to real data and some of the conclusions that can be reached based on the findings, let us consider a numerical example in some detail. The example deals with the problem of testing independence in a 2×2 contingency table. The example illustrates the exact meaning of the mixture index of fit in this very simple case, and that the property "measured" by the mixture index of fit is very different from the property "measured" by the chi-squared statistic. Notice that the mixture index of fit can be applied to measuring the fit of any statistical model, not just independence, and not only in the setup of contingency tables. The real applications later in this section will consider fitting models other than independence to contingency tables, and in the next section details of some of the applications to other statistical problems will be described.

Suppose the size of the sample available is 100, and the data given in Table 5 were observed.

The Pearson chi-squared statistic for the hypothesis of independence is 0.794 on 1 degree of freedom. With the use of the chi-squared distribution

Table 5. A Hypothetical Set of Observations

10	20
30	40

Table 6. Mixture Representation of the Distribution in Table 5

$$0.933 \times \begin{array}{cc} 0.1072 & 0.1426 \\ 0.3215 & 0.4287 \end{array} + 0.067 \times \begin{array}{cc} 0.00 & 1.00 \\ 0.00 & 0.00 \end{array}$$

as reference, this value of the test statistic does not suggest the rejection of the hypothesis of independence at the usual levels of significance.

The maximum-likelihood estimation procedure for π^* is identical to computing π^* for the observed distribution. The observed distribution is exactly the one considered in the previous section. The value of π^* for this distribution can be obtained as $0.2 - (0.1 \times 0.4)/0.3 = 0.067$. For a proof of this fact, see Clogg et al. (1995); it is based on the observation that the value of π^* is obtained when the smallest number of observations is removed from the table, so that the remaining table has an odds ratio equal to 1. The decomposition related to π^*, that is, the one given in Equation (2), is represented in Table 6.

The residual distribution $\hat{R}^*$ in Table 6 has the somewhat surprising property of being concentrated to one cell only. In fact, the number of positive entries in the table of residual distribution cannot exceed the number of degrees of freedom of model H. In the present case, independence for a 2×2 table has 1 degree of freedom. Any possible deviation a distribution may have from H (independence, for example) can be attributed to cells, the number of which does not exceed the number of degrees of freedom of $H\,[(I-1)(J-1)$ for an $I \times J$ table and independence]. This fact sheds new light on the meaning of the degrees of freedom associated with a statistical model.

The conclusion is that we estimate that independence cannot describe more than 93.33% of the population, or, at least 6.67% of the population lies outside of independence. Whether this is interpreted as a good or bad fit may depend on several factors. These percentages are estimated values, and as such they are subject to deviation from their respective true values as a result of sampling. To assess the influence sampling has on the estimated values, one may want to compute confidence bounds for the true value of π^*. For these data, the 95% confidence interval for π^* contains zero, implying that a formal test of the fit of the hypothesis that $\pi^* = 0$, that is, independence describes the entire population, would not lead to rejection. Notice that there is no simple formula available to compute confidence bounds for π^*. Rudas et al. (1994) described an iterative procedure to determine the value of a lower confidence bound. It follows

Table 7. Another Hypothetical Set of Observations with Frequencies Proportional to Those in Table 5

100	200
300	400

from Equation (1) that if the mixture representation is possible with a certain value of π, then it is also possible with any other π value bigger than this. The confidence interval contains all π values that "perform nearly as well" as the estimated value, in terms of their ability to represent the data as a mixture (1). Because π values bigger than the estimate are just as good as the estimate itself, they will all be contained in the confidence interval, and therefore the upper confidence bound is always 1. If the value $\pi = 0$ is "not much worse" than the estimate, just like in the present case, the value $\pi = 0$ is also contained in the 95% confidence interval or, more precisely, it is the lower confidence bound at a level less than or equal to 95%.

Things look quite different when, instead of the data in Table 5, the observations in Table 7 are to be analyzed.

For many researchers, these data provide the same evidence for or against independence as the data in Table 5. This viewpoint is supported by the fact that the estimated value of the odds ratio (see Rudas, 1998a), as a measure of association, is the same for both sets of data. The impression gained by looking at the Pearson chi-squared statistic is very different: its value is equal to ten times its previous value, that is, nearly 8. One still has 1 degree of freedom; therefore, at any usual level of significance, the test would suggest the rejection of the hypothesis of independence. The rationale behind this is, of course, that with larger sample sizes one would expect the properties of the population (independence under the present hypothesis) to show more clearly in the sample than with smaller sample sizes. In the present case, one observes the same deviation from independence in this sample as in the smaller sample (as measured, e.g., by the estimated value of the odds ratio), but this deviation from independence is deemed significant with 1,000 observations, but it is not deemed significant based on 100 observations.

The value of the maximum-likelihood estimate of the mixture index of fit is the same here as with the previously analyzed set of data: 0.667. It is not affected by the sample size. The gain associated with the larger

Table 8. Citation Practices of Two Operations Research Journals

	Cited Journal			
Citing Journal	**Neither**	**MS Only**	**OR Only**	**Both**
MS	61	63	20	59
OR	42	3	47	41

Source: Fienberg (1980a). Reprinted by permission of the Applied Probability Trust.

sample size is a shortened confidence interval. In fact, the 95% confidence interval does not contain the value of zero any more; the approximate 95% confidence interval for the true value of the mixture index of fit is (0.029, 1). Therefore, a formal test of fit for the original null hypothesis (independence of the two variables for the entire population) would have to be rejected at the 95% level.

The mixture index of fit is not a function of the Pearson chi-squared statistic. It can happen that when two tables are compared; one has the greater value of the Pearson chi-squared statistic, and the other has the greater value of the mixture index of fit. It is less obvious that the mixture index of fit is also not a function of the odds ratio. They measure deviation from independence in different senses.

To illustrate the analyses that are possible by using the mixture index of fit, we first reconsider data (Table 8) from Fienberg (1980a) regarding the citation practices of two operations research journals, *Management Science* (MS) and *Operations Research* (OR). Data were collected from 336 papers from the 1969 and 1970 issues of these journals. The papers were classified into four categories according to whether other papers from these two journals were referred to.

The homogeneity of the citation practices is technically equivalent to the independence of the two variables forming the table. The data also appear in Fienberg (1980b, pp. 25–6) in a problem, where he suggests "viewing the data as if they were a sample from a hypothetically infinite population," and testing the independence of the two variables. In fact, *any* test of a statistical hypothesis is only meaningful under the assumption that the available data are a sample from some underlying population, finite or infinite. When this assumption is made, no particular difficulties are encountered in testing independence. The value of the Pearson chi-squared statistic is 60.20 on 3 degrees of freedom, which suggests the rejection of the hypothesis of independence.

The correctness of an assumption similar to the one mentioned earlier may be dubious in certain social science applications. When, rather than the observation of randomly selected units of a well-defined population, complete data are collected from certain parts of the population defined by variables that may not be independent from the characteristic under investigation (like time, or geographic location), the question whether there is any population to which the data are "representative" always arises. In spite of this problem, such data-collecting procedures may be useful and sometimes are unavoidable. In these cases it may be necessary to consider the data as complete observations.

If for this, or for other reasons, one is not willing to make the assumption suggested by Fienberg (1980b) as to the existence of an underlying population, the question of independence takes a very different form. Viewed as population data, the two variables *are not* independent, and one may want to assess the magnitude of deviation from independence. The value of the Pearson chi-squared statistic can, of course, be considered as a measure of association, but in this context it lacks any calibration and therefore is not very informative. The associated probability levels, applied by many researchers, also should not be used to assess strength of association (see Schervish, 1996). Thus, one must choose another measure of association, and consequently the analyses under the sample and population assumptions are carried out in *conceptually* very different frameworks.

The advantage of the mixture index of fit in this case is that whether the data are considered a sample from a larger population, or the population itself, the same technique can be used, with different interpretations, of course.

The value of π^* is 0.2746 for the data. This can be interpreted as the maximum-likelihood estimate of the true (population) value if the data are considered a sample, and this is the actual population value if the data are considered to describe the entire population. Independence, that is, homogeneous citation practices, can describe at most 72.54% of the articles, and this is the population value under the latter assumption, or its maximum-likelihood estimate under the former assumption. The decomposition in Equation (2), obtained for the data, is as given in Table 9.

Again, the decomposition in Table 9 is either the "optimal" decomposition of the population, or the maximum-likelihood estimate of it. The

Table 9. Mixture Representation of the Distribution in Table 8

$$0.7254 \times \begin{array}{cccc} 0.2503 & 0.0179 & 0.0821 & 0.2421 \\ 0.1723 & 0.0123 & 0.0565 & 0.1667 \end{array} + 0.2746 \times \begin{array}{cccc} 0 & 0.6276 & 0 & 0 \\ 0 & 0 & 0.3682 & 0.0042 \end{array}$$

first table is our estimate for the distribution that describes the largest possible independent fraction of the population, and the second distribution is the estimate for the smallest fraction where independence, that is, homogenous citation practices, does not hold true. It is estimated that at most, 72.54% of the population can be described by homogenous citation practices, and in at least 27.46% of the population the citation practices of the two journals are different. The estimate for the distribution in this part of the population reveals that the deviation from homogenous citation practices is due to a strong tendency of self-reference in both journals. In fact, more than 99% of the papers in the "residual class" refer only to other papers in the same journal where they appeared. One may conclude that the citation practices of the two journals are not homogenous because in addition to a fairly large fraction of papers (72.54%) in which the citation practices can be considered homogenous, there are a number of other papers containing only self-references. The lack of fit of the model of homogenous citation practices (for the entire population) can be attributed to an excess number of papers containing references to other papers in the same journal only.

When the data are considered a sample, one may want to carry out statistical inference regarding the true value of π^*. In this case, the asymptotic 95% confidence interval for π^* is (0.223, 1), indicating that at the 95% level one has strong evidence against independence (i.e., against $\pi^* = 0$).

A further example that we consider here is the death penalty data published by Radelet (1981). The additional features here include the presence of a zero observed frequency and two nested models being of interest. The data are from 20 counties of Florida during the period 1976–7. The data (Table 10) are in the form of a $2 \times 2 \times 2$ table, and the three variables are *race of victim* (white, black), *race of defendant* (white, black), *death penalty verdict* (yes, no). The research problem of interest was

Table 10. The Death Penalty Data

		Death Penalty	
Defendant	**Victim**	**Yes**	**No**
White	White	19	132
	Black	0	9
Black	White	11	52
	Black	6	97

Source: Radelet (1981). Reprinted by permission of the American Sociological Association.

whether the data suggest the presence of racial bias in imposing the death penalty.

To operationalize the research problem of interest, it appears to be natural to consider the odds of the death penalty being imposed as opposed to not being imposed, for the various groups of victims and defendants, or the related odds ratios. Therefore, the handling of the zero frequency in one of the cells, a feature shared by several real data sets, plays an important role in the analysis.

Agresti (1990, pp. 135–8) gave a detailed analysis of the death penalty data. Disregarding the victim's race, white defendants have a higher chance of receiving the death penalty. When the victim's race is held fixed, for both white and black victims, black defendants received the death penalty more often than white defendants. That is, the *marginal* and *conditional* associations among the defendant's race and whether the death penalty was imposed are in opposite directions. In order to be able to compute the relevant odds, Agresti (1990) added 0.5 to every frequency.

Adding 0.5 to the observed frequencies to remove the adverse effects that observed zeros have on the analysis is widely applied in practice but lacks general justification. When the observed zero is a random one (i.e., can be attributed to the sampling process) rather than a structural one (i.e., nonexistent category in the population), then an argument is possible in terms of the simultaneous analysis of several samples that are not very far from each other. A more precise formulation of this argument is outside of the scope of this chapter; we mention only that it is closely related to ideas underlying bootstrap methods. The analysis justified by this argument would, however, pay close attention to the extent to which the results depend on the choice of the constant added (0.5, usually).

There are two hypotheses that are relevant in the present problem: the victim's and the defendant's race jointly are independent from the penalty, and given the victim's race, the defendant's race and the penalty are conditionally independent. For the first hypothesis, Agresti (1990) found that the likelihood ratio statistic was 8.1 on 3 degrees of freedom (probability level, 0.04), and for the second hypothesis the likelihood ratio statistic was 1.9 on 2 degrees of freedom (probability level, 0.39). From the p values one may conclude that the first model gives a poor fit, whereas the second one gives a very good fit. It would, however, be difficult to answer the question of how much better the second model fits than the first one.

It will be shown later how the latent class approach to measuring the fit of a model can be used to complement these analyses. In particular, it will be illustrated how the effect of adding a constant (and of the choice of its value) on model fit can be quantified; and how the gain in fit, when one goes from the more restrictive model to the less restrictive one, can be measured.

When 0.5 is added to every cell frequency, one finds for the first model that $\hat{\pi}^*$ is 0.063. That is, we estimate that at least slightly more than 6% of the population lies outside of the model of independence (of penalty from combined race of victim and defendant). In other words, the model of independence can account for nearly 94% of the population. When, rather than 0.5, the value added to every cell is 0.1, $\hat{\pi}^*$ is 0.080; that is, we estimate that at most 92% of the population can be described by the independence model. As smaller and smaller quantities are added to the cell frequencies, then the estimate of the largest possible fraction of the population that can be described by the model of interest becomes smaller and smaller. These fractions have a limiting value (as the quantity added goes to zero), and this limiting value is 91.4%. This value does not depend on any particular choice of the quantity added to the cell frequencies.

For the model of conditional independence (of defendant's race and penalty, given the victim's race) the maximum-likelihood estimate of the smallest possible fraction of the data that cannot be described by the model is 0.016, when 0.5 is added to every cell frequency. In other words, we estimate that at most slightly more than 98% of the population can be accounted for by the conditional independence model. When 0.1 is added, one estimates that at most 97.4% of the population can be accounted for by the model, and the limiting value of the estimates (when the quantity added approaches zero) is 97.1%.

The gain in estimated ability of the model to account for the population when, instead of the independence model, the less restrictive conditional independence model is assumed, is the difference of these values. When 0.5 is added to every frequency, the gain is 98.4% − 93.7% = 4.7%; when 0.1 is added, the gain is 97.4% − 92% = 5.4%; and in the limiting case, the gain is estimated to be 97.1% − 91.4% = 5.7%. These values are, of course, subject to rounding error. One may conclude that the model of conditional independence is able to describe a fraction of the population that is approximately 5% larger than the maximal fraction the independence model may be able to describe.

Note that although in order to be able to calculate the mixture index of fit, one has to add a constant to the cell frequencies, the results depend on the quantity added only very little, and by considering the limiting values as earlier, the effect of adding a constant to the (zero) cell frequencies can be removed entirely. The results obtained have clear and natural interpretation and make the comparison of the relative fits of nested models straightforward.

Xi (1996) considered extensions of the mixture index of fit that do not require this special treatment of zero observed frequencies. See also Formann (2000) for a discussion on handling empty cells.

5. GENERALIZATIONS TO OTHER STATISTICAL PROBLEMS

The idea underlying the mixture index of fit applies to any kind of data, and to any statistical model. The application of the π^* approach may lead to appealing interpretations of well-known statistical quantities. Rudas et al. (1994) considered the relationship between the mixture index of fit and the correlation coefficient.

When two variables have a joint normal distribution, their correlation coefficient can be used as a measure of the strength of their association. If the correlation is equal to any of the values of −1, 0, 1, then its interpretation is straightforward. The correct assessment of the amount of association when the correlation coefficient takes on other values is difficult, because of the lack of an intuitive interpretation. As the correlation coefficient is a measure of association, one may expect that the smaller its absolute value, the more similar the joint distribution of the two variables to independence. Or, the larger the absolute value of the correlation coefficient, the lesser the ability of the model of independence to describe the joint distribution. This suggests the application of the mixture index of fit.

In fact, π^* is the following function of the correlation coefficient:

$$\pi^* = 1 - \sqrt{(1 - |\rho|)/(1 + |\rho|)},$$

where $|\rho|$ is the absolute value of the correlation coefficient. For example, when the correlation is 0.6, at most 50% of the population can be described by independence. This is an intuitively clear interpretation of the meaning of the given correlation.

The mixture index of fit can also be applied to several problems in multivariate statistics, including parameter estimation (Rudas, 1998b). It provides a natural measure of model fit and a coherent model selection criterion in regression analysis (Rudas, 1999). In fact, the mixture index of fit is optimized whenever a minimax estimation procedure is applied, in the case of normal or uniform error structures. A detailed account of these results, however, lies outside of the scope of the present chapter.

ACKNOWLEDGMENTS

The research reported here was supported in part by Grants OTKA T-016032 and OTKA T-032213 from the Hungarian National Science Foundation.

APPENDIX: ALGORITHMS TO COMPUTE THE MIXTURE INDEX OF FIT π^*

Emese Verdes
Kossuth Lajos University, Debrecen

There are three algorithms available to compute π^* for contingency tables. The EM algorithm was suggested by Rudas et al. (1994). Xi and Lindsay (1997) suggested the use of the SQP (sequential quadratic programming) algorithm within a table decomposition approach. The third method is based on the minimax theory. The EM algorithm is completely general, but it exhibits very slow convergence and requires an inner optimization routine in every second step; therefore only the more efficient other two algorithms will be presented here.

Loglinear models, as generalized linear models, can be written in the form

$$X\beta = \log(m),$$

where X is a design matrix, β is a parameter vector, and m is the vector of the expected counts. In these terms, the problem of finding π^* can be written in the form

$$\text{maximize } \sum_i \exp\left[X(i,.)\,\beta\right]$$
$$\text{subject to } X\beta \leq \log(n),$$

where n is the vector of observed counts and $X(i,.)$ refers to the ith row of the matrix X. The above problem is a constrained maximization problem, which can be solved by the SQP algorithm available in the MATLAB package (http://www.mathworks.com). This algorithm can be activated by calling the *constr* function. Details of this procedure for the examples can be found at http://www.klte.hu/~vemese/pistar.htm. Note that according to our experience, for a loglinear model one obtains several local optima, and it is important to start from a "good" starting point (see Xi, 1996).

A more generally applicable algorithm can be based on the following result (Rudas, 1999).

Let H be any generalized linear model, and let g denote the observed density (or a smoothed version of it). Then

$$1 - \pi^* = \sup_{h \in H} inf_{\text{supp}\, h} \frac{g}{h},$$

where h is a density in H and supp h is its support. In the case of loglinear models, the values of the function g are n_i, $i = 1, \ldots, t$ and the values of the function f are m_i, $i = 1, \ldots, t$, where $X\beta = \log m$. The support of h consists of t points and so this expression is the minimum of t fractions depending on x when h is selected to maximize this minimum:

$$1 - \pi^* = \max_\beta \min_i \left\{ \frac{n_i}{\exp[X(i, .)\beta]}, \quad i = 1, \ldots, t \right\}.$$

By reciprocating the elements in this expression, one obtains the minimax problem

$$\frac{1}{1 - \pi^*} = \min_i \max_\beta \left\{ \frac{\exp[X(i, .)\beta]}{n_i}, \quad i = 1, \ldots, t \right\}.$$

This problem can also be solved by the MATLAB package, using the function *minimax* available in the Optimization Toolbox. For details of this, see again http://www.klte.hu/~vemese/pistar.htm.

REFERENCES

Agresti, A. (1990). *Categorical Data Analysis*. New York: Wiley.

Clogg, C. C., Rudas, T., & Xi, L. (1995). “A new index of structure for the analysis of models for mobility tables and other cross classifications,” *Sociological Methodology*, **25**, 197–222.

Clogg, C. C., Rudas, T., Matthews, S. (1998). “Analysis of model misfit, structure, and local structure in contingency tables using graphical displays based on the mixture index of fit.” In M. Greenacre & J. Blasius (eds.), *Visualization of Categorical Data*. San Diego: Academic Press, pp. 425–39.

Cramer, H. (1946). *Mathematical Methods of Statistics.* Princeton, NJ: Princeton University Press.

Dempster, A. P., Laird, N. M., Rubin, D. B. (1977). Maximum likelihood from incomplete data via the EM algorithm. *Journal of the Royal Statistical Society, Series B*, **39**, 1–38.

Fienberg, S. E. (1979). “The use of chi-squared statistics for categorical data problems,” *Journal of the Royal Statistical Society, Series B*, **41**, 54–64.

Fienberg, S. E. (1980a). “Using loglinear models to analyze cross-classified categorical data,” *Mathematical Scientist*, **5**, 13–30.

Fienberg, S. E. (1980b). *The Analysis of Cross-Classified Categorical Data*. 2nd ed. Cambridge: MIT Press.

Fisher, R. A. (1941). *Statistical Methods for Research Workers*, 8th ed. Oliver and Boyd.

Formann, A. K. (2000). "Rater agreement and the generalized Rudas-Clogg-Lindsay index of fit," *Statistics in Medicine*, **19**, 1881–8.

Radelet, M. (1981). "Racial characteristics and the imposition of the death penalty," *Americal Sociological Review*, **46**, 918–27.

Read, T. R. C., Cressie, N. A. C. (1988). *Goodness-of-fit Statistics for Discrete Multivariate Data*. New York: Springer.

Rudas, T. (1984). "Testing goodness-of-fit of loglinear models based on small samples: a Monte Carlo study," *Colloquia Mathematica Societas János Bolyai*, **45**. *Goodness-of-fit*, 467–83.

Rudas, T. (1986). "A Monte Carlo comparison of the small sample behaviour of the Pearson, the likelihood ratio and the Cressie-Read statistics," *Journal of Statistical Computation and Simulation*, **24**, 107–20.

Rudas, T. (1998a). *Odds Ratios in the Analysis of Contingency Tables*. Quantitave Application on the Social Sciences, Vol. 119. Thousand Oaks, CA: Sage.

Rudas, T. (1998b). "Minimum mixture estimation and regression analysis." In B. Marx & H. Friedl (eds.), *Proceedings of the 13th International Workshop on Statistical Modeling*. Louisiana State University, pp. 340–7.

Rudas, T. (1999). "The mixture index of fit and minimax regression," *Metrika*, **50**, 163–72.

Rudas, T., Clogg, C. C., Lindsay, B. G. (1994). "A new index of fit based on mixture methods for the analysis of contingency tables," *Journal of the Royal Statistical Society, Series B*, **56**, 623–39.

Rudas, T., & Zwick, R. (1997). "Estimating the importance of differential item functioning," *Journal of Educational and Behavioral Statistics*, **22**, 31–45.

Schervish, M. J. (1996). "*P* values: What are they and what are they not," *The American Statistician*, **50**, 203–06.

Xi, L. (1996). "The mixture index of fit," Ph.D. Thesis, Department of Statistics, The Pennsylvania State University.

Xi, L., & Lindsay, B. G. (1997). "A note on calculating the π^* index of fit for the analysis of contingency tables," *Sociological Methods and Research*, **25**, 248–59.

THIRTEEN

Mixture Regression Models

Michel Wedel and Wayne S. DeSarbo

1. INTRODUCTION

The development of mixture models dates back to the nineteenth century (Newcomb, 1886). In finite mixture models, it is assumed that the observations of a sample arise from two or more unobserved classes, of unknown proportions, that are mixed. The purpose is to unmix the sample and to identify the underlying classes. Mixture models present a model-based approach to clustering. They allow for hypothesis testing and estimation within the framework of standard statistical theory. The mixture model approach to clustering moreover presents an extremely flexible class of clustering algorithms that can be tailored to a very wide range of substantive problems. Mixture models are statistical models, which involve a specific form of the distribution function of the observations in each of the underlying populations (which is to be specified). The distribution function is used to describe the probabilities of occurrence of the observed values of the variable in question. The normal distribution, for example, is the most frequently used distribution for continuous variables that take values in the range of minus infinity to infinity. The binomial distribution describes the probabilities of occurrence of binary (0/1) variables, and the Poisson distribution the probabilities of occurrence of discrete (count) variables. Certain classes of mixture models based on the latter two distributions have become known in the literature as *latent class models*. Lazarsfeld and Henry (1968) provide one of the first extensive treatments of this topic. Major contributions to the development of latent class models were made by Clogg (see, e.g., Clogg and Goodman, 1984, 1986; Clogg, 1988).

In this chapter, we review developments in a particular area of mixture modeling: mixture regression models. These methods extend the traditional mixture approach in that they simultaneously allow for the classification of a sample into classes, as well as for the estimation of a regression model within each of these classes. A general framework for mixture regression models is provided, based on the work of Wedel and DeSarbo (1995) and Wedel and Kamakura (1999). Within that general framework we review (1) standard mixture models, (2) mixture regression models, and (3) concomitant variable mixture regression models that allow for a simultaneous description of the underlying classes with background variables (comparable with passive variables in a cluster analysis). For each of the latter two models an application is provided. We start by describing the foundations of the standard mixture model approach. This discussion is based on the books by Titterington, Smith, and Makov (1985), McLachlan and Basford (1988), and McLachlan and Peel (2000).

In describing those mixture models, we deal with several types of variables. First we have dependent variable, or K-type variables, which may constitute K repeated measures on the same variable for each subject, or alternatively K different variables for each subject. The observed values of those variables are denoted by y. Then there are independent predictor variables, P-type variables, that may be used to predict the outcomes of the dependent variables. These are denoted by x. Both y and x may be either discrete or continuous. Then there is an unobserved latent variable that can take on a finite number of T values, corresponding to T unobserved classes in the data. This T-type variable is indicated by ξ. Finally, there are L observed variables that are related to the latent variable. These L-type variables are denoted by z.

2. FINITE MIXTURE MODELS

In order to formulate the finite mixture model, assume that a sample of N subjects is drawn. On each subject K variables $y_n = (y_{nk}, n = 1, \ldots, N; k = 1, \ldots, K)$ are measured. These subjects are assumed to arise from a population that is a mixture of T unobserved classes, in (unknown) proportions $\pi_1, \ldots, \pi_T$. It is not known in advance from which class a particular subject arises. The probabilities π_t obey the following constraints:

$$\sum_{t=1}^{T} \pi_t = 1, \quad \pi_t > 0. \tag{1}$$

Given that y_{nk} comes from class t, the distribution function of the vector of measurements y_n is represented by the general form $f_t(y_n|\theta_t)$. Here,

θ_t denotes the vector of all unknown parameters for class t. For example, in the case that the y_{nk} within each class is independent normally distributed, θ_t contains the means, μ_{kt}, and variances, σ_t^2, of the normal distribution within each of the T classes. The basic idea behind mixture distributions is that the unconditional distribution is obtained from the conditional distributions as

$$f(y_n|\phi) = \sum_{t=1}^{T} \pi_t f_t(y_n|\theta_t), \tag{2}$$

where $\phi = (\pi, \theta)$ denotes all parameters of the model. This can easily be derived from the basic principles of probability theory: the unconditional probability is equal to the product of the conditional probability given t, times the probability of t, and this expression summed over all values of t.

The conditional density function, $f_t(y|\theta_t)$, can take many forms, including the previously mentioned normal, Poisson, and binomial distribution functions, as well as other well-known distribution functions such as the negative binomial, exponential, gamma, and inverse Gaussian. All of these more commonly used distributions present specific members of the so-called exponential family of distributions. This family is a general family of distributions that encompasses both discrete and continuous distributions. The exponential family is a very useful class of distributions. The common properties of these distributions enable them to be studied simultaneously, rather than as a collection of unrelated cases. These distributions are characterized by their means, μ_{kt}, and possibly a dispersion parameter λ_t. In mixtures these parameters are typically assumed to be constant over observations within each class t.

Table 1 presents several characteristics of a number of the most well-known univariate and multivariate distributions in the exponential family. The table lists a short notation for each distribution, the form of the distribution as a function of the parameters, and the canonical link function, which will be discussed more extensively in the next section.

Often, the K repeated measurements (or K variables) for each subject are assumed independent. In the latent class literature this assumption is called the *assumption of local independence*. This implies that the joint distribution function for the K observations factors into the product of the marginal distributions:

$$f_t(y_n|\theta_t) = \prod_{k=1}^{K} f_t(y_{nk}|\theta_{tk}). \tag{3}$$

Table 1. Some Distributions in the Exponential Family

Distribution	Notation	Distrib. Function	Link
Binomial	$B(K, \mu)$	$\binom{K}{y}\left(\frac{\mu}{K}\right)^{y}\left(1-\frac{\mu}{K}\right)^{(K-y)}$	$\ln[\mu/(K-\mu)]$
Poisson	$P(\mu)$	$\frac{e^{-\mu}\mu^{y}}{y!}$	$\ln(\mu)$
Negative binomial	$\mathrm{NB}(\mu, v)$	$\left(\frac{v}{v+\mu}\right)^{v}\frac{\Gamma(v+y)}{y!\Gamma(v)}\left(\frac{\mu}{v+\mu}\right)^{y}$	$\ln[\mu/(v+\mu)]$
Multinomial	$M(\mu)$	$\prod_{k=1}^{K}\mu_k^{y_k}$	$\ln(\mu_k/\mu_K)$
Normal	$N(\mu, \sigma)$	$\frac{1}{\sqrt{2\pi}\sigma}\exp\left[\frac{-(y-\mu)^2}{2\sigma^2}\right]$	μ
Multivariate normal	$\mathrm{MVN}(\mu, \sum)$	$\frac{1}{(2\pi)^{K/2}\lvert\sum\rvert^{1/2}}$	
		$\times\exp\left[-\frac{1}{2}(y_n-\mu)'\sum^{-1}(y_n-\mu)'\right]$	μ
Exponential	$E(\mu)$	$\frac{1}{\mu}\exp\left(-\frac{y}{\mu}\right)$	$1/\mu$
Gamma	$G(\mu, v)$	$\frac{1}{\Gamma(v)}\left(\frac{yv}{\mu}\right)^{v-1}\exp\left(-\frac{vy}{\mu}\right)$	$1/\mu$
Dirichlet	$D(\mu)$	$\frac{\Gamma\left(\sum_{k=1}^{K}\mu_k\right)\prod_{k=1}^{K}y_{\kappa}^{\mu_k-1}}{\prod_{k=1}^{K}\Gamma(\mu_k)}$	$\ln(\mu_k/\mu_K)$

If, given the knowledge of the classes or mixture components, the observations cannot be assumed independent, then one of the members of the multivariate exponential family may be appropriate. The two most important and most frequently used distributions in this family are the multinomial distribution and the multivariate normal distribution (see Table 1). In the latter case, the distribution of y takes the well-known form shown in Table 1, with μ_t as the $(K \times 1)$ vector of expectations, and $\sum_t$ as the $(K \times K)$ covariance matrix of the vector y, given class t.

The parameter vector ϕ is usually estimated by using the method of maximum likelihood, as described in the introduction of this book. Often, the expectation–maximization (EM) algorithm is applied for that purpose (cf. Wedel and Kamakura, 1999). Once estimates of ϕ are obtained, estimates of the posterior probability, p_{nt}, that subject n comes from class t can be calculated by using these parameter estimates and Bayes' Theorem:

$$p_{nt} = \pi_t f_t(y_n|\theta_t) \Bigg/ \sum_{t=1}^{T} \pi_t f_t(y_n|\theta_t). \tag{4}$$

A. Identification

A potential problem associated with mixture models is that of identification. Identification of a model refers to the situation in which thereis only one set of parameter values that uniquely maximizes the likelihood. When there exists more than one set that provide a maximum, the model is not identified. The interpretation of parameters from nonidentified models is thus useless because an infinite set of parameters yields the same solution. Throughout the exposition shown earlier, it was assumed that ϕ is identifiable. Titterington et al. (1985) show that, in general, mixtures involving members of the exponential family, including the univariate binomial, normal, Poisson, exponential, and gamma distributions, are identified.

3. MIXTURE REGRESSION MODELS

In the previous sections the general mixture models have been described. *Unconditional* refers to the situation in which there are no exogenous or explanatory variables to predict a dependent variable, y_n. For example, in the unconditional mixtures of normal distributions, the mean and variance of the underlying classes are estimated; in "conditional" mixture models, the class means are constrained in the form of regression models. These regression models relate a dependent variable to a set of independent variables. Whereas the majority of applications of mixture regression models has been in marketing and business research, a potential for substantive applications exist in all the social sciences. In this section, we describe a general framework for mixture regression models. The material in this section is based on the work of Wedel and DeSarbo (1994, 1995) and Wedel and Kamakura (1999).

The number of applications of generalized linear models (which include as special cases linear regression, logit and probit models, loglinear, and multinomial models) in the social sciences is enormous. Generalized linear models (Nelder and Wedderburn, 1972) are regression models in which the dependent variable is specified to be distributed according to one of the members of the exponential family (see Table 1). Generalized linear models deal with continuous variables, which can be specified to follow a normal, gamma, or exponential distribution; for discrete variables, the binomial, multinomial, Poisson, or negative binomial distributions can be utilized. The expectation of the dependent variable is modeled as a function of a set of explanatory variables as in standard multiple regression models (which are a special case of generalized linear models). However, the estimation of a single aggregate regression equation across all subjects in a sample may be inadequate if the subjects arise from a number of unknown classes in which the regression coefficients differ.

Where the behavior of different subjects is studied, it is not very difficult to come up with potential reasons for the existence of such heterogeneous classes. It is, therefore, no surprise that the application of the mixture regression approach has proved to be of great use. An alternative motivation for mixture regression models comes from random coefficient specifications. In random coefficient models, the coefficients of a generalized linear model are assumed to follow some distribution across the population to account for heterogeneity. Often, the normal distribution is assumed, but a discrete distribution can be assumed instead. In that case, a finite number of support points with accompanying probability mass are used to approximate the continuous heterogeneity distribution of the coefficients. This is the finite mixture formulation, which is often more convenient than the continuous heterogeneity approximation, because of the interpretation of the coefficients for each class, and because of ease of estimation, where the evaluation of high-dimensional integration required in the continuous case is alleviated.

In describing the mixture regression framework, we extend the unconditional mixture approach described in the previous sections. As earlier, assume that the vector of observations (on the dependent variable) of subject n, y_n, arises from a population that is a mixture of T unknown classes in proportions. The distribution of y_n, given that y_n comes from class t, $f_t(y_n|\theta_t)$, is assumed to be one of the distributions in the exponential family, or the multivariate exponential family (see Table 1). The exponential family is a very useful family of distributions in this context that enables the formulation of a general framework for the class of mixture regression models. In addition to the dependent variables, a set of P nonstochastic explanatory variables $X_1, \ldots, X_P[X_p = (X_{nkp});\ p = 1, \ldots, P]$ is specified.

The development of the class of mixture regression models is very similar to that of the mixture models described earlier. A major difference, however, is that the means of the observations in each class are to be predicted from a set of explanatory variables. To this end, the mean of the distribution, μ_t, is written as

$$\eta_{nt} = g(\mu_{nt}), \tag{5}$$

where $g(\cdot)$ is some function called a link function, and η_{nkt} is the called linear predictor. Convenient link functions, called canonical links, are respectively the identity, log, logit, inverse, and squared inverse functions for the normal, Poisson, binomial, gamma, and inverse Gaussian distributions (see Table 1). The linear predictor in class t is a linear combination of the P explanatory variables:

$$\eta_{nkt} = \sum_{p=1}^{P} X_{nkp}\,\beta_{tp}, \tag{6}$$

where $\beta_t = (\beta_{tp})$ is a set of regression parameters to be estimated for each class.

A. Identification

The same remarks on identification made in the previous sections apply to mixture regression models. However, for the mixture regression model, an additional identification problem presents itself concerning the conditioning of the X matrix and the size of P. Collinearity between the predictors within classes may lead to instable estimates of the regression coefficients and large standard errors. In mixture regression models, this situation is compounded by the fact that there are fewer observations for estimating the regression model in each class than at the aggregate level. Therefore the condition of the X variables is an important issue in applications mixture regression models. Comments made in Chapter 12 on local optima and possible remedies apply here as well, as do the comments concerning convergence of the algorithm. Problems of model identification can be examined empirically by investigating the Hessian matrix of second derivatives of the likelihood with respect to the parameters. For a well-identified model, all the eigenvalues of the Hessian should be greater than zero.

B. Application: Trade Show Performance

DeSarbo and Cron (1988) first proposed a mixture regression model that enables the estimation of separate regression functions (and corresponding object memberships) in a number of classes using maximum likelihood. The model is a finite mixture of univariate normal densities. The expectations of these densities are specified as linear functions of a set of explanatory variables. The model was used to analyze the factors that influence perceptions of trade show performance and to investigate the presence of classes that differ in the importance attributes to these factors in evaluating trade show performance.

In their study, 129 marketing executives were asked to rate their firms' trade show performance on eight performance factors, as well as on overall trade show performance. The performance factors included the following: 1, identifying new prospects; 2, servicing current customers; 3, introducing new products; 4, selling at the trade show; 5, enhancing corporate image; 6, testing of new products; 7, enhancing corporate moral; and 8, gathering competitive information.

An aggregate level regression analysis of overall performance on the eight performance factors, the results of which are depicted in Table 2, revealed that identifying new prospects and new product testing were significantly related to trade show performance. These results are derived by a standard OLS regression of the overall performance ratings on the ratings of the eight factors. However, a mixture regression model revealed two classes (on the basis of the Akaike information criterion), composed of 59 and 70 marketing executives, respectively. The effects of the performance factors in the two classes are markedly different from those at the aggregate level, as shown in Table 2.

Managers in Class 1 primarily evaluate trade shows in terms of nonselling factors, including servicing current customers, and enhancing corporate image and moral. Managers in Class 2 evaluate trade shows

Table 2. Aggregate and Segment Level Results of the Trade Show Performance Study

Parameter	Aggregate	Class 1	Class 2
Intercept	3.03[a]	4.093[a]	2.218[a]
1. New prospects	0.15[a]	0.126	0.242[a]
2. New customers	−0.02	0.287[a]	−0.164[a]
3. Product introduction	0.09	−0.157[a]	0.204[a]
4. Selling	−0.04	−0.133[a]	0.074[a]
5. Enhancing image	0.09	0.128[a]	0.072
6. New product testing	0.18[a]	0.107	0.282[a]
7. Enhancing morale	0.07	0.155[a]	−0.026
8. Competitive information	0.04	−0.124	0.023
Size (%)	1	0.489	0.511

[a] $p < .05$.

primarily on selling factors, including identifying new prospects, introducing new products, selling at the shows, and new product testing. Neither of the two classes considers gathering competitive information important.

Whereas the percentage of variance explained by the aggregate regression was 37%, the percentages of explained variance in overall trade show performance in Classes 1 and 2 were, respectively, 73% and 76%. Table 3 presents an overview of a number of mixture regression applications. For

Table 3. Mixture Regression Applications

Authors	Year	Application	*D*
Quandt	1972	Switching regression for wage prediction	N
Quandt and Ramsey	1978	Switching regression of house construction	N
DeSarbo and Cron	1988	Trade show performance	N
Ramaswamy et al.	1993	Marketing mix effects	MVN
Helsen et al.	1993	Country segmentation	MVN
DeSarbo et al.	1992	Conjoint analysis, consumer products	MVN
Wedel and DeSarbo	1994	Conjoint analysis, service quality	MVN
Wedel and DeSarbo	1995	Customer satisfication	MVN
Jedidi et al.	1996	Structural equation model of price–quality relation	MVN
Kamakura and Russell	1989	Brand choice analysis	M
Kamakura and Mazzon	1991	Value segmentation	M
Bucklin and Gupta	1992	Purchase incidence and brand choice	M
Russell and Kamakura	1994	Linking microlevel and macrolevel data	M
Kamakura et al.	1994	Conjoint analysis & segment description	M
DeSoete and DeSarbo	1990	Pick-any choices	B
Wedel and DeSarbo	1993	Paired comparison choices	B
Dillon and Kumar	1994	Paired comparison choices & segment description	B
Wedel et al.	1993	Direct mail	P
Wedel and DeSarbo	1995	Coupon usage	P
Wedel et al.	1995	Hazard model of brand switching	P
Ramaswamy et al.	1994	Purchase frequency analysis	NB
DeSarbo et al.	1995	Multiple criteria decision-making	D
Böckenholt	1993	Brand choice	DG, MG, DP, MP

Note: The mixture distributions used are as follows: N, normal; MVN, multivariate normal; M, multinomial; B, binomial; P, Poisson; TP, truncated Poisson; NB, negative binomial; D, Dirichlet; DM, Dirichlet multinomial; DG, Dirichet gamma; MG, multinomial gamma; DP, Dirichlet–Poisson; MP, multnomial Poisson.

each application, the substantive area of application is provided, as well as the distribution used to describe the dependent variable.

4. CONCOMITANT VARIABLE MIXTURE REGRESSION MODELS

In mixture and mixture regression models, the classes of subjects identified are often described by background variables of the subjects (i.e., demographics; in this context these variables are often called *concomitant variables*) to obtain insights into the composition of the classes. Such profiling of classes is typically performed on the basis of the posterior class membership probabilities [see Equation (4)]. These posterior memberships, p_{nt}, provide the probability that a particular subject belongs to a certain class. The classes are frequently profiled in a second step of the analyses: a logit transform of the posterior membership probabilities, $\log[p_{nt}/(1 - p_{nt})]$, is regressed on designated exogenous variables. The coefficient of a specific concomitant variable for a certain class represents the effects of the variable in question on the relative probabilities of belonging to that class. However, the two-step procedure has several disadvantages. First, the logit regression is performed independently from the estimation of the mixture model, and it optimizes a different criterion – the sum of squared errors in the posterior probabilities rather than the likelihood. Thereby, the classes derived in the first stage do not possess an optimal structure with respect to their profile on the concomitant variables. Second, this procedure does not take into account the estimation error of the posterior probabilities. Therefore, several authors have proposed models that simultaneously profile the derived classes with descriptor variables (see Wedel and DeSarbo, 1994). These models are based on the earlier concomitant variable latent class model proposed by Dayton and MacReady (1988).

In order to develop the concomitant variable mixture regression model, let $l = 1, \ldots, L$ index concomitant variables, and z_{nl} be the value of the lth concomitant variable for subject n, $Z = [(z_{nl})]$. It is again assumed that the y_n is distributed in the exponential family, conditional upon unobserved classes, as earlier. The starting point for the development of the model is the general mixture regression model defined earlier. The unconditional distribution for the concomitant variable mixture is formulated as

$$f(y_n|\phi) = \sum_{t=1}^{T} \pi_{t|Z} f_t(y_n|\beta_t, \lambda_t). \tag{7}$$

Note that Equation (7) is similar to Equation (2), but that here the prior probabilities, π_t, have been replaced by $\pi_{t|Z}$. This is the core of the concomitant variable model approach: the prior probabilities of class membership are explicitly parameterized as functions of the concomitant variables. For this purpose, the logistic formulation is mostly used:

$$\pi_{t|Z} = \exp\left(\sum_{l=1}^{L} \gamma_{lt} z_{nl}\right) \bigg/ \sum_{t=1}^{T} \exp\left(\sum_{l=1}^{L} \gamma_{lt} z_{nl}\right). \tag{8}$$

Equation (8) is called the submodel of the concomitant variable mixture and relates the prior probabilities to the concomitant variables. In order to include an intercept for each class, $z_n^1 = 1$ is specified for all t. The parameter γ_{lt} denotes the impact of the lth consumer characteristic on the prior probability for class t. For example, a positive value of γ_{lt} implies that a higher value of variable l increases the prior probability that consumer n belongs to class t. Another way of interpreting the concomitant variable model is based on the fact that the prior probabilities are equivalent to the class sizes. Thus, this model allows the sizes of the classes to vary across demographic (and other) variables. The concomitant variable model designates what the class sizes are according to each of an a priori specified number of variables. Note that the mixture regression model of the previous section arises as a special case if the matrix Z consists of one column of ones so that for each class, only a constant, γ_t, is specified in the submodel.

The posterior probabilities are again calculated by updating the priors according to the Bayes rule:

$$p_{nt} = \pi_{t|Z} f_t(y_n|\beta_t, \lambda_t) \bigg/ \sum_{t=1}^{T} \pi_{t|Z} f_t(y_n|\beta_t, \lambda_t) \tag{9}$$

A. Identification

For the concomitant variable mixture regression models, the same identification constraints hold as for the corresponding mixture regression models without concomitant variables. In addition, in the submodel, the coefficients γ_{lT} for the last class T are usually set to zero for identification. In the concomitant variable mixture model an additional identification problem presents itself concerning the conditioning of the Z matrix and the size of L.

B. Application: Conjoint Study on Banking

We provide the results of the application of a concomitant variable mixture model by Kamakura et al. (1994). The study pertained to consumers' preferences for bank services. A large U.S. bank wanted information on consumers' concern with four attributes and was interested in the existence of classes of consumers with different importance weights for these attributes. These four attributes (with their levels in parentheses) were as follows: the minimum balance required to exempt the customer from a monthly service fee (MINBAL: \$0, \$500, \$1,000); the amount charged per check issued by the customer (CHECK: 0¢, 15¢, 35¢); the monthly service fee charged if the account balance falls below the minimum (FEE: \$0, \$3, \$6); the availability and cost of automatic teller machines in a network of supermarkets (ATM: not available, free, 75¢ per transaction). Conjoint analysis was used to investigate the importance weights of these attributes. In conjoint analysis, profiles of hypothetical products are constructed from the attributes and corresponding levels of the attributes, typically using fractional factorial designs.

In this study, two equivalent but distinct sets of nine profiles were created from the attributes. These profiles were presented to a random sample of 269 of the bank's customers in the form of "peeling stickers" in a mail survey. The respondents were instructed to peel off their first choice and stick it to a designated place, then their second choice and so on, until a full ranking was obtained for the profiles in each of the two sets. The purpose of conjoint analysis is to derive the importance weights of the attribute levels from the overall rankings of the profiles by using regression techniques. For that purpose, the effects of the first three attributes were assumed to be linear and modeled by using a single dummy variable, whereas the effects of the availability and cost of automatic teller machines were modeled by using two dummy variables. The regression model used to estimate the attribute-level importance weights was a multinomial logit model that describes the rank orders as a series of successive choices of highest preference among the set of remaining profiles.

In addition to the conjoint data, the following information for each respondent was obtained from the bank: average balance kept in the account during the previous 6 months, earning 5.5% interest (BALANCE); number of checks issued per month in the previous 6 months (NCHECK); and the number of ATM transactions per month (NATM). These variables represent the actual past banking behavior of each respondent, and

Table 4. Parameter Estimates of the Rank-Order Concomitant Variable Model

Parameter	Class 1	Class 2	Class 3	Class 4
Attributes				
MINBAL	−0.320	−1.978[a]	−2.633[a]	−10.117[a]
CHECK	−7.856[a]	−3.984[a]	−2.479[a]	−4.167[a]
FEE	−0.336[a]	−0.172[a]	−0.220[a]	−0.274[a]
ATM	0.515[a]	2.414[a]	0.818[a]	0.681[a]
ATM 75	−0.025	0.736[a]	−0.707[a]	−0.357[a]
Concomitant variables				
BALANCE	0.048[a]	0.017	0.028[a]	−0.093[a]
NCHECK	0.023[a]	−0.03	−0.001	0.008
NATM	−0.049[a]	0.075[a]	−0.001	−0.025
Size	0.211	0.209	0.267	0.314

[a] $p < .05$.

constituted the concomitant variables included in the model to explain class membership.

The rank orders of the nine stimuli were analyzed with the multinomial concomitant variable mixture regression model, where the rank orders for each subject were expanded into choices, using the rank-explosion rule. The CAIC statistic indicated $S = 4$ classes to be appropriate. Table 4 presents the estimated coefficients for the four-class solution (the estimated coefficients of the model with and without concomitant variables are quite similar). A likelihood ratio test with a chi-square value of 51.2 with 12 degrees of freedom showed that the contribution of the concomitant variables is statistically significant.

Table 4 shows that customers in Class 1 (21.1% of the sample) are concerned mainly with the amount charged per check and the monthly service fee charged. The parameters of the submodel show that customers with high average balances are more likely to belong to this class than to others, which is consistent with the low concern for minimum balance of this class. Furthermore, these customers issue a relatively large number of checks per month, which explains their sensitivity for the amount charged per check. The relatively low usage of ATM machines is consistent with their low concern of the availability of these machines. In Class 2 (20.9% of the sample) consumers are most sensitive to the availability of automatic teller machines in supermarkets. This is consistent with the effects of the concomitant variables in the submodel: these show that that subjects in Class 2 are relatively heavy users of these teller machines. The remain-

Table 5. Applications of Concomitant Variable Mixture Regression Models

Authors	Year	Application	*D*
Dayton and McReady	1988	Analysis of test items in testing theory	B
Dillon et al.	1993	Analysis of paired comparisons of food products	B
Dillon and Kumar	1994	Analysis of paired comparisons	B
Formann	1992	Analysis social mobility tables	M
Kamakura et al.	1994	Conjoint segmentation for bank services	M
Gupta and Chintagunta	1994	Consumer durable brand choice segmentation, using scanner data	M
Peng et al.	1996	Speech recognition	M

Note: The mixture distributions used are as follows: N, normal; M, multinomial; B, binomial; P, Poisson.

ing attributes, although significant, appear to have less impact as compared with the other classes. Customers in Class 3 (26.7% of the sample) are somewhat sensitive to minimum balance and ATM availability. Customers with high average balances have a higher prior probability to belong to this class than to Classes 2 or 4. In Class 4 (31.4% of the sample), the coefficient for the minimum balance required is nearly four times as large as that in Class 3. The coefficients in the submodel explain this result: Consumers with low average balances have the highest probability of belonging to this class, and thus they are predominantly concerned with the minimum balance required to prevent a monthly service fee. The authors showed that the predictive validity of the concomitant variable mixture regression model on the holdout set ($R^2 = 0.265$) was somewhat better than that of the mixture regression model without concomitant variables ($R^2 = 0.260$).

Table 5 lists some other applications of concomitant variable mixture and mixture regression models and presents both the nature of the application and the distribution function used.

5. CONCLUSION

We have attempted to provide an integrative overview of mixture regression models. These models hold great promise in accounting for subject heterogeneity. It is therefore not surprising that a large volume of literature has accumulated on this topic, and it is expected that these models will attract further academic interest in the future. In future research, a number of problems must be resolved. Mixture models assume that subjects within classes are homogeneous. This assumption may be too restrictive and may be the cause of the fact that in certain situations mixture models show relatively poor predictive validity as compared with models estimated at the individual level (Vriens et al., 1996). Promising attempts to include within-class heterogeneity in mixture models include those by Böckenholt (1993) and by Ramaswamy et al. (1993), who used compound distributions such as the negative binomial and the Dirichlet multinomial to describe within-class behavior of subjects. In addition, Bayesian estimation procedures show promise in that they allow one to specify a prior distribution of the parameters and to evaluate the posterior predictive distribution (e.g., Peng et al., 1996). An additional topic for future research, and one that has received the least satisfactory statistical treatment in mixture models, is the selection of the appropriate number of underlying classes (Titterington, 1990). This issue of the selection of the appropriate number of classes may be related to the lack of predictive validity and the negligence of accounting for within-class heterogeneity, as was demonstrated by Böckenholt (1993).

REFERENCES

Böckenholt, U. (1993). "A latent class regression approach for the analysis of recurrent choice data," *British Journal of Mathematical and Statistical Psychology*, **46**, 95–118.

Bucklin, R. E., & Gupta, S. (1992). "Brand choice, purchase incidence and segmentation: an integrated approach," *Journal of Marketing Research*, **29**, 201–16.

Clogg, C. C., & Goodman, L. A. (1984). "Latent structure analysis of a set of multidimensional contingency tables," *Journal of the American Statistical Association*, **79**, 762–71.

Clogg, C. C., & Goodman, L. A. (1986). "On scaling models applied in several groups," *Psychometrika*, **51**, 123–35.

Clogg, C. C. (1988). "Latent class models for measuring." In R. Langeheine & J. Rost (eds.), *Latent Trait and Latent Class Models.* New York: Plenum, pp. 173–206.

Dayton, M. C., & MacReady, G. B. (1988). "Concomitant variable latent class models," *Journal of the American Statistical Association*, **83** (401), 173–9.

DeSarbo, W. S., & Cron, W. L. (1988). "A maximum likelihood methodology for clusterwise linear regression," *Journal of Classification*, **5**, 249–82.

DeSarbo, W. S., Ramaswamy, V., & Chatterjee, R. (1995). "Analyzing constant-sum multiple criterion data: a segment level approach," *Journal of Marketing Research*, **32**, 222–32.

DeSarbo, W. S., Wedel, M., Vriens, M., & Ramaswamy, V. (1992). "Latent class metric conjoint analysis," *Marketing Letters*, **3**, 273–88.

DeSoete, G., & DeSarbo, W. S. (1990). "A latent class probit model for analyzing pick any *N*/data," *Journal of Classification*, **8**, 45–63.

Dillon, W. R., & Kumar, A. (1994). "Latent structure and other mixture models in marketing: an integrative survey and overview." In Richard P. Bagozzi (ed.), *Advanced Methods for Marketing Research*. Cambridge: Blackwell, pp. 295–351.

Dillon, W. R., Kumar, A., & Smith de Borrero, X. (1993). "Capturing individual differences in paired comparisons: an extended BTL model incorporating descriptor variables," *Journal of Marketing Research*, **30**, 42–51.

Formann, A. K. (1992). "Linear logistic latent class analysis for polytomous data," *Journal of the American Statistical Association*, **87**, 476–86.

Gupta, S., & Chintagunta, P. K. (1994). "On using demographic variables to determine segment membership in logit mixture models," *Journal of Marketing Research*, **31**, 128–36.

Helsen, K., Jedidi, K., & DeSarbo, W. S. (1993). "A new approach to country segmentation utilizing multinational diffusion patterns," *Journal of Marketing*, **57**, 60–71.

Jedidi, K., Ramaswamy, V., DeSarbo, W. S., & Wedel, M. (1996). "The disaggregate estimation of simultaneous equation models: an application to the price-quality relationship," *Journal of Structural Equation Modelling*, **3**, 266–89.

Kamakura, W. A., & Mazzon, J. A. (1991). "Values segmentation: a model for the measurement of values and value systems," *Journal of Consumer Research*, **18**, 208–19.

Kamakura, W. A., & Russell, G. J. (1989). "A probabilistic choice model for market segmentation and elasticity structure," *Journal of Marketing Research*, **26**, 379–90.

Kamakura, W. A., Wedel, M., & Agrawal, J. (1994). "Concomitant variable latent class models for conjoint analysis," *International Journal for Research in Marketing*, **11**, 451–64.

Lazarsfeld, P. F., & Henry, N. W. (1968). *Latent Structure Analysis.* New York: Houghton Mifflin.

McLachlan, G. J., & Basford, K. E. (1988). *Mixture Models: Inference and Applications to Clustering*. New York: Marcel Dekker.

McLachlan, G. J., & Peel, G. (2000). *Finite Mixtures.* New York: Wiley.

McCullagh, P., & Nelder, J. A. (1989). *Generalized Linear Models*. New York: Chapman & Hall.

Nelder, J. A., & Wedderburn, R. W. M. (1972). "Generalized linear models," *Journal of the Royal Statistical Society A*, **135**, 370–84.

Newcomb, S. (1986). "A Generalized theory of the combination of observations so as to obtain the best result," *American Journal of Mathematics*, **8**, 343–66.

Peng, F., Jacobs, R. A., & Tanner, M. A. (1996). "Bayesian inference in mixtures of experts and hierarchical mixtures of experts models with an application to speech recognition," *Journal of the American Statistical Association*, **91**, 953–62.

Quandt, R. E. (1972). "A new approach to estimating switching regressions," *Journal of the American Statistical Association*, **67**, 306–10.

Quandt, R. E., & Ramsey, J. B. (1978). "Estimating mixtures of normal distributions and switching regressions," *Journal of the American Statistical Association*, **73**, 730–8.

Ramaswamy, V., DeSarbo, W. S., Reibstein, D. J., & Robinson, W. T. (1993). "An empirical pooling approach for estimating marketing mix elasticities with PIMS data," *Marketing Science*, **12**, 103–24.

Ramaswamy, V., Anderson, E. W., & DeSarbo, W. S. (1994). "A disaggregate negative binomial regression procedure for count data analysis," *Management Science*, **40**, 405–17.

Russell, G. J., & Kamakura, W. A. (1994). "Understanding brand competition using micro and macro scanner data," *Journal of Marketing Research*, **31**, 289–303.

Titterington, D. M. (1990). "Some recent research in the analysis of mixture distributions," *Statistics*, **4**, 619–41.

Titterington, D. M., Smith, A. F. M., & Makov, U. E. (1985). *Statistical Analysis of Finite Mixture Distributions.* New York: Wiley.

Vriens, M., Wedel, M., & Wilms, T. J. (1996). "Segmentation for metric conjoint analysis: a Monte Carlo comparison," *Journal of Marketing Research*, **33**, 73–85.

Wedel, M., & DeSarbo, W. S. (1994). "A review of latent class regression models and their applications." In Richard P. Bagozzi (ed.), *Advanced Methods for Marketing Research.* Cambridge: Blackwell, pp. 353–88.

Wedel, M., & DeSarbo, W. S. (1995). "A mixture likelihood approach for generalized linear models," *Journal of Classification*, **12**, 1–35.

Wedel, M., & DeSarbo, W. S. (1993). "A latent class binomial logit methodology for the analysis of paired comparison choice data: an application reinvestigating the determinants of perceived risk," *Decision Sciences*, **24**, 1157–70.

Wedel, M., DeSarbo, W. S., Bult, J. R., & Ramaswamy, V. (1993). "A latent class Poisson regression model for heterogeneous count data," *Journal of Applied Econometrics*, **8**, 397–411.

Wedel, M., Kamakura, W. A., DeSarbo, W. S., & ter Hofstede, F. (1995). "Implications for asymmetry, nonproportionality and heterogeneity in brand switching from piece-wise exponential mixture hazard models," *Journal of Marketing Research*, **32**, 457–62.

Wedel, M., & Kamakura, W. A. (1999). *Market Segmentation, Methodological and Conceptual Foundations*, 2nd ed. Dordrecht: Kluwer.

FOURTEEN

A General Latent Class Approach to Unobserved Heterogeneity in the Analysis of Event History Data

Jeroen K. Vermunt

1. INTRODUCTION

In the context of the analysis of survival and event history data, the problem of unobserved heterogeneity, or the bias caused by not being able to include particular important explanatory variables in the regression model, has received a great deal of attention (see, e.g., Vaupel et al., 1979; Heckman and Singer, 1982, 1984; Chamberlain, 1985; Trussell and Richards, 1985; Yamaguchi, 1986; Mare, 1994; Guo and Rodriguez, 1994). The reason for this is that this phenomenon has a much larger impact in hazard models than in other types of regression models: Unobserved heterogeneity may introduce, among other things, downward bias in the time effects, spurious effects of time-varying covariates, spurious time-covariate interaction effects, and dependence among competing risks and repeatable events. This may be true even if the unobserved heterogeneity is uncorrelated with the values of the observed covariates at the start of the process under study.

The models that have been proposed to correct for unobserved heterogeneity differ mainly with respect to the assumptions made about the distribution of the latent variable capturing the unobserved heterogeneity. Heckman and Singer (1982, 1984) proposed a nonparametric random-effects approach that is strongly related to latent class analysis (LCA). Vermunt (1996a, 1997a) proposed extending their latent class (LC) approach by specifying simultaneously with the event history model a system of logit models (or causal loglinear model) for the covariates. By an explicit modeling of the relationships among the observed and unobserved covariates, it becomes possible to relax and test some of the assumptions that are generally made about the nature of the unobserved heterogeneity. Models can be specified in which the unobserved heterogeneity is related to observed (time-varying) covariates, in which the unobserved heterogeneity itself is time varying, and in which there are several mutually related latent covariates.

The next section explains event history analysis by means of hazard models, paying special attention to situations in which unobserved heterogeneity may distort the results. Subsequently, methods for dealing with unobserved heterogeneity are presented, including the standard and the extended LC approach. The presented LC methodology is illustrated by two empirical examples.

2. EVENT HISTORY ANALYSIS

The purpose of event history analysis is to explain differences in the time at which individuals experience the events under study. The best way to define an event is as a transition from a particular origin state to a particular destination state. The time variables indicating the duration of nonoccurrence of the events of interest may either be discrete or continuous variables. Although, because of space limitations, here we deal solely with continuous-time methods, the results can easily be generalized to discrete-time situations (Vermunt, 1997). Textbooks on event history and survival analysis are those by Kalbfleisch and Prentice (1980), Allison (1984), Tuma and Hannan (1984), Lancaster (1990), Yamaguchi (1991), Blossfeld and Rohwer (1995), and Vermunt (1997a).

Suppose that we are interested in explaining individual differences in women's timing of their first birth. In that case, the event is having a first child, which can be defined as the transition from the origin state of no children to the destination state of one child. This an example of what is called a *single nonrepeatable event*. The term *single* refers to the fact that the origin state of no children can only be left by one type of transition. The term *nonrepeatable* indicates that the event can occur only once. Later, situations in which there are several types of events (multiple risks) and in which events may occur more than once (repeatable events) are presented. In the first birth example, it seems most appropriate to assume the time variable to be a continuous variable, although it is, of course, measured discrete, for instance, in days, months, or years after a woman's 15th birthday.

A. Basic Concepts

Suppose T is a continuous random variable indicating the duration of nonoccurrence of the first birth. Let $f(t)$ be the probability density function of T, and $F(t)$ the distribution function of T. As always, the following relationships exist between these two quantities:

$$f(t) = \lim_{\Delta t \to 0} \frac{P(t \leq T < t + \Delta t)}{\Delta t} = \frac{\partial F(t)}{\partial t},$$

$$F(t) = P(T \leq t) = \int_0^t f(u)d(u).$$

The survival probability or survival function, indicating the probability of nonoccurrence of an event until time t, is defined as

$$S(t) = 1 - F(t) = P(T \geq t) = \int_t^{\infty} f(u)d(u).$$

Another important concept is the hazard rate or hazard function, $h(t)$, expressing the instantaneous risk of experiencing an event at $T = t$, given that the event did not occur before t. The hazard rate is defined as

$$h(t) = \lim_{\Delta t \to 0} \frac{P(t \leq T < t + \Delta t | T \geq t)}{\Delta t} = \frac{f(t)}{S(t)},$$

in which $P(t \leq T < t + \Delta t | T \geq t)$ indicates the probability that the event will occur during $[t \leq T < t + \Delta t]$, given that the event did not occur before t. The hazard rate is equal to the unconditional instantaneous probability of having an event at $T = t$, $f(t)$, divided by the probability of not having an event before $T = t$, $S(t)$. It should be noted that the hazard rate itself cannot be interpreted as a conditional probability. Although its value is always nonnegative, it can take values greater than one. However, for small Δt, the quantity $h(t)\Delta t$ can be interpreted as the approximate conditional probability that the event will occur between t and $t + \Delta t$.

Earlier, $h(t)$ was defined as a function of $f(t)$ and $S(t)$. It is also possible to express $S(t)$ and $f(t)$ completely in terms of $h(t)$, that is,

$$S(t) = \exp\left[-\int_0^t h(u)d(u)\right],$$

$$f(t) = h(t)S(t) = h(t)\exp\left[-\int_0^t h(u)d(u)\right].$$

This shows that the functions $f(t)$, $F(t)$, $S(t)$, and $h(t)$ give mathematically equivalent specifications of the distribution of T.

B. Loglinear Models for the Hazard Rate

When one is working within a continuous-time framework, the most appropriate method for regressing the time variable T on a set of covariates is through the hazard rate. This makes it straightforward to assess the effects of time-varying covariates – including the time dependence itself and time-covariate interactions – and to deal with censored observations (Yamaguchi, 1991, pp. 10–11; Vermunt, 1997a, pp. 91–92). Censoring is a form of missing data that is explained in more detail later.

Let $h(t|\mathbf{x}_i)$ be the hazard rate at $T = t$ for an individual with covariate vector $\mathbf{x}_i$. Because the hazard rate can take on values between 0 and infinity, most hazard models are based on a log transformation of the hazard rate, which yields a regression model of the form

$$\log h(t|\mathbf{x}_i) = \log h(t) + \sum_j \beta_j x_{ij}. \tag{1}$$

This hazard model is not only loglinear but also proportional. In proportional hazard models, the time dependence is multiplicative and independent of an individual's covariate values (Lancaster, 1990, pp. 42–43). Later, it will be shown how to specify nonproportional loglinear hazard models by including time-covariate interactions.

The various types of loglinear continuous-time hazard models are defined by the functional form that is chosen for the time dependence, that is, for the term $\log h(t)$. In Cox's model, the time dependence is treated in a nonparametric way (Cox, 1972). Exponential models assume the hazard rate to be constant over time, whereas piecewise exponential models assume the hazard rate to be a step function of T, that is, constant within time periods. Other examples of parametric loglinear hazard models are Weibull, Gompertz, and polynomial models (Blossfeld and Rohwer, 1995).

As was demonstrated by several authors (Holford, 1980; Laird and Oliver, 1981; Clogg and Eliason, 1987; Vermunt, 1997a, pp. 106–117), loglinear hazard models can also be defined as loglinear Poisson models, which are also known as *log-rate models*. Assume that we have – besides the event history information – two categorical covariates denoted by A and B. In addition, assume that the time axis is divided into a limited number of time intervals in which the hazard rate is postulated to be constant. In the first birth example, this could be 1-year intervals. The discretized time variable is denote by Z. Let h_{abz} denote the constant hazard rate in the zth time interval for an individual with $A = a$ and $B = b$. To see the similarity with standard loglinear models, note that the log hazard rate can also be written as

$$\log h_{abz} = \log(m_{abz}/E_{abz}),$$

where m_{abz} denotes the expected number of events and E_{abz} denotes the total exposure time in cell (a, b, z). Using the notation of hierarchical log-linear models, we can write the saturated loglinear model for the hazard rate h_{abz} as

$$\log h_{abz} = u + u_a^A + u_b^B + u_z^Z + u_{ab}^{AB} + u_{az}^{AZ} + u_{bz}^{BZ} + u_{abz}^{ABZ}, \tag{2}$$

in which the u terms are loglinear parameters that are constrained in the usual way, for instance, by means of ANOVA-like restrictions. It should be noted that this is a nonproportional model as a result of the presence of time-covariate interactions.

As the standard loglinear model, the log-rate model can be formulated in a more general way by using a design matrix, that is,

$$\log h_{abz} = \sum_j \beta_j x_{abzj}. \tag{3}$$

These log-rate models can be estimated by using standard programs for loglinear analysis that use E_{abz} as a weight vector (Vermunt, 1997a, p. 112).

Restricted variants of the (saturated) model described in Equation (**??**) can be obtained by omitting some of the higher-order interaction terms. For example,

$$\log h_{abz} = u + u_a^A + u_b^B + u_z^Z$$

yields a model that is similar to the proportional loglinear hazard model described in Equation (B). Different types of hazard models can be obtained by the specification of the time dependence. Setting the u_z^Z terms equal to zero yields an exponential model. Unrestricted u_z^Z parameters yield a piecewise exponential model. Other parametric models can be approximated by defining the u_z^Z terms to be some function of Z (or T; Yamaguchi, 1991, pp. 75–77). An approximate Gompertz model, for instance, is obtained by linearly restricting the effect of Z (or T) in a similar way as in loglinear association models (Clogg, 1982). Finally, if there are as many time intervals as observed survival times and if the time dependence of the hazard rate is not restricted, one obtains a Cox regression model (Laird and Olivier, 1981; Vermunt, 1997a, pp. 113–15).

The presence of unobserved heterogeneity may bias the results obtained from the hazard models discussed so far in various ways (Yamaguchi, 1986; Vermunt, 1997a, pp. 140–1). The best-known phenomenon is the downward bias of the time dependence that occurs even if – at $T = 0$ – the unobserved factors are uncorrelated with the values of covariates included in the model (Vaupel et al., 1979; Heckman and Singer, 1982, 1984). In such situations, covariate effects will be biased as well because the unobserved variables and the observed variable become correlated after $T = 0$. When the unobserved factors are related with some of the covariates at $T = 0$, that is, when there is some form of selection bias, not only the effects of the covariates concerned are biased, but also spurious time-covariate interactions will be found.

C. Censoring

A subject that always receives a great amount of attention in discussions on event history analysis is the problem of censoring (Yamaguchi, 1991, pp. 3–9; Guo, 1993; Vermunt, 1997a, pp. 117–30). An observation is called *censored* if it is known that it did not experience the event of interest during some time, but it is not known when it experienced the event. In fact, censoring is a specific type of missing data. In the first-birth example, a censored case could be a woman who is 30 years of age at the time of interview (and has no follow-up interview) and does not have children. For such a woman, it is known that she did not have a child until age 30, but it is not known whether or when she will have her first child.

This is, actually, an example of what is called *right censoring*. *Left censoring* means that it is known that the event did not occur, but the exact length of the period that the case concerned did not experience the event is unknown. Left censoring is more difficult to deal with than right censoring (Guo, 1993).

As long as it can be assumed that the censoring mechanism is not related to the process under study, dealing with censored observations in the maximum-likelihood estimation of the parameters of hazard models is straightforward. Let δ_i be a censoring indicator taking the value 0 if observation i is censored and 1 if it is not censored. The contribution of case i to the likelihood function that must be maximized when there are censored observations is

$$\mathcal{L}_i = h(t_i|\mathbf{x}_i)^{\delta_i} S(t_i|\mathbf{x}_i) = h(t_i|\mathbf{x}_i)^{\delta_i} \exp\left[-\int_0^{t_i} h(u|\mathbf{x}_i)du\right].$$

This likelihood function is, however, only valid if the censoring mechanism can be ignored for likelihood-based inference. The presence of unobserved heterogeneity that is shared by the process of interest and the censoring process will lead to a violation of the assumption that censoring is ignorable.

D. Time-Varying Covariates

A strong point of hazard models is that one can use time-varying covariates. These are covariates that may change their value over time. Examples of interesting time-varying covariates in the first-birth example are a woman's marital and work status. It should be noted that, in fact, the time variable and interactions between time and time-constant covariates are time-varying covariates as well (Kalbfleish and Prentice, 1980, pp. 121–2). The saturated log-rate model described in Equation (**??**) contains both time effects and time-covariate interaction terms. Inclusion of ordinary time-varying covariates does not change the structure of this hazard model. Suppose, for instance, that covariate B is time varying rather than time constant. This has only implications for the matrix with exposure times E_{abz}. When computing this matrix, one must take into account that individuals can change their value on B (Vermunt, 1997a, pp. 144–5).

The presence of unobserved heterogeneity may seriously bias the effects of time-varying covariates. Earlier, we already mentioned the effect of unobserved risk factors on the time dependence and on time-covariate interactions. The effects of time-varying covariates may be partially spurious as a result of the presence of unobserved risk factors, which influence both the covariate process and the dependent process (Yamaguchi, 1991, pp. 134–9; Vermunt, 1997a, pp. 140–1).

E. Multiple Risks

Thus far, only hazard rate models for situations in which there is only one destination state were considered. In many applications it may, however, prove necessary to distinguish between different types of events or risks. In the analysis of the first-union formation, for instance, it may be relevant to make a distinction between marriage and cohabitation. In the analysis of death rates, one may want to distinguish different causes of death. And in the analysis of the length of employment spells, it may be of interest to make a distinction between the events of voluntary job change, involuntary job change, redundancy, and leaving the labor force.

The standard method for dealing with situations in which – as a result of the fact that there is more than one possible destination state – individuals may experience different types of events is the use of a multiple-risk or competing-risk model. For a discussion of the various types of multiple-risk situations, see Allison (1984, pp. 42–4) or Vermunt (1997a, pp. 146–50). A multiple-risk variant of the general log-rate model described in Equation (3) is

$$\log h_{abzd} = \sum_j \beta_{dj} x_{abzdj}.$$

Here, D is a variable indicating the destination state or the type of event that one has experienced. As can be seen, this variable can be used in the hazard model in the same way as the other variables. Note that the index d is used in the parameters and the elements of the design matrix to indicate that the covariate and time effects may be risk dependent.

Again the presence of unobserved heterogeneity may distort the results obtained from the hazard model. More precisely, if the different types of events have shared unmeasured risks factors, the results for each of the types of events is only valid under the observed hazard rates for the other risks. In fact, the resulting dependence among risks is comparable with what in the field of discrete choice modeling is known as the violation of the assumption of independence of irrelevant alternatives (Hill et al., 1993).

F. Multivariate Hazard Models

Most events studied in social sciences are repeatable, and most event history data contain information on repeatable events for each individual. This is in contrast to biomedical research, in which the event of greatest interest is death. Examples of repeatable events are job changes, having children, arrests, accidents, promotions, and residential moves.

Often, events are not only repeatable but also of different types; that is, we have a multiple-state situation. When people can move through a sequence of states, events cannot only be characterized by their destination state, as in competing risks models, but they may also differ with respect to their origin state. An example is an individual's employment history: an individual can move through the states of employment, unemployment, and out of the labor force. In that case, six different kinds of transitions can be distinguished that differ with regard to their origin and destination states. Of course, all types of transitions can occur more than once. Other examples are people's union histories, with the states of living with parents, living alone, unmarried cohabitation, and married cohabitation, or people's residential histories, with different regions as states. Special multiple-state models are the well-known Markov and semi-Markov chain models (Tuma and Hannan, 1984, pp. 91–115).

Hazard models for analyzing data on repeatable events and multiple-state data are special cases of the general family of multivariate hazard rate models. Another application of these multivariate hazard models is the simultaneous analysis of different life-course events, or as Willekens (1989) calls it, *parallel careers*. For instance, it can be of interest to investigate the relationships between women's reproductive, relational, and employment careers, not only by means of the inclusion of time-varying covariates in the hazard model, but also by explicitly modeling their mutual interdependence.

Another application of multivariate hazard models is the analysis of dependent or clustered observations. Observations are clustered, or dependent, when there are observations from individuals belonging to the same group or when there are several similar observations per individual. Examples are the occupational careers of spouses, educational careers of brothers (Mare, 1994), child mortality of children in the same family (Guo and Rodriguez, 1994), or in medical experiments, measures of the sense of sight of both eyes or measures of the presence of cancer cells in different parts of the body. In fact, data on repeatable events can also be classified under this type of multivariate event history data because in that case there is more than one observation of the same type for each observational unit as well.

The log-rate model can easily be generalized to situations in which there are several origin and destination states and in which there may be more than one event per observational unit (Vermunt, 1997a, p. 169). The only thing we need to do is to define – besides the covariates and the time variables – variables indicating the origin state (O), the destination state (D), and the rank number of the event (M). The general log-rate model for such a situation is of the form

$$\log h_{abzodm} = \sum_j \beta_{odmj} x_{abzodmj}. \tag{4}$$

Of course, notation could be simplified by replacing the variables D, O, and M by a single variable indicating the type of event defined by origin, destination, and rank number.

The different types of multivariate event history data have in common that there are dependencies among the observed survival times. These dependencies may take several forms. The occurrence of one event may influence the occurrence of another event. Events may be dependent as a result of common antecedents. Furthermore, survival times may be correlated because they are the result of the same causal process, with the same antecedents and the same parameters determining the occurrence or nonoccurrence of an event. If these common risk factors are not observed, the assumption of statistical independence of observation is violated, which may seriously distort the results. This is, actually, the same type of problem that motivated the development of multilevel models (Goldstein, 1987; Bryk and Raudenbush, 1992).

3. DEALING WITH UNOBSERVED HETEROGENEITY

As described in the previous section, unobserved heterogeneity may have different types of consequences in hazard modeling. The best-known phenomenon is the downward bias of the duration dependence. In addition, it may bias covariate effects, time-covariate interactions, and effects of time-varying covariates. Other possible consequences are dependent censoring, dependent competing risks, and dependent observations. There are two main types of methods that have been proposed for correcting for unobserved heterogeneity: random-effects and fixed-effects methods. Later, these two methods are described and a more general nonparametric random-effects method is presented.

A. Random-Effects Approach

The random-effects approach is based on the introduction of a time-constant latent covariate in the hazard model (Vaupel et al., 1979). The latent variable is assumed to have a multiplicative and proportional effect on the hazard rate, that is,

$$\log h(t|\mathbf{x}_i, \theta_i) = \log h(t) + \sum_j \beta_j x_{ij} + \log \theta_i.$$

Here, θ_i denotes the value of the latent variable for subject i. In the parametric random-effects approach, the latent variable is postulated to have a particular distributional form. The amount of unobserved heterogeneity is determined by the size of the standard deviation of this distribution: The larger the standard deviation of θ, the more unobserved heterogeneity there is.

Vaupel et al. (1979) proposed using a gamma distribution for θ, with a mean of 1 and a variance of $1/\gamma$, where γ is the unknown parameter to be estimated. Several other authors have proposed incorporating a gamma-distributed multiplicative random term in hazard models (Tuma and Hannan, 1984, pp. 177–179; Lancaster, 1990, pp. 65–70). According to Vaupel et al. (1979), the gamma distribution was chosen because it is analytically tractable and readily computational. Moreover, it is a flexible distribution that takes on a variety of shapes as the dispersion parameter γ varies: When $\gamma = 1$, it is identical to the well-known exponential distribution; when γ is large, it assumes a bell-shaped form reminiscent of a normal distribution.

Heckman and Singer (1982, 1984) demonstrated by an analysis of one particular data set that the results obtained from continuous-time hazard models can be very sensitive to the choice of the functional form of the mixture distribution. Therefore, they proposed using a nonparametric characterization of the mixing distribution by means of a finite set of so-called mass points, or points of support, whose number, locations, and weights are empirically determined. In this approach, the continuous mixing distribution of the parametric approach is replaced by a discrete density function defined by a set of empirically identifiable mass points that are considered adequate to characterize fully the form of the heterogeneity. Often, two or three points of support suffice (Guo and Rodriguez, 1994).

It should be noted that Heckman and Singer's arguments against the use of parametric mixing distributions have been criticized by other authors who claimed that the sensitivity of the results to the choice of the mixture distribution was caused by the fact that Heckman and Singer misspecified the duration dependence in the hazard model they formulated for the data set they used to demonstrate the potentials of their nonparametric approach. Trussell and Richards (1985) demonstrated that the results obtained with Heckman and Singer's nonparametric mixing distribution can severely be affected by a misspecification of the functional form of the distribution of T.

The nonparametric unobserved heterogeneity model proposed by Heckman and Singer (1982, 1984) is strongly related to LCA. As in LCA, the population is assumed to be composed of a finite number of exhaustive and mutually exclusive groups formed by the categories of a latent variable. Suppose W is a categorical latent variable with W^* categories, and w is a particular value of W. The nonparametric hazard model with unobserved heterogeneity can be formulated as follows:

$$\log h(t|\mathbf{x}_i, \theta_w) = \log h(t) + \sum_j \beta_j x_{ij} + \log \theta_w.$$

Here, θ_w denotes the (multiplicative) effect on the hazard rate for latent class w. It should noted that this loglinear hazard model can also be written in the form of a log-rate model by using the parameter notation from hierarchical loglinear models. This is demonstrated below in the examples.

The contribution of the ith subject to the likelihood function that must be maximized in the case of a single nonrepeatable event is

$$\mathcal{L}_i = \sum_{w=1}^{W^*} \pi_w h(t_i|\mathbf{x}_i, \theta_w)^{\delta_i} S(t_i|\mathbf{x}_i, \theta_w),$$

where π_w is the proportion of the population belonging to latent class w. In the terminology used by Heckman and Singer (1982), the number of latent classes (W^*), the latent proportions (π_w), and the effects of W (θ_w) are called the number of mass points, the weights, and the mass points locations, respectively.

The most important drawback of the parametric and nonparametric random-effects approaches to unobserved heterogeneity is that the mixture distribution is assumed to be independent of the observed covariates. This is, in fact, in contradiction to the omitted variables argument, which is often used to motivate the use of these types of mixture models. If one assumes that particular important variables are not included in the model, it is usually implausible to assume that they are completely unrelated to the observed factors. In other words, by assuming independence among unobserved and observed factors, the omitted variable bias, or selection bias, will generally remain (Chamberlain, 1985; Yamaguchi, 1986, 1991, p. 132). Later, a nonparametric random-effects approach is presented that overcomes this weak point of the standard approaches.

The use of parametric mixture distributions is relatively simple in models for a single nonrepeatable event. However, when there is more than one (latent) survival time per observational unit, that is, when there is a model for competing risks, a model for repeatable events, or another type of multivariate hazard model, this is generally not true anymore. It is not so easy to include several possibly correlated parametric latent variables in a hazard model because that makes it necessary to specify the functional form of the multivariate mixture distribution. Therefore, in such cases, most applications use either several mutually independent cause, spell, or transition-specific latent variables, or one latent variable that may have a different effect on the several cause, spell, or transition-specific hazard rates. The former approach was adopted, for instance, by Tuma and Hannan (1984, pp. 177–83), and the latter, for instance, by Flinn and Heckmann (1982) in a study in which they used a normal mixture distribution. These two specifications of the unobserved factors have also been used in models with nonparametric unobserved heterogeneity (Heckman and Singer, 1985; Moon, 1991). The general LC approach that is presented later makes it possible to specify models with several mutually related latent variables without the necessity of specifying a distributional form for their joint distribution. The joint distribution of the latent variables is nonparametric as well and can, if necessary, be restricted by means of a loglinear parameterization of the latent proportions.

B. Fixed-Effects Approach

A second method for dealing with unobserved heterogeneity involves adding cluster-specific effects, or incidental parameters, to the model (Chamberlain, 1985; Yamaguchi, 1986). In fact, a categorical variable is included in the hazard model that indicates to which cluster a particular observation belongs: Observations belonging to the same cluster have the same value for this "observed" variable, whereas observations belonging to different clusters have different values. Thus, besides the time and covariate effects, we have to estimate a separate parameter for each cluster.

This approach to unobserved heterogeneity, which is called the *fixed-effects approach*, can only be applied with multivariate survival data, that is, when there is more than one observation for the largest part of the observational units. Note that actually the unobserved heterogeneity is transformed into a form of observed heterogeneity capturing the similarity among observations belonging to the same cluster.

The advantage of using fixed-effects methods to correct for unobserved heterogeneity is that they circumvent two objections against random-effects methods that were presented earlier: No functional form must be specified for the (joint) distribution of the unobserved heterogeneity, and the unobserved heterogeneity is automatically related to both the initial state and the time-constant covariates.

The major limitation of the fixed-effects approach is that because each cluster has its own incidental parameter, no parameter estimates can be obtained for the effects of covariates that have the same value for the different observations belonging to the same cluster: Only the effects of observation-specific or of time-varying covariates can be estimated. Another problem is that the incidental parameters cannot be estimated consistently, because by definition they are based on a limited number of observations regardless of the sample size. This inconsistency may be carried over to the other parameters if the parameters are estimated by means of maximum-likelihood methods (Yamaguchi, 1986).

C. General Nonparametric Random-Effects Approach

To overcome the limitations of the fixed- and random-effects approaches, Vermunt (1997a, pp. 189–256) proposed a more general nonparametric latent variable approach to unobserved heterogeneity. The main difference between this latent variable method and Heckman and Singer's method is that in the former different types of specifications can be used for the joint distribution of the observed covariates, the unobserved covariates, and the initial state. This means that it becomes possible to specify hazard models in which the unobserved factors are related to the observed covariates and to the initial state. A special case is, for instance, the Mover–Stayer model proposed by Farewell (1982), in which the probability of belonging to the class of stayers is regressed on a set of covariates by means of a logit model. Moreover, when hazard models are specified with several latent covariates, different types of specifications can be used for the relationships among the latent variables, one of which leads to a time-varying latent variable as proposed by Böckenholt and Langeheine (1996). By means of a multivariate hazard model, the latent covariates can also be related to observed time-varying covariates.

The model that is used for dealing with unobserved heterogeneity consists of two parts: a loglinear path model in which the relationships among the time-constant observed covariates, the initial state, and unobserved covariates are specified, and an event history model in which the determinants of the dynamic process under study are specified.

Suppose there is a model with two time-constant observed covariates denoted by A and B and two unobserved covariates denoted by W and Y. In the first part of the model, the relationships between these four variables are specified by means of a loglinear path model (Goodman, 1973; Hagenaars, 1990; Vermunt 1996a, 1996b, 1997a). Let π_{abwy} denote the probability that an individual belongs to cell (a, b, w, y) of the contingency table formed by the variables A, B, W, and Y. Specifying a modified path model for π_{abwy} involves two things, namely, decomposing π_{abwy} into a set of conditional probabilities on the basis of the assumed causal order among A, B, W, and Y, and specifying loglinear or logit models for these conditional probabilities. At least three meaningful specifications for the causal order among A, B, W, and Y are possible, namely, all the variables are of the same order, the latent variables are posterior to the observed variables, and the observed variables are posterior to the latent variables. In the first specification, π_{abwy} is not decomposed in terms of conditional probabilities. The second specification is obtained by

$$\pi_{abwy} = \pi_{ab}\,\pi_{wy|ab},$$

and the third one by

$$\pi_{abwy} = \pi_{wy}\,\pi_{ab|wy}.$$

Note that the second type of specification in which the latent variable(s) depend on covariates was used by Clogg (1981) in his LC model with external variables.

Suppose that the second specification is chosen. In that case, π_{ab} and $\pi_{wy|ab}$ can be restricted by means of a nonsaturated (multinomial) logit model. A possible specification of the dependence of the unobserved covariates on the observed covariates is

$$\pi_{wy|ab} = \frac{\exp\left(u_w^W + u_y^Y + u_{aw}^{AW} + u_{bw}^{BW}\right)}{\sum_{wy}\exp\left(u_w^W + u_y^Y + u_{aw}^{AW} + u_{bw}^{BW}\right)}.$$

Here, W depends on A and B, whereas Y is assumed to be independent of W and the observed covariates. Other specifications are, for instance,

$$\pi_{wy|ab} = \frac{\exp\left(u_w^W + u_y^Y + u_{wy}^{WY}\right)}{\sum_{wy}\exp\left(u_w^W + u_y^Y + u_{wy}^{WY}\right)} = \pi_{wy},$$

where the joint latent variable is assumed to be independent of the observed variables, and

$$\pi_{wy|ab} = \frac{\exp\left(u_w^W + u_y^Y\right)}{\sum_{wy}\exp\left(u_w^W + u_y^Y\right)} = \pi_w\pi_y,$$

in which the two latent variables are mutually independent and independent of the observed variables.

The second part of the model with nonparametric unobserved heterogeneity consists of an event history model for the dependent process to be studied. The hazard model that is used here is a log-rate model that, as was shown in Equation (4), in its most general form involves specifying a regression model for the hazard rate $h_{abzwyodm}$. As before, z serves as index for the time variable, o for the origin state, d for the destination state, and m for the rank number of the event.

When maximum-likelihood estimates are obtained of the parameters of a hazard model with the observed covariates A and B and latent covariates W and Y, the contribution of case i to the likelihood function is

$$\mathcal{L}_i = \sum_{wy} \pi_{abwy} \mathcal{L}_i^*(h),$$

in which $\mathcal{L}_i^*(h)$ denotes the contribution of person i to the complete data likelihood function for the hazard part of the model, and a and b are the observed values of A and B for person i. Vermunt (1997a, pp. 186–8) proposed estimating the parameters of a general class of event history models with missing data by means of the expectation–maximization (EM) algorithm. A computer program called *lEM* (Vermunt, 1997b) has been developed in which this nonparametric random-effects approach is implemented.

4. EXAMPLES

This section presents two applications of the general random-effects approach presented earlier. The first application concerns a single nonrepeatable event. In the second example, there is information on the occurrence of four events for each observational unit.

A. Example I: First Interfirm Job Change

In his textbook on event history analysis, Yamaguchi (1991) presented an example on employees' first interfirm job change to illustrate the use of log-rate models. He reported the data taken from the 1975 Social Stratification and Mobility Survey in Japan in Table 4.1. Here, the same data set is reanalyzed to demonstrate a number of possible specifications of unobserved heterogeneity.

The event of interest is the first interfirm job separation experienced by the sample subjects. The time variable is measured in years. In the analysis, the last 1-year time intervals are grouped together in the same way as Yamaguchi did, which results in 19 time intervals. It should be noted that contrary to Yamaguchi, we do not apply a special formula for the computation of the exposure times for the first time interval.

Besides the time variable denoted by Z, there is information on one observed covariate: firm size (F). The first five categories range from small firm (1) to large firm (5). Level 6 indicates government employees. The most general log-rate model that will be used is of the form

$$\log h_{fzw} = u_f^F + u_z^Z + u_{zf}^{ZF} + u_w^W,$$

Table 1. Test Results for Example I

Model	Log-Likelihood	No. of Parameters	BIC
Without unobserved heterogeneity			
A. { }	−7261	1	14,530
B. $\{ZF\}$	−6956	114	14,665
C. $\{Z, F\}$	−7012	24	14,203
D. $\{F\}$	−7183	6	14,410
E. $\{Z\}$	−7072	19	14,286
F. $\{Z_1, Z_2, F\}$	−7123	8	14,306
G. $\{Z_1, Z_2, Z_{lin}, F\}$	−7030	9	14,129
With unobserved heterogeneity			
H. $\{Z_1, Z_2, F, W\}$	−7051	10	14,215
I. $\{Z_1, Z_2, F, W\} + \{FW\}$	−7027	15	14,166
J. $\{Z_1, Z_2, W\} + \{FW\}$	−7028	10	14,132
K. $\{Z_1, Z_2, W\} + \{F_{lin}W, F_6W\}$	−7031	7	14,114

Note: The BIC we use here is computed with the log-likelihood value and the number of parameters rather than with the L^2 and the degrees of freedom.

where W is the label of the latent variable assumed to capture the unobserved heterogeneity. The log-likelihood values, the number of parameters, and the Bayesian information criterion (BIC) values for the estimated models are reported in Table 1.

We will first describe the results for the models without unobserved heterogeneity. Models A and B serve as reference points. Model A postulates that the hazard rate does not depend on Z and F. Model B is a saturated log-rate model: It includes the main effects of Z and F and the ZF interaction term. The comparison of these two models shows that the log likelihood is 305 points higher in Model B, using 113 additional parameters. A Cox-like model is obtained by omitting the two-variable interaction between Z and F (Model C). By a *Cox-like model*, we mean a model that contains a separate parameter for each time point and that does not contain time-covariate interactions. This model with 90 parameters less than Model B does not perform so badly. The much lower log-likelihood values of Models D and E compared with Model C indicate that neither the time effect nor the effect of firm size can be omitted from Model C. The estimates of time parameters of Model C showed that the hazard rate rises in the first time periods and subsequently starts decreasing slowly. Models F and G were estimated to test whether it is possible simplify the time dependence of the hazard rate on the basis of this information. Model F contains only time parameters for the first and second time points, which means that the hazard rate is assumed to be constant from time point 3 to 19. Model G is the same as Model F, except for that it contains a linear term to describe the negative time dependence after the second time point. The comparison between Models F and G shows that this linear time dependence of the log hazard rate is extremely important: The log-likelihood increases 97 points, using only one additional parameter. A comparison of Model G with the less restricted Model C and the more restricted Model D shows that Model G captures the most important part of the time dependence. Although according to the likelihood-ratio statistic the difference between Models C and G is significant, Model G is the preferred model according to the BIC criterion.

The observed negative time dependence after the second time point may be caused by the presence of unobserved heterogeneity: Individuals with a lower rate of job changing remain with their first employer and therefore the observed hazard rate declines. For this explanation of the negative time dependence to be tested, a model is specified with an unobserved heterogeneity component. Model H contains a two-class latent variable that is assumed to be independent of the observed covariate. This is, in fact, Heckman and Singer's nonparametric model. The other part of the specification is the same as in Model F. As can be seen from the test results, the log-likelihood value of Model H is 72 points higher that of Model F, using only two additional parameters. This provides evidence that the negative time dependence is at least partially the result of the selection mechanism resulting from unobserved heterogeneity. The comparison of the log-likelihood values of Models G and H indicates that the two-class latent variable W does not fully describe the observed negative time dependence. Inclusion of a third latent class does not, however, lead to an improvement of the log-likelihood.

Model I relaxes one of the main assumptions that is generally made in models with unobserved heterogeneity: the unobserved covariate W is allowed to be related to the observed covariate F. The relationship between W and F is modeled by treating W as posterior to F, which implies that a logit model is specified for $\pi_{w|f}$. From a substantive point of view, this means that it is assumed that firm size determines the (unobserved) individual characteristics of employees. As can be seen from the increase of the log-likelihood of 24 points using five additional parameters, including the direct of effect F on W improves the model a great deal. In Model J, the direct effect of F on the hazard rate is omitted. The fact that Model J performs almost as well as Model I with five fewer parameters indicates that – given the postulated causal order between F and W – there is only an indirect effect of firm size on the hazard rate by means of the latent variable W.

It should be noted that in this case, the same fit is obtained irrespective of the assumed causal order between W and F. The interpretation of the results is, however, quite different. If F were assumed to be posterior to W, the results from Model J would indicate that the effect of F on the hazard rate is spurious.

Finally, to simplify the model even more, in Model K, the loglinear parameters describing the effect of firm size on W is restricted to be linear for levels 1 to 5 of F. In fact, the log odds of belonging to the high-risk class is assumed to decrease linearly with the size of the firm. Because the result of Model J showed that level 6 (government) is situated between levels 4 and 5, a separate parameter is included for this category. The resulting Model K performs very well.

The parameter estimates of Model K indicate that Class 1 – containing more than 60% of the sample – has an eleven times higher rate of an interfirm job change than Class 2. The probability of belonging to Class 1 is higher for employees of small firms than for employees of large firms: The membership probabilities of class one for the first five levels of the variable firm size are 0.87, 0.79, 0.67, 0.53, and 0.38, respectively. Government employees take a position in between the two largest firm sizes (0.46).

The main substantive conclusion is that there are two types of employees: stayers and movers. When this unobserved heterogeneity is controlled for, the association between firm size and the risk of a job change disappears. However, there is a strong association between firm size and being a mover or a stayer. Two plausible explanations for this phenomenon are that (1) firms of different sizes select and contract different types of employees, or (2) firms of different sizes offer different job conditions making employees stayers or movers. These explanations assume that the causal effect goes from firm size to type of employee. The effect could, however, also be reversed: people belonging to the stayer type prefer working at larger firms or in government whereas movers prefer working at smaller firms.

Another substantive conclusion is that the declining risk of a job change after the second year of employment can be attributed to unobserved heterogeneity. This is a good illustration of the well-known phenomenon of spurious negative time dependence.

B. Example II: First Experience with Relationships

The second example concerns adolescents' first experiences with relationships. The data are taken from a small-scale two-wave panel survey of 145 cases (Vinken, 1998). In this application, we use the information on the age at which the sample members experienced for the first time the events of "sleeping with someone," "having a steady friend," "being very much in love," and "going out." A substantial part of the sample is censored; that is, many of the individuals did not experience one or more of these events. The model that is used for each of the hazard rates has the form of a Cox model:

$$\log h_{zwm} = u_{zm}^{ZM} + u_{wm}^{WM}.$$

Here, M is the label for the type of event, Z for the time variable, and W for the latent variable. As can be seen, both the time dependence and the effect of W is assumed to be event specific.

The purpose of the analysis is to use the dependence between the four events to construct a latent typology. This example can either be seen as the application of LCA as a clustering method, in which the complication is that the indicators are censored survival times, or as a method for dealing with correlated survival times. In the former case, we could use the label LC cluster analysis, and in the latter, LC regression analysis (see Vermunt and Magidson, 2000; Vermunt and Magidson, this volume; and Wedel and DeSarbo, this volume). In a second step of the analysis, it is checked whether it is possible to explain differences in class membership by means of three observed covariates.

Table 2 reports the log-likelihood values and the number of parameters for the one-, two-, three-, and four-class models. Although the inclusion of the second class gives the largest improvement in log likelihood (46 points), the inclusion of a third and a fourth class also improves the

Table 2. Test Results for Example II

Model	Log Likelihood	No. of Parameters	BIC
A. 1 class	−1279	4	2,995
B. 2 classes	−1233	9	2,929
C. 3 classes	−1220	15	2,929
D. 4 classes	−1209	20	2,930

Note: The BIC we use here is computed with the log-likelihood value and the number of parameters rather than with the L^2 and the degrees of freedom.

Table 3. Parameter Estimates for Example II: Three-Class Model

Parameter	Class 1	Class 2	Class 3
Class size	0.17	0.42	0.41
Loglinear hazard parameters			
Event 1	−1.79	−0.24	2.03
Event 2	−2.14	0.20	1.92
Event 3	−0.87	0.24	0.63
Event 4	−0.45	−0.16	0.61
Median ages			
Event 1	23	19	16
Event 2	23	18	15
Event 3	19	15	14
Event 4	15	14	14
Loglinear covariate effects			
Youth centristic	−0.33	0.18	0.15
Boy	1.03	−0.34	−0.68
Low education	−0.62	0.44	0.18

log-likelihood. According to the BIC, the two- and three-class models perform equally well.

The results for the three-class model are presented in Table 3. The loglinear hazard parameters show that the three identified latent classes differ clearly with respect to the timing of the four events of interest. Class 1 has the lowest hazard rates for each of the four events, whereas Class 3 has the highest hazard rates. This implies that the first class, which contains 17% of the sample, experiences the events later than Classes 2 and 3; whereas Class 3, which contains 41% of the sample, experiences the events earlier than Classes 1 and 2. Table 3 also reports the estimates of the median ages at which the three types experienced the events of interest. These indicate that the largest differences between the three types occur in the timing of events 1 (sleeping with someone) and 2 (having a steady relationship).

Finally, a three-class model was estimated in which class membership was regressed on three dichotomous covariates by means of multinomial logit model. These covariates are youth centrism (Y), sex (S), and educational level (E). Youth centrism is one of the central concepts in youth studies; it indicates the extent to which young people perceive their peers as a positively valued ingroup and perceive adults as a negatively valued outgroup. The dichotomous youth-centrism scale that is used here was construct by Vinken (1998), using LCA.

The covariate part of the model is a logit model of the form

$$\pi_{w|yse} = \frac{\exp\left(u_w^W + u_{yw}^{YW} + u_{sw}^{SW} + u_{ew}^{EW}\right)}{\sum_w \exp\left(u_w^W + u_{yw}^{YW} + u_{sw}^{BW} + u_{ew}^{EW}\right)}.$$

The other part of the model consists of the same multivariate hazard model as we used earlier. The covariate effects are reported in Table 3, where the omitted level is used as reference category. It can be seen that youth-centristic adolescents, girls, and lower educated belong less often to the late type and more often to the early types. In addition, sex has a much stronger effect on class membership than the other two covariates.

In this example, we showed that the information on the timing of four life events can be summarized into an easy to interpret typology. The typology identified subgroups that experience the relational events at different (median) ages. We also showed how the latent typology can be related to covariates. It turns out that girls experience the events of interest earlier than boys, low educated earlier than high educated, and youth-centristic earlier than non-youth-centristic adolescents.

5. FINAL REMARKS

This paper described a general nonparametric random-effects approach for dealing with unobserved heterogeneity in the analysis of event histories. This approach, which is based on specifying simultaneously with the hazard model a loglinear path model for the observed and unobserved covariates, overcomes most of the weak points of the standard random- and fixed-effects approaches. Two applications were presented to demonstrate the potentials of the new method. Many other applications can be found in Vermunt (1996a, 1997a), for instance, on models for dependent competing risks, multiple-state models, models with several latent variables, and models with spurious effects of time-varying covariates.

It must, however, be stressed that substantive arguments must determine the specification of the nature of the unobserved heterogeneity. In the job change example, the latent covariate was postulated to be posterior and related to the observed covariate, whereas in the relationship example, the latent variable was assumed to describe the observed correlations between four survival times. The obtained results make sense only if these are the correct specifications. The LCA approach described here is a special case of a more general framework for dealing with missing data problems in event history analysis proposed by Vermunt (1996a, 1997a).

Other forms of missing data that can be dealt with are measurement error in covariate values, measurement error in occupied states, partially missing information in covariate values, and partially missing information in occupied states.

REFERENCES

Allison, P. D. (1984). *Event History Analysis: Regression for Longitudinal Event Data*. Beverly Hills: Sage.

Blossfeld, H. P., & Rohwer, G. (1995). *Techniques of Event History Modeling*. Mahwah, NJ: Erlbaum.

Bryk, A. S., & Raudenbush, S. W. (1992). *Hierarchical Linear Models: Applications and Data Analysis Methods*. Newbury Park: Sage.

Böckenholt, U., & Langeheine, R. (1996). "Latent change in recurrent choice data," *Psychometrika*, **61**, 285–302.

Chamberlain, G. (1985). "Heterogeneity, omitted variable bias, duration dependence." In J. J. Heckman & B. Singer (eds.), *Longitudinal Analysis of Labor Market Data*. Cambridge: Cambridge University Press.

Clogg, C. C. (1981). "New developments in latent structure analysis." In D. J. Jackson & E. F. Borgotta (eds.), *Factor Analysis and Measurement in Sociological Research*. Beverly Hills: Sage, pp. 215–45.

Clogg, C. C. (1982). "Some models for the analysis of association in multiway cross-classifications having ordered categories," *Journal of the American Statistical Association*, **77**, 803–15.

Clogg, C. C., & Eliason, S. R. (1987). "Some common problems in loglinear analysis," *Sociological Methods and Research*, **16**, 8–14.

Cox, D. R. (1972). "Regression models and life tables," *Journal of the Royal Statistical Society B*, **34**, 187–203.

Farewell, V. T. (1982). "The use of mixture models for the analysis of survival data with long-term survivors," *Biometrics*, **38**, 1041–6.

Flinn, C. J., & Heckman, J. J. (1982). "New methods for analysing individual event histories." In S. Leinhardt (ed.), *Sociological Methodology 1982*. San Francisco: Jossey-Bass, pp. 99–140.

Goldstein, H. (1987). *Multilevel Models in Educational and Social Research*. London: Charles Griffin.

Goodman, L. A. (1973). "The analysis of multidimensional contingency tables when some variables are posterior to others: a modified path analysis approach," *Biometrika*, **60**, 179–92.

Guo, G. (1993). "Event-history analysis for left-truncated data." In P. V. Marsden (ed.), *Sociological Methodology 1993*. Oxford: Basil Blackwell, pp. 217–43.

Guo, G., & Rodriguez, G. (1994). "Estimating a multivariate proportional hazards model for clustered data using the EM algorithm, with an application to child survival in Guatemala," *Journal of the American Statistical Association*, **87**, 969–76.

Hagenaars, J. A. (1990). *Categorical Longitudinal Data – Loglinear Analysis of Panel, Trend, and Cohort Data*. Newbury Park: Sage.

Heckman, J. J., & Singer, B. (1982). "Population heterogeneity in demographic models." In K. Land & A. Rogers (eds.), *Multidimensional Mathematical Demography*. New York: Academic, pp. XXX–XXX.

Heckman, J. J., & Singer, B. (1984). "The identifiability of the proportional hazard model," *Review of Economic Studies*, **51**, 231–41.

Heckman, J. J., & Singer, B. (1985). "Social sciences duration analysis." In J. J. Heckman & B. Singer (eds.), *Longitudinal Analysis of Labour Market Data*. Cambridge: Cambridge University Press, pp. 38–110.

Hill, D. H., Axinn, W. G., & Thornton, A. (1993). "Competing hazards with shared unmeasured risk factors." In P. V. Marsden (ed.), *Sociological Methodology 1993*. Oxford: Basil Blackwell, pp. 245–77.

Holford, T. R. (1980). "The analysis of rates and survivorship using loglinear models," *Biometrics*, **32**, 299–306.

Kalbfleisch, J. D., & Prentice, R. L. (1980). *The Statistical Analysis of Failure Time Data*. New York: Wiley.

Laird, N., & Oliver, D. (1981). "Covariance analysis of censored survival data using loglinear analysis techniques," *Journal of the American Statistical Association*, **76**, 231–40.

Lancaster, T. (1990). *The Economic Analysis of Transition Data*. Cambridge: Cambridge University Press.

Mare, R. D. (1994). "Discrete-time bivariate hazards with unobserved heterogeneity: a partially observed contingency table approach." In P. V. Marsden (ed.), *Sociological Methodology 1994*. Oxford: Basil Blackwell, pp. 341–85.

Moon, C. G. (1991). "A grouped data semiparametric competing risks model with nonparametric unobserved heteogeneity and mover-stayer structure," *Economics Letters*, **37**, 279–85.

Trussell, J., & Richards, T. (1985). "Correcting for unobserved heterogeneity in hazard models using the Heckman-Singer procedure." In N. Tuma (ed.), *Sociological Methodology 1985*. San Francisco: Jossey–Bass, pp. 242–76.

Tuma, N. B., & Hannan, M. T. (1984). *Social Dynamics: Models and Methods*. New York: Academic.

Vaupel, J. W., Manton, K. G., & Stallard, E. (1979). "The impact of heterogeneity in individual frailty on the dynamics of mortality," *Demography*, **16**, 439–54.

Vermunt, J. K. (1996a). "Loglinear event history analysis: a general approach with missing data, unobserved heterogeneity, and latent variables," Tilburg, The Netherlands: Tilburg University Press.

Vermunt, J. K. (1996b). "Causal loglinear modeling with latent variables and missing data." In U. Engel & J. Reinecke (eds.), *Analysis of Change: Advanced Techniques in Panel Data Analysis*. Berlin/New York: Walter de Gruyter, pp. 35–60.

Vermunt, J. K. (1997a). "Loglinear models for event history histories." In *Advanced Quantitative Techniques in the Social Sciences Series*, Vol. 8. Thousand Oaks, CA: Sage.

Vermunt, J. K. (1997b). *LEM: A General Program for the Analysis of Categorical Data. User's Manual*. Tilburg, The Netherlands: Tilburg University.

Vermunt, J. K., & Magidson, J. (2000). *Latent GOLD's User's Guide*. Boston: Statistical Innovations.

Vinken, H. (1998). *Political Values and Youth Centrism. Theoretical and Empirical Perspectives on the Political Value Distinctiveness of Dutch Youth Centrists.* Tilburg, The Netherlands: Tilburg University Press.

Willekens, F. J. (1991). "Understanding the interdependence between parallel careers." In J. J. Siegers, J. De Jong-Gierveld, & E. Van Imhoff (eds.), *Female Labour Market Behavior and Fertility, a Rational Choice Approach.* The Hague: NIDI, pp. 2–31.

Yamaguchi, K. (1986). "Alternative approaches to unobserved heterogeneity in the analysis of repeatable events." In N. B. Tuma (ed.), *Sociological Methodology 1986.* Washington, DC: American Sociological Association, pp. 213–49.

Yamaguchi, K. (1991). "Event history analysis." In *Applied Social Research Methods*, Vol. 28. Newbury Park: Sage.

FIFTEEN

Latent Class Models for Contingency Tables with Missing Data

Christopher Winship, Robert D. Mare, and John Robert Warren

1. INTRODUCTION

Missing data is a common problem in many types of data analysis. In this paper, we show how to deal with missing data in loglinear analyses of frequency tables.[1] Our approach is based on two ideas: (1) Latent class models can be adapted to contingency tables with missing data by defining variables that are latent (missing) for some cases and are manifest (observed) for others; and (2) latent class models can be viewed as loglinear models for tables in which some cells are unobserved or partially observed. Using our approach, we can retain the loglinear model framework and notation and deal with missing data through a modest extension of the standard model. Flexible software for latent class models, such as DNEWTON (Haberman, 1989) and LEM (Vermunt, 1996) is required, but the conceptual extension of elementary loglinear models is straightforward.[2]

By explicitly incorporating missing data into the analysis of a contingency table, one can address two concerns. First, a researcher may be worried about the possible loss of statistical power or precision of estimation that results when observations with missing data are excluded from an analysis. If many cases have missing data on at least one variable, exclusion of these cases from the analysis may substantially reduce the sample and create unacceptably large standard errors. One may want to incorporate cases with missing data into the analysis so that the information associated with these cases can be used to obtain more precise estimates. When the loss of statistical power is the only problem, incorporating missing data into a loglinear analysis is usually straightforward.

Second, one may be concerned that the exclusion of missing data may result in inconsistent parameter estimates in loglinear models. This is likely to occur if there is a systematic mechanism producing the missing data. In this case one should develop a model of the missing data process jointly with the substantive model of interest. Later, we discuss more precisely the types of missing data processes that lead to inconsistent estimates. Although correcting problems of this type can be difficult, it may be essential if one is to make appropriate substantive conclusions.

The next section of the paper presents an example that illustrates that how one deals with missing data affects both the precision of parameter estimates and the substantive conclusions that one draws. The subsequent section briefly discusses one conventional approach to dealing with missing data, namely, adding a category to a variable for missing data. We show that this procedure typically leads to inconsistent estimates. Next we show how a contingency table can be extended to incorporate missing data so that loglinear models for partially observed data can be applied. We then consider alternative assumptions about missing data and the models that these assumptions imply. We examine different models for our first example, and then present a more complex empirical example. We then discuss various problems in estimation and identification.

2. EXAMPLE 1: PRENATAL CARE AND INFANT MORTALITY

Panel (a) of Table 1 presents data on the relationship between prenatal care and infant mortality in two clinics. These data were first analyzed in Bishop, Fienberg, and Holland (1975). Little and Rubin (1987)

Table 1. Contingency Table with Partially Classified Observations

		Survival (S)	
Clinic (C)	**Prenatal Care (P)**	**Died**	**Lived**
	(a) Completely Classified Cases		
Clinic A	less	3	176
	more	4	293
Clinic B	less	17	197
	more	2	23
	(b) Partially Classified Cases (Clinic Missing)		
	less	10	150
	more	5	90

Source: (a) Bishop et al. (1975), Table 2.4-2; (b) artificial data from Little and Rubin (1987, Table 9.8, p. 187).

supplement these data with the hypothetical data in Panel (b) of the table, which contain 255 infants whose clinic ID is missing. A researcher who wishes to analyze the combined data in the two panels faces two problems. First, data are missing for 255 out of 970 cases. If all cases with missing data were omitted, this would substantially reduce statistical power. This is a particularly serious issue because the response variable, infant mortality, measures a rare event.

Second, assumptions about the true values of the missing data may markedly affect the estimated effect of prenatal care on infant mortality. Table 2 illustrates a range of possible outcomes under various extreme assumptions. All other possible assignments of the missing data are less

Table 2. Mortality Rates for Data in Table 1 Under Alternative Assumptions about Missing Data

		Level of Prenatal Care		
Assumption	**Clinic**	**Less**	**More**	**Difference**
1. Observed data collapsed over clinic		5.4	2.6	2.8
2. Complete data	A	1.7	1.3	0.4
	B	7.9	8.0	0.1
Missing are all:				
3. Clinic = A	A	3.8	2.3	0.5
4. Clinic = B	B	7.2	5.8	1.4
5. If care = less, clinic = A	A	3.8	1.3	2.5
If care = more, clinic = B	B	7.9	5.8	2.1
6. If care = more, clinic = A	A	1.7	2.3	−0.6
If care = less, clinic = B	B	7.2	8.0	−0.8
7. If survival = died, clinic = A	A	6.9	2.3	4.3
If survival = lived, clinic = B	B	6.4	1.8	4.6
8. If survival = lived, clinic = A	A	1.3	1.0	0.3
If survival = died, clinic = B	B	12.6	23.3	−10.7
9. If care = less & survival = died, or care = more & survival = lived, clinic = A	A	6.9	1.0	5.9
If care = more & survival = died, or care = less & survival = lived, clinic = B	B	11.8	23.3	−11.5
10. If care = less & survival = died, or care = more & survival = lived, clinic = B	A	0.9	3.0	−2.1
If care = more & survival = died, or care = less & survival = lived, clinic = A	B	12.0	1.7	10.3

extreme in that they yield estimates that fall within the range of those reported in Table 2.

Estimates of the effect of prenatal care vary across a wide range. If clinic status is ignored (assumption 1), the estimated difference in mortality rates by level of prenatal care is large (5.4% – 2.6% = 2.8%). However, if only complete data are used and estimates are computed within clinics (assumption 2), the estimated differences are very small (0.4% for Clinic A and 0.1% for Clinic B). Assuming that all missing data are from Clinic A (assumption 3) or Clinic B (assumption 4) produces somewhat higher estimates of the effect of the difference (0.5% and 1.4%, respectively). Assuming more complex patterns of missing data increases the range of possible effects. Under assumptions 5–10, the pattern of missing data is the opposite for the two clinics. For example, assumption 5 is that those who in fact were in Clinic A but are missing information on their clinic received less prenatal care, and that those who in fact were in Clinic B but are missing this information had more prenatal care. Across these assumptions for Clinic A, the difference in the mortality rates between the two levels of prenatal care ranges from −2.1% to +5.9%, and for Clinic B it ranges from −11.5% to +10.3%. The estimates of the percentage difference effects are greatest for the most complex assumed patterns for the missing data, that is, assumptions 9 and 10, where the missing data pattern is a function of both level of care and survival status and the effect of level of prenatal care is the opposite in the two clinics. Given the wide range of estimates in Table 2, it is impossible to infer for either clinic whether more prenatal care is beneficial. What this exercise does show, however, is that the most extreme estimates of the effect of prenatal care occur when we assume that the mechanism generating missing data differs between the two clinics.

3. CONVENTIONAL APPROACHES TO MISSING DATA

The most common approach to incorporating missing data into a loglinear analysis is to add a "missing" category to the variables with missing data. This is analogous to creating a dummy variable in a regression analysis to indicate that respondents are missing on a variable. In both cases, inconsistent estimates generally result (Little and Rubin, 1987). We consider the loglinear case here. The inclusion of a missing data category may affect both estimates of associations among the variables that contain no missing data and also estimates of associations involving variables that have missing data. The top panel of Table 3 presents a hypothetical table

Table 3. Hypothetical Data for Showing the Effects of Using a Missing Data Category

No Missing Data ($N = 3{,}600$)

Marginal Associations

	$X=0$	$X=1$
$Z=0$	1,000	800
$Z=1$	800	1,000
Odds Ratio	1.56	

Conditional Associations

	$Y=0$		$Y=1$	
	$X=0$	$X=1$	$X=0$	$X=1$
$Z=0$	800	400	200	400
$Z=1$	400	200	400	800
Odds Ratio	1.0		1.0	

With Missing Data

Conditional Associations

	$Y=0$		$Y=1$		$Y=$ Missing	
	$X=0$	$X=1$	$X=0$	$X=1$	$X=0$	$X=1$
$Z=0$	640	320	160	320	200	160
$Z=1$	320	160	320	640	160	200
Odds Ratio	1.0		1.0		1.56	

with three variables, *X*, *Y*, and *Z*, in which no data are missing. In this table *X* and *Z* are marginally associated with an odds ratio of 1.56, but, within levels of variable *Y*, *X*, and *Z* are independent.

Now suppose that 20% of the data on *Y* are missing purely at random. The lower panel of Table 3 arrays these data and distinguishes between missing and nonmissing data by adding a third category to *Y*. If we exclude the missing data from the sample, we get the same result as with the original data – *X* and *Z* are conditionally independent. However, the standard errors of any estimates will be larger because of the smaller sample. Consider the association between *X* and *Z* for those cases in which *Y* is missing. For these cases, *X* and *Z* are conditionally dependent with an odds ratio of 1.56, the same odds ratio as in the marginal table between *X* and *Y* in the original data. This is because the missing category contains cases in which both $Y = 0$ and $Y = 1$. In fact, because the data are missing completely at random, it contains the same proportion of cases with $Y = 0$ and $Y = 1$ as in the full sample. Thus, the conditional association of X and Z for those with missing data is equal to the marginal association between X and Z in the original table, not their conditional relationship. If one uses both the missing and nonmissing cases and estimates a common value for the conditional association between X and Z, the estimate will lie between the marginal and the true conditional association between X and Z. Because it is impossible to control properly for Y for the cases with missing values on Y, including a separate category for cases that are missing on Y will not give a consistent estimate for the association of X and Z conditional on Y.[3]

4. LATENT CLASS MODELS

The most common use of latent class models is to account for the associations among observed categorical variables with one or more latent categorical variables. One examines alternative models for a cross-classification of the observed variables and typically assumes that the observed variables are conditionally independent within categories of the latent variable(s) (Goodman, 1974; Dayton and Macready, 1980; McCutcheon, 1987; Clogg, 1988; Hagenaars, 1988, 1993). By definition, the latent dimensions of tables in latent class models have missing data for every observation. In contrast, in latent class models for missing data, data are observed for some cases and unobserved for others. Thus, in the approach discussed here, the latent class models have latent *cells* rather than latent dimensions.

In our discussion and analysis of contingency tables with missing data, we use several configurations of the data. We distinguish among three configurations: (1) observed table, (2) complete data table, and (3) expanded table. In the observed table, each variable that has missing data contains an additional category for missing observations. The complete data table is the cross-classification of observations that have no missing data. The expanded table, which is only partially observed, consists of the substantive variables cross-classified by a set of dichotomous variables that indicate whether observations are missing on each of the original variables. We term these variables *missing indicators*. The expanded table represents not only how the substantive variables in the data are related to each other, but also how the missing indicators relate to each other and to the substantive variables.

We illustrate the relationships among these tables using the infant mortality data shown in Table 1. The data contain four variables: clinic (C), prenatal care (P), survival status (S), and a missing on clinic indicator (M) coded not missing $= 0$ and missing $= 1$. The 2^4 table of C by P

by S by M is the expanded table. It contains two 2^3 subtables. One subtable table consists of the cross-classification of C by P by S on the data with no missing values on C. This is the complete data table and is equivalent to subtable (a) in Table 1. The other subtable is a 2^3 subtable of the incomplete data. This is subtable (b) in Table 1 cross-classified by the unobserved clinical status of these individuals. Thus, this 2^3 table for the incomplete data is latent. The union of the complete data table with the two-way margin for the incomplete data is the observed table, which is simply Table 1.

5. LOGLINEAR–LATENT CLASS MODELS FOR MISSING DATA

Missing data can be incorporated into an analysis by applying loglinear models to the expanded table. Because neither table is directly observed, we require a latent class model. Assuming that observations are obtained under a multinomial sampling scheme, we see that a loglinear model for the expanded table in the infant mortality example is

$$\begin{aligned}\log \pi_{ijkl} = {} & \lambda + \lambda_i^M + \lambda_j^C + \lambda_k^P + \lambda_l^S + \lambda_{ij}^{MC} + \lambda_{ik}^{MP} + \lambda_{il}^{MS} \\ & + \lambda_{jk}^{CP} + \lambda_{jl}^{CS} + \lambda_{kl}^{PS} + \lambda_{ijk}^{MCP} + \lambda_{ijl}^{MCS} + \lambda_{ikl}^{MPS} \\ & + \lambda_{jkl}^{CPS} + \lambda_{ijkl}^{MCPS},\end{aligned}$$

where π_{ijkl} denotes the probability that an individual falls into the cell for the ith level of M ($i = 0,1$), the jth level of C ($j = 0,1$), the kth level P ($k = 0,1$), and the lth level of the S ($l = 0,1$). Here, λ is determined by the constraint that $\sum_{ijkl} \pi_{ijkl} = 1$, and the remaining λ's are parameters, some of which may be zero. (For example, λ_{jk}^{CP} denotes the clinic by prenatal care association for a contrast involving the jth clinic and the kth prenatal care categories.) We assume that these parameters satisfy the typical constraint that the sum of all parameters of a particular type is zero.

Let f_{ijkl} represent the observed frequency in the $ijkl$th cell. When this cell is not directly observed, we denote the hypothetical frequency by f_{ijkl}^*. If $i = 0$ when clinic is known and $i = 1$ when clinic is missing, then we observe the collapsed tables $f_{1+kl} = f_{10kl}^* + f_{11kl}^*$, but not the f_{10kl}^* or f_{11kl}^* separately. The log-likelihood function for the model is

$$\log L = \sum_{jkl} f_{0jkl} \log(\pi_{0jkl}) + \sum_{kl} f_{1+kl} \log(\pi_{10kl} + \pi_{11kl}),$$

where the first summation is for cells that are not missing and the second summation is for cells that contain missing data. The terms contained in the first summation follow the form for likelihood terms in a loglinear model (e.g., Agresti, 1990, p. 166), whereas the terms in the second summation follow the form for likelihood terms in a latent class model (Andersen, 1980, p. 260). Our missing data model, therefore, combines elements of loglinear and latent class models.

6. TYPES OF MISSING DATA MODELS

The discussion of missing data patterns in Table 2 showed that estimates of the effect of prenatal care depend critically on assumptions about how the missing data are distributed. Having shown how to include missing data into a loglinear analysis through the use of the expanded table, we can examine specific models.

A. MCAR Models

The most restrictive model assumes that data are missing completely at random (MCAR) (Little and Rubin, 1987). This means that the missing indicators are assumed to be independent of all other variables. Consider the hypothetical data in Table 3 where there are three variables, X, Y, and Z, and there is missing information on Y for some cases. Let M be the missing data indicator for Y. Then the MCAR model for these data is equivalent to (M) (XYZ), where we have used the parentheses to indicate an arbitrary set of associations among X, Y, and Z. The models $(M)(XY)(YZ)$, $(M)(XZ)(Y)$, and $(M)(XYZ)$ are all examples of models in which the data are assumed to be MCAR. In each of these cases, whether data are missing, as indicated by M, is assumed to be independent of the other variables in the model. When data are MCAR, they can be omitted from the analysis without affecting parameter estimates. However, because omitting missing data reduces the sample size, it reduces the precision of estimated parameters involving variables that have no missing data.

B. MAR Models

A less restrictive and usually more realistic assumption is that whether a variable is missing is a function of the values of other observed variables; that is, data are missing at random (MAR), conditional on the observed data (Little and Rubin, 1987). With MAR data, whether a variable is missing for a particular case is random conditional on the observed values on the other variables. In terms of our hypothetical data in Table 3, MAR models are of the form $(MXZ)\,(XYZ)$, where, as before, the associations among the variables inside the parentheses are arbitrary. These models are MAR because M is conditionally independent of the variable it pertains to, Y. Typically, the inclusion of MAR data has very small effects on the estimated relationships between the variable that has missing data and the other variables. However, dropping cases with missing data may affect the estimated associations among the other variables in the model. Thus, incorporating cases with missing data into the analysis may affect both the precision and consistency of the estimates for the relationships among variables that do not have missing data (Winship and Mare, 1989).

C. NINR Models

When the probability that a variable has missing data is associated with the variable itself conditional on the other variables in the model, this requires a model for nonignorable nonresponse, or NINR. NINR models are required when it is likely that some survey respondents refuse to answer a question. For example, persons with unusually high or low incomes may be less willing to divulge their incomes than persons with incomes in the middle of the income distribution. Elsewhere (Winship and Mare, 1989) we examined data on whether individuals had ever been arrested. Here, one would expect that persons who had been arrested would be less likely to answer the question than those who had not been arrested.[4] In our hypothetical data, any model that assumes dependence between M and Y is a NINR model. For example $(MY)\,(XY)\,(YZ)$, $(MY)\,(MX)\,(XY)\,(YZ)$, and $(MYZ)\,(XY)\,(YZ)$ are all NINR models.

When data are in fact subject to nonignorable nonresponse, omitting observations with missing values results in estimates that are not only inefficient but inconsistent as well. Often the bias can be considerable. For example, in Table 2, assumptions 3 through 10 are consistent with a variety of NINR models. Depending on what assumptions one makes about the true values of the missing data, one gets quite different estimates for the effect of Prenatal Care on Survival. Unfortunately, NINR models can often be difficult to estimate.

7. FITTING MISSING DATA MODELS TO INFANT MORTALITY DATA

We can illustrate the fitting of alternative missing data models with the data in Table 1. Table 4 presents the estimated likelihood ratio G^2 and

Table 4. Goodness of Fit of Selected Models for Infant Mortality Data

Model	G^2	df	BIC
Complete Data			
1. *SC PC*	0.08	2	−13.06
MCAR Models			
2. *M S PC*	24.81	7	−23.33
3. *M SP PC*	20.02	5	−14.37
4. *M SC PC*	7.98	5	−26.41
5. *M SC PC SP*	7.84	4	−19.67
MAR Models			
6. *MP S PC*	20.13	5	−14.26
7. *MP PC SP*	15.33	4	−12.18
8. *MP SC PC*	3.30	4	−24.21
9. *MP SC PC SP*	3.16	3	−17.47
10. *MS PC SP*	17.83	4	−9.68
11. *MS SC PC*	5.79	4	−21.72
12. *MS SC PC SP*	5.65	3	−14.98
13. *MS MP PC*	17.94	4	−9.57
14. *MS MP SP PC*	13.55	3	−7.08
15. *MS MP SC PC*	1.57	3	−19.06
16. *MS MP SC PC SP*	1.38	2	−12.37
NINR Models			
17. *MC S PC*	20.13	5	−14.26
18. *MC PC SP*	15.33	4	−12.18
19. *MC SC PC*	2.26	4	−25.25
20. *MC SC PC SP*	2.00	3	−18.63
21. *MC MP S PC*	20.13	4	−7.38
22. *MC MP PC SP*	15.33	3	−5.30
23. *MC MP SC PC*	0.39	3	−20.24
24. *MC MP SC PC SP*	0.39	2	−13.36
25. *MC MS PC*	17.94	4	−9.60
26. *MC MS PC SP*	13.55	3	−7.08
27. *MC MS SC PC*	1.64	3	−18.99
28. *MC MS SC PC SP*	1.40	2	−12.35

Bayesian information criterion (BIC) (Raftery, 1995) statistics for the complete data and selected missing data models. In this analysis, we regard infant mortality as the response variable and clinic and prenatal care as the explanatory variables. Thus, we consider only models that include the association between prenatal care and clinic (*PC*). The G^2 values here should be viewed with caution. Because several of the frequencies in Table 1 are very small, the G^2 statistics may not follow a X^2 distribution (Agresti, 1990).

For the complete data, model 1, which includes all one-way effects and the two-way associations between Survival Status and Clinic (*SC*), and between Prenatal Care and Clinic (*PC*), fits the data extremely well ($G^2 = 0.08$, df = 1), implying that Prenatal Care and Survival Status are conditionally independent within clinics. The data, therefore, suggest that the level of prenatal care does *not* affect the probability that an infant will survive (Bishop et al., 1975; Little and Rubin, 1987). If we include the missing data, then a much larger class of models is available. Among MCAR models, the (*M*) (*SC*) (*PC*) model (model 4), which is analogous to the best-fitting complete data model, fits the data well ($G^2 = 7.98$, df = 5). This is an adequate fit, assuming that G^2 does in fact follow a X^2 distribution for these data.

Although the fit of the MCAR model is reasonable, it is nonetheless of interest to examine MAR and NINR models for these data. Considering these models as a whole, one can draw a number of general conclusions. First, models that do not contain both the (*SC*) and (*PC*) associations fit the data poorly. Second, for all three types of missing data models, the (*SP*) association appears to be statistically insignificant. Thus, for these data, our analysis is consistent with the analysis of the complete data alone. The data support the assumption that prenatal care and survival status are conditionally independent.

Across the three types of missing data models, however, there are some interesting differences. MAR models fit better than their corresponding MCAR models by the G^2 criterion. Whereas the (*M*) (*SC*) (*PC*) model has a G^2 of 7.98 on 5 degrees of freedom, the two comparable MAR models where *M* is assumed to be a function of Level of Prenatal Care (8) or Survival Status (11) have G^2 of 3.30 and 5.79, respectively (df = 4). Differencing these amounts from the MCAR G^2, we get G^2 statistics of 4.68 and 2.19 (df = 1). Assuming that G^2 follows a X^2 distribution, we see that the first of these differences is significant at the 0.05 level, indicating that a significant improvement in fit is achieved by assuming that *M* is associated with level of prenatal care (*P*). Adding the (*MS*) term to the (*MP*) (*SC*) (*PC*) model lowers the G^2 to 1.57 (Model 15), but this decrement is not statistically significant. The NINR models also fit the data better by the G^2 criterion. A model in which the likelihood of having missing data is simply a function of the clinic, that is, (*MC*) (*SC*) (*PC*), gives a G^2 of 2.26 on 4 degrees of freedom (Model 19). This model has a lower G^2

with the same degrees of freedom as the MAR model (*MP*) (*SC*) (*PC*) (Model 8), although a nested comparison between these two models is obviously impossible. The (*MC*) (*SC*) (*PC*) model implies that the likelihood of missing data on clinic is only associated with which clinic a mother was served by. Adding an (*MS*) term to the model lowers the G^2 to 1.64 (df = 3) (Model 27). Alternatively, adding an (*MP*) term to the (*MC*) (*SC*) (*PC*) model lowers the G^2 to 0.39 (Model 24), an extremely good fit. Neither of the changes, however, is significant at the 0.05 level, assuming a X^2 distribution for G^2. With considerations of fit, parsimony, and plausibility, we conclude that Model 19 is the most satisfactory model for the data.

It is certainly possible to conceive of more complex NINR models. Table 2 illustrated how various assumptions about the pattern of missing data lead to substantial differences in estimates for the effect of Level of Prenatal Care on Survival. Cases 5 through 10 in Table 2 are all equivalent to saturated NINR models, that is, models that fit the data perfectly and that have zero degrees of freedom. Many of these models are also unidentified and, by definition, cannot be tested against the data. All of these models assume that the missing mechanism has a three-way or higher interaction with the clinic and another variable. For example, cases 5 and 6 assume a three-way interaction among *M*, *C*, and *P*. Cases 9 and 10 assume a four-way interaction among *M*, *C*, *P*, and *S*. These models all assume that in the two clinics different types of data are likely to produce missing data on clinic. Without some assumptions we cannot rule out the estimates associated with these allocations. If, however, we assume that the missing data mechanism is the same in the two clinics except for the rate at which missing data occurs, then this rules out these possibilities. The (*MC*) (*SC*) (*PC*) model represents this assumption.

8. EXAMPLE 2: INTERGENERATIONAL EDUCATIONAL MOBILITY

A common use for loglinear models is the analysis of intergenerational social mobility. These analyses typically focus on cross-classifications of parents' and offsprings' socioeconomic characteristics derived from retrospective reports by the offspring. A problem for these analyses is missing data on the parents' characteristics. In this example, we examine the association between father's schooling and offspring's educational attainment, using data from the 1994 General Social Survey (GSS) and the Survey of American Families (SAF). The 1994 GSS, a cross-section survey of the

Table 5. Offspring's Educational Attainment by Father's Educational Attainment

	Offspring's Schooling		
Father's Schooling	**<12**	**12**	**>12**
<12	46	135	113
12	12	75	145
>12	4	30	206
Missing	23	29	18

Source: 1994 General Social Survey, respondents age 18 and older with a sibling interviewed in the Survey of American Families.

U.S. population, included a module on the socioeconomic characteristics of persons related to GSS respondents, including parents, children, spouses, and siblings. The SAF was a telephone survey administered to one randomly selected sibling of the GSS respondents (Mare and Hauser, 1994). Table 5 cross-classifies father's and respondent's schooling as reported by respondents in the GSS. These data were restricted to persons age 18 and older who had a sibling who was interviewed in the SAF. Of the 836 persons included in the table, 70, or 8.4%, failed to report their father's educational attainment. This may be construed as a relatively modest amount of missing data, and some researchers would simply omit observations in which father's schooling is missing. This decision, however, may have a big effect on one's estimates and inferences. One way of seeing the consequences of omitting missing data is to examine the estimated distribution of father's schooling and association between father's and offspring's schooling under several alternative hypothetical patterns of missing data. Table 6 shows the four local odds ratios for the 3×3 table of father's schooling by offspring's schooling under four scenarios: (1) data missing completely at random; (2) missing observations on father's schooling are all drawn from respondents whose fathers have less than 12 years of schooling; (3) missing observations are all drawn from respondents whose fathers have more than 12 years of schooling; and (4) missing observations are all drawn from the same level of schooling as the respondent. Under these alternative assumptions, the proportions of persons whose fathers have less than 12 years of schooling ranges from approximately 33% to approximately 40%. The estimated local odds ratios under alternative assumptions about missing data vary substantially. For example, the local cross-product ratio between whether a father is a high school graduate versus a dropout and the corresponding contrast

Table 6. Local Odds Ratios and Distributions of Father's Schooling Under Alternative Assumptions about Missing Data

	Local Odds Ratios		**Distribution of Father's Schooling**		
	<12/12	**12/>12**	**<12**	**12**	**>12**
1.			Observed Data		
<12/12	2.13	2.31			
			0.384	0.303	0.313
12/>12	1.20	3.55			
2.			Missing Data All Taken from Diagonal		
<12/12	4.43	1.67			
			0.354	0.293	0.352
12/>12	0.87	5.36			
3.			Missing Data All Taken from <12		
<12/12	2.62	2.42			
			0.409	0.257	0.332
12/>12	1.20	3.55			
4.			Missing Data All Taken from >12		
<12/12	2.13	2.31			
			0.328	0.257	0.413
12/>12	0.35	1.96			

Source: Table 5.

for offspring varies between 2.13 for data missing at random and 4.43 for data missing exclusively from persons who have the same schooling level as their fathers.

We can examine the association between father's and offspring's schooling while taking account of missing data by using variants of the missing data models discussed earlier. Although data may be missing completely at random, it is more likely that whether data are missing on father's educational attainment is associated with a respondent's own educational attainment and possibly father's educational attainment itself. The former problem may arise because better educated respondents are more likely to be cooperative and conscientious in answering survey questions. The latter problem may arise if respondents perceive that it is desirable to have a better educated father or if individuals with more poorly educated fathers are less likely to know their father's schooling. Some offspring of fathers with low levels of educational attainment may misreport their father's schooling, but others may simply not report it.

By itself, Table 5 provides limited information with which to investigate the impact of missing data on our estimates of the distribution of father's education or of the association between father's and offspring's education. That better educated respondents may be more likely to report their father's schooling can be investigated with this table. If this is the only systematic source of missing data, and if father's and offspring's educational attainments are associated, this idea can be represented as a saturated MAR model for this table. However, the idea that whether data are missing on father's schooling is associated with father's schooling itself requires a model of nonignorable nonresponse (NINR), which is not identified from these data. To investigate nonignorable nonresponse requires a variable that is associated with father's schooling but not with whether father's schooling is missing. Such variables are difficult to find because most characteristics of an individual that are associated with father's schooling are also associated with the individual's propensity to report father's schooling. A solution to this problem is to use the responses to the same item in an independent interview conducted with a person related to the original respondent. The SAF asked a sibling of each GSS respondent to report on father's schooling. Table 7 cross-classifies GSS respondent's report of father's schooling, GSS respondent's report of his or her own schooling, and the SAF respondent's – that is, GSS respondent's sibling's – report of father's schooling. This table includes categories for missing data on both reports of father's educational attainment and can be used to examine a variety of models for missing data.

These models can be regarded as applying to an *expanded table* that includes separate dimensions for the substantive variables of interest and for whether these variables are missing. The $3 \times 3 \times 3 \times 2 \times 2$ expanded table has the following five dimensions: GSS respondent's educational attainment (O), GSS respondent's report of father's educational attainment (F_G), SAF respondent's report of father's educational attainment (F_S), whether F_G is missing (M_G), and whether F_S is missing (M_S). For example, from Table 7 we can identify GSS respondents who have missing data on father's schooling classified by their own schooling and their siblings' reports of father's schooling. We do not know the educational attainment of these individuals' fathers (although a large fraction of these individuals have the same father as their sibling and, for them, their father's education can be inferred). Thus, for a given level of own schooling and sibling's report of father's schooling, GSS respondents who have missing data on their own report of father's schooling are distributed in an unknown way across categories of their father's schooling. Our models for missing data are based on selected interactions among the five dimensions of this expanded table.

Table 7. GSS Respondent's Educational Attainment by Father's Educational Attainment Reported by GSS Respondent by Father's Educational Attainment Reported by SAF Respondent

	Father's Schooling (SAF)											
	<12 Offspring's Schooling			12 Offspring's Schooling			>12 Offspring's Schooling			Missing Offspring's Schooling		
Father's Schooling (GSS)	<12	12	>12	<12	12	>12	<12	12	>12	<12	12	>12
<12	33	104	88	1	11	13	0	2	2	12	18	10
12	2	11	16	7	49	109	1	13	16	2	2	4
>12	0	0	3	1	9	23	3	21	177	0	0	3
Missing	12	17	5	4	6	6	0	3	2	7	3	5

Many logically possible models may be fit to the frequencies in Table 7. We limit the range of possible models through the following substantive considerations. First, because of the well-known correlation between the socioeconomic positions of parents and offspring, father's and offspring's educational attainment are associated. Indeed, this association provides the substantive interest in this table. Inasmuch as the GSS and SAF respondents' reports of father's educational attainment (F_G and F_S) apply to the same individual in most families, they are two reports of the same trait and thus both of these measures are associated with offspring's schooling. Thus, all models should include the $F_G O$ and the $F_S O$ associations.[5] Second, inasmuch as most siblings share the same father, their reports of father's schooling are likely to be strongly associated. Thus, all of our models include the $F_G F_S$ association. Third, siblings' propensities to fail to report father's schooling may be associated, either because of a shared reluctance to provide this information or a shared ignorance of their father's schooling. For most models, therefore, we include the $M_G M_S$ association. Fourth, it is an empirical question whether data are missing on father's schooling is associated with respondent's own educational attainment and with father's schooling itself. Thus, we examine alternative models with and without the OM_G, OM_S, $F_G M_G$, and $F_S M_S$ associations. Finally, we assume that whether a person reports father's schooling is *conditionally* independent of his or her sibling's reported level of father's schooling, given the association between each sibling's reported level of father's schooling. Thus, we assume the absence of the $F_G M_S$ and the $F_S M_G$ associations. These are the key restrictions for identifying NINR models for these data.

Table 8 presents goodness-of-fit statistics for selected models fit to the observed data in Table 7. Model 1 is an MCAR model in that it

Table 8. Goodness of Fit of Selected Models for Educational Mobility Table

Model	G^2	df	BIC
1. $F_G F_S$, $F_G O$, $F_S O$, M_G, M_S (MCAR)	1,470.6	27	1,288.9
2. $F_G F_S$, $F_G O$, $F_S O$, $M_G M_S$ (MAR)	117.9	26	−57.0
3. $F_G F_S$, $F_G O$, $F_S O$, OM_G, $M_G M_S$ (MAR)	73.6	24	−87.9
4. $F_G F_S$, $F_G O$, $F_S O$, OM_S, $M_G M_S$ (MAR)	87.1	24	−74.4
5. $F_G F_S$, $F_G O$, $F_S O$, OM_G, OM_S, $M_G M_S$ (MAR)	51.5	22	−96.5
6. $F_G F_S$, $F_G O$, $F_S O$, $F_G M_G$, $F_S M_S$, $M_G M_S$ (NINR)	48.0	22	−100.0
7. $F_G F_S$, $F_G O$, $F_S O$, OM_G, OM_S, $F_G M_G$, $M_G M_S$ (NINR)	46.2	20	−88.4
8. $F_G F_S$, $F_G O$, $F_S O$, OM_G, OM_S, $F_S M_S$, $M_G M_S$ (NINR)	28.5	20	−106.1
9. $F_G F_S$, $F_G O$, $F_S O$, OM_G, OM_S, $F_G M_G$, $F_S M_S$, $M_G M_S$ (NINR)	20.3	18	−100.8

assumes conditional independence of whether data are missing on the two measures of father's schooling from any of the other dimensions of the table. This model includes parameters for the marginal distributions of whether data are missing on F_G and F_S (M_G and M_S, respectively), but no association between whether data are missing on these two variables. As indicated by both the likelihood ratio G^2 and the BIC statistics (Raftery, 1995), this model fits very poorly.

Models 2–5 are MAR models. Model 2 includes a parameter for the association between M_G and M_S and fits the data much better than Model 1. This suggests that GSS respondents and their siblings both fail to report their father's schooling at a much higher rate than one would expect if their rates of nonresponse were statistically independent. Common family circumstances may determine whether offspring know their father's educational attainment. Models 3, 4, and 5 incorporate parameters for the association between GSS respondent's schooling and whether GSS and SAF respondents' have missing data on father's schooling. Inclusion of both of these associations significantly improves the fit of the model. The OM_G association implies that better educated respondents differ from more poorly educated respondents in their level of cooperation with the survey or their knowledge of their father's schooling. The OM_S association may arise because the table does not include a dimension for SAF respondent's own educational attainment. Given a strong correlation between siblings' educational attainments, we observe an OM_S association when the schooling of SAF respondents is not taken into account.

Models 6–9 are NINR models that include associations between F_G and F_S on the one hand and M_G and M_S on the other. Model 6 includes the $F_G M_G$ and $F_S M_S$ associations, but it excludes the associations between M_G and M_S and GSS respondent's educational attainment (O). This model fits much better than Model 2, the corresponding MAR model ($G_2{}^2 - G_6{}^2 = 69.9$, 4 df, $p < .001$), and it provides provisional evidence of nonignorable nonresponse. Model 6, however, does not fit the data well by the G^2 criterion. A more stringent test for NINR is to estimate the $F_G M_G$ and $F_S M_S$ associations in the presence of the OM_G and OM_S associations. Models 7 and 8 include the $F_G M_G$ and $F_S M_S$ associations, respectively, as well as OM_G and OM_S. Model 7 fits marginally better than Model 5, the corresponding MAR model ($G_5^2 - G_7^2 = 5.3$, 2 df; $p = .071$), whereas Model 8 fits much better than Model 5 ($G_5^2 - G_8^2 = 23.0$, 2 df, $p < .001$). This provides strong evidence for NINR for F_S and somewhat weaker evidence for F_G. Model 9 includes the OM_G, OM_S, $F_G M_G$, and $F_S M_S$ associations and fits significantly better than Model 5 as well as the three

Table 9. Estimated Association Parameters for NINR Model (Model 9)

Model Terms	λ	SE(λ)	exp(λ)
$M_G M_S$	0.670	0.358	1.954
$F_{G12} F_{S12}$	3.868	0.294	47.847
$F_{G12} F_{S>12}$	3.967	0.570	52.826
$F_{G>12} F_{S12}$	4.476	0.646	87.882
$F_{G>12} F_{S>12}$	8.056	0.778	3,152.654
$F_{G12} O_{12}$	0.075	0.501	1.078
$F_{G12} O_{>12}$	0.704	0.488	2.022
$F_{G>12} O_{12}$	−0.285	0.876	0.752
$F_{G>12} O_{>12}$	1.562	0.836	4.768
$F_{S12} O_{12}$	0.645	0.514	1.906
$F_{S12} O_{>12}$	0.917	0.504	2.502
$F_{S>12} O_{12}$	1.325	0.858	3.762
$F_{S>12} O_{>12}$	1.606	0.831	4.983
$F_{G12} M_G$	−0.197	0.462	0.821
$F_{G>12} M_G$	−2.292	1.469	0.101
$F_{S12} M_S$	−2.156	1.233	0.116
$F_{S>12} M_S$	−2.672	0.951	0.069
$O_{12} M_G$	−0.924	0.326	0.397
$O_{>12} M_G$	−1.620	0.281	0.198
$O_{12} M_S$	−0.883	0.351	0.414
$O_{>12} M_S$	−0.865	0.378	0.421

other NINR models. By the likelihood ratio criterion, Model 9 is the best fitting model ($G_9^2 = 20.3$, 18 df, $p = .316$), although the BIC indicates that Model 8 is slightly more satisfactory.

Table 9 presents estimates of two-way association parameters and partial odds ratios from Model 9, which reveal the systematic nature of missing data on father's educational attainment. First, the estimated partial odds ratio for the two missing data indicators ($M_G M_S$) indicates a positive association between siblings' propensities to fail to report father's educational attainment. The SAF respondent is almost twice as likely to have missing data on father's schooling if his or her sibling in the GSS fails to report father's schooling than if the sibling reports father's schooling. Second, more highly educated GSS respondents are much more likely to report their father's schooling than their more poorly educated counterparts. Relative to GSS respondents who were high school dropouts, for example, those who had more than a high school degree had odds of not reporting father's schooling that are only one fifth as great. Finally, missing data on father's educational attainment is associated with the level of father's education. For example, for SAF respondents whose fathers had more than a high school degree, the odds of missing data are only approximately 7% of the odds for SAF respondents whose fathers were high school dropouts. Both the parameter estimates and the goodness-of-fit tests provide more evidence of nonignorable nonresponse for SAF respondents' reports of fathers' schooling than for GSS respondents' reports. This may occur because GSS respondents' own educational attainments are controlled in our models, whereas SAF resondents' attainments are not.

9. ESTIMATION AND IDENTIFICATION

A. Estimation

The estimates reported in this paper were calculated by using DNEWTON, a flexible program for the estimation of latent class models, including models with latent cells such as the ones discussed in this paper (Haberman, 1989). Further discussion of DNEWTON is provided in the Technical Appendix. A good alternative to DNEWTON is LEM (Vermunt, 1996), which has similar capabilities. The development of several user-friendly latent class programs has substantially reduced the burden of estimating latent class models. Although modern software makes the estimation of loglinear models for missing data feasible, several estimation problems nonetheless are common. These include (1) failure of the program to converge; (2) estimates on the boundary of the parameter space; (3) cells with small expected frequencies (<5), and (4) poor model fit. These important issues, which are problems for latent class models more generally, are discussed in the introductory chapter of this volume.

B. Identification

Consider the hypothetical data shown in Table 3 consisting of variables X, Y, and Z but collapsed over Z. If we add to this a variable M indicating whether information on Y is missing for a case, we then have a 2^3 table of X by Y by M. This is shown in Table 10 both as a 2×3 subtable of the observed table and as a 2^3 expanded table that is partially latent. In a complete eight-cell table, a fully saturated model has seven parameters

Table 10. Hypothetical Data from Table 3 Collapsed Over *Z*

(a) Observed Table

	$Y=0$	$Y=1$	$Y=$ Missing
$X=0$	960	480	360
$X=1$	480	960	360

(b) Expanded Table

	$M=0$		$M=1$		
	$Y=0$	$Y=1$	$Y=0$	$Y=1$	Total
$X=0$	960	480	?	?	360
$X=1$	480	960	?	?	360

(plus a grand mean). However, Table 10 has only six cells – four for the *X*–*Y* complete data subtable and two for the categories of *X* among respondents with missing data on *Y*. Therefore, a model with at most five parameters is identified. If one fits a marginal parameter for each of the dimensions *X*, *Y*, and *M*, then, among hierarchical models, the most complex five-parameter model can include at most two two-way interactions. The potential models are as follows: (1) (*MX*) (*XY*) (MAR Model); (2) (*MY*) (*XY*) (NINR Model); and (3) (*MX*) (*MY*), which is not identified.

To see that the (*MX*) (*MY*) model is not identified, note that in panel (b) of Table 10 one can observe the association between *X* and *Y* for the complete data. The (*MX*) (*MY*) model assumes that the partial *X*–*Y* association (conditional on *M*) is zero, which can be tested by using the complete data. That we can test that the *X*–*Y* interaction is zero means that a parameter for this association is identified, irrespective of whether other parameters of the (*MX*) (*MY*) model are not identified. Such a test would use 1 degree of freedom, leaving only 4 degrees of freedom for estimating the five parameters of the (*MX*) (*MY*) model. Thus the (*MX*) (*MY*) model cannot be identified.

MAR models are usually identified, even if data are missing on several variables (Little and Rubin, 1987, pp. 171–94). A sufficient condition for identification of MAR models is that some observations are complete on all variables and that these observations represent all possible combinations of the variables in the model (Fuchs, 1982).

General rules for the identification of NINR models have not yet been developed. Some guidance is available, however, from results on the identifiability of NINR models for two-way tables and from rules for identification of latent class models. Little and Rubin (1987, pp. 238–9) summarize the identifiability of models for two-way ($J \times K$) tables in which one dimension of the table has missing observations. Denote the dimensions of the table by X and Y. Let all observations be present for X, but some observations on Y be missing. Let a third variable, M, denote whether data are missing on Y. Among the several loglinear models that can be fit to the X, Y, M table, NINR models are those that include the M–Y interaction. The only NINR model that is potentially identifiable is (MY) (XY), that is, a model in which the fully observed variable is associated with the partially observed variable, but is independent of whether data are missing on the partially observed variable. This model is identified if $J \geq K$, that is, if the number of categories of the fully observed variable is at least as large as the number of categories of the partially observed variable. In a 2×2 table, $J = K$ and the model is just identified.

These results suggest that NINR models are identified if (1) for every partially observed variable, there exists a fully observed variable that is conditionally independent of the missing data indicator; and (2) the number of categories of the partially observed variable does not exceed the number of categories of the fully observed variable. The fully observed variable plays a role in identifying the model that is analogous to an instrumental variable in a structural equation model, and the condition is equivalent to assuming that parameter for the two-way interaction between the fully observed variable and the missing indicator is zero. This restriction can sometimes also be met by assuming that higher-order interaction terms are zero. Baker and Laird (1988), for example, estimate a model in which two fully observed variables both affect the missing indicator, but the model is identified because the three-way interaction among the variables is assumed to be zero.

An intuitive way of understanding the identifiability of some NINR models is to note that they are often similar to standard latent class models that are analogous to factor models (e.g., Goodman, 1974). Consider the NINR (MC) (SC) (PC) model for the Little–Rubin data. If C were missing on *every* observation, the model would be a standard latent class model in which C is a latent "factor" and M, S, and P are its observed "indicators." The model assumes that M, S, and P are conditionally independent given C. If C were missing on every case, however, the model would not be identified without either another indicator for C or additional restrictions on the parameters. In the pure latent variable case, we observe only the two-way relationship between P and S, which is insufficient for estimating the associations between both S and C and also P and C.

But when data are missing for only some of the cases, we observe the relationship between S, P, and C in the complete data. Thus we can estimate both the S–C and P–C associations simply from the complete data. Only the relationship between M and C has to be estimated indirectly. When C is partially observed, therefore, this NINR model is identified.

10. CONCLUSION

Conventional methods of dealing with missing data in multivariate models run serious risks. Omitting observations with missing data from an analysis certainly reduces sample size (and thus the precision of estimates) and, at worst, may lead to severe biases in parameter estimates. Simply incorporating missing data by adding categories for missing data to the observed variables generally results in inconsistent estimates. Fortunately, alternatives to the conventional approaches provide powerful methods for investigating the degree to which data are missing systematically and for carrying out appropriate substantive analyses.

The approach illustrated in this chapter[6] is to extend standard analyses of categorical data by recognizing that when some data are missing, the resulting contingency tables have cells that are partially observed. We suggest that one analyze such data by using latent class loglinear models for tables that have a mixture of fully and partially observed cells. This approach enables one to investigate hypotheses about the mechanisms by which missing data come about as well as the substantive relationships of interest. Inasmuch as these models are simply variants of standard loglinear models, one can incorporate missing data by using well-known procedures for specifying multiway interactions in contingency tables. Unlike standard loglinear models for fully observed data, however, these models typically require that the analyst incorporate additional data or make simplifying assumptions to identify the relationships between substantive variables of interest and indicators of whether data are missing.

REFERENCES

Agresti, A. (1990). *Categorical Data Analysis*. New York: J Wiley.

Andersen, E. B. (1980). *Discrete Statistical Models with Social Science Applications*. Amsterdam: North-Holland.

Baker, S. G., & Laird, N. M. (1988). "Regression analysis for categorical variables with outcome subject to nonignorable nonresponse," *Journal of the American Statistical Association*, **83**, 62–9.

Bishop, Y. M., Fienberg, S., & Holland, P. W. (1975). *Discrete Multivariate Analysis: Theory and Practice*. Cambridge: MIT Press.
Clogg, C. C. (1988). "Latent class models for measuring." In R. Langeheine and J. Rost (eds.), *Latent Trait and Latent Class Models*. New York: Plenum, pp. 173–205.
Davis, J. A., & Smith, T. W. (1986). *General Social Surveys, 1972–1986* (MRDF), NORC ed. Chicago: NORC; distributed by Roper Public Opinion Research Center, New Haven, CN.
Dayton, C. M., & Macready, G. B. (1980). "A scaling model with response errors and intrinsically unscalable respondents," *Psychometrika*, **45**, 343–56.
Fuchs, C. (1982). "Maximum likelihood estimation and model selection in contingency tables with missing data," *Journal of the American Statistical Association*, **77**: 270–8.
Goodman, L. A. (1974). "The analysis of systems of qualitative variables when some of the variables are unobservable. Part I: a modified path analysis approach," *American Journal of Sociology*, **79**, 1179–1259.
Haberman, S. J. (1989). "A stabilized Newton–Raphson algorithm for log-linear models for frequency tables derived by indirect observation." In C. C. Clogg (ed.), *Sociological Methodology 1988*, Vol. 18. **City: pub**, pp. 193–211.
Hagenaars, J. (1988). "Latent structure models with direct effects between indicators," *Sociological Methods and Research*, **16**, 379–405.
Hagenaars, J. (1993). *Loglinear Models with Latent Variables*. Newbury Park, CA: Sage.
Little, R. J. A., & Rubin, D. B. (1987). *Statistical Analysis with Missing Data*. New York: Wiley.
Mare, R. D., and Hauser, R. M. (1994). "A Survey Module on Families and Social Mobility," unpublished grant proposal to National Science Foundation. Madison, WI: Dept. of Sociology, University of Wisconsin.
McCutcheon, A. L. (1987). *Latent Class Analysis*. Quantitative Applications in the Social Sciences Paper No. 64. Beverly Hills, CA: Sage.
Park, Taesung, & Brown, M. B. (1994). "Models for categorical data with non-ignorable nonresponse," *Journal of the American Statistical Association*, **89**, 44–52.
Raftery, A. E. (1995). "Bayesian model selection in social research." *Sociological Methodology 1995*, Vol. 25, 111–63.
Vermunt, J. K. (1996). *LogLinear Event History Analysis*. Tilburg, The Netherlands: Tilburg University Press.
Winship, C., & Mare, R. D. (1989). "Loglinear Models with Missing Data: A Latent Class Approach." *Sociological Methodology 1989*, Vol. 19, 331–68.
Winship, C., & Mare, R. D. (1992). "Models for sample selection bias," *Annual Review of Sociology*, **18**, 327–50.

NOTES

1. Fuchs (1982), Little and Rubin (1987), Baker and Laird (1988), Winship and Mare (1989), and Park and Brown (1994) provide a technical discussion of

how missing data can be incorporated into loglinear models. In this chapter, we discuss the key ideas needed for the researcher to incorporate missing data in a loglinear analysis.

2. These programs can estimate loglinear models with arbitary patterns of missing data. As a result, we do not need to consider the issues found in earlier literature as to whether the missing data pattern is monotone and/or whether the likelihood can be factored (Little and Rubin, 1987).
3. Winship and Mare (1989) provide a more extensive analysis of this example. In particular, we show that the estimated effect for the association between X and Z is biased when we add a missing data category to Table 3.
4. Econometric models for sample selection bias are examples of NINR models (Winship and Mare, 1992).
5. A related set of models explicitly distinguishes between the "true" father's schooling and the fallible reports of father's schooling provided by each sibling. This distinction can be incorporated into latent class models. We do not consider these models in this chapter.
6. This research was supported by National Science Foundation Grants SBR-94-11875 and SBR-94-11670, and by the Graduate School of the University of Wisconsin-Madison.

Appendix A: Notational Conventions

The purpose of this appendix is to indicate the similarities among what might, at first reading, appear as highly disparate forms of notation used by the various contributors to this volume. The following discussion will be most accessible to those who are familiar with the basic latent class model. Thus, it is recommended that readers familiarize themselves with the notational styles in the two introductory chapters, Chapters 1 and 2, before reading through this appendix.

As with nearly all statistical models, the latent class model can be expressed by using a variety of different notational conventions. The problem of differing notation is exacerbated with the latent class model because the LCM can be parameterized in two seemingly different, though equivalent, parameterizations: probabilistic and loglinear (see McCutcheon, Chapter 2).

The probabilistic parameterization – the most commonly used parameterization – was first introduced to a broader audience by Goodman, who used the iterative proportional fitting estimation method (1974a, 1974b). Goodman adopted a horizontal "overbar" notational style to represent conditional probabilities; thus, in this notational style, the basic latent class model is represented as

$$\pi_{ijkl}^{ABCD} = \sum_{t=1}^{T} \pi_t^X \pi_{it}^{\bar{A}X} \pi_{jt}^{\bar{B}X} \pi_{kt}^{\bar{C}X} \pi_{lt}^{\bar{D}X},$$

where the symbol π_t^X represents the latent class, or mixing, proportion $\mathrm{P}(X = t)$. This is a relatively widespread and common representation usage, although some authors occasionally elect to use some other character than X to represent the latent variable, and some choose to use some other letter than $t = 1, \ldots, T$ to represent the specific classes of the latent variable. The conditional probabilities are represented by the horizontal overbar symbols such as $\pi_{it}^{\bar{A}X}$ and $\pi_{jt}^{\bar{B}X}$. Thus, for example, the symbol $\pi_{it}^{\bar{A}X}$ represents the probability that a randomly selected observation will be at level i of variable A, given that it is in class t of the latent variable X; that is, $P(A = i \mid X = t)$.

Nearly all of the authors in this volume begin with a close variant of this basic latent class model (see, e.g., Wedel & DeSarbo, Chapter 13), so we will use this four-indicator variable model as our common "rosetta stone" with which to examine the alternative notational conventions. This will, it is hoped, prove useful to those readers who initially find these differing notational conventions confusing. It is important to remember that the probabilistic parameterization of the basic LCM has two fundamental types of parameters – conditional probabilities and latent class probabilities – and that although some of our contributors use alternative notational conventions, each of these two types of parameters will be found in all alternative forms. It is also important to recognize that most of the authors chose their alternative notation in order to better streamline the presentation of their own modifications and extensions of the latent class model.

Although several of the authors use the notation with the horizontal overbar for conditional probabilities (see, e.g., Chapters 1, 4, 8, and others), a common alternative to this notation is the vertical bar notation. This variation on the horizontal overbar notation is also frequently used by the contributing authors (e.g., Chapters 2, 3, 5, 9, 14, and others):

$$\pi_{ijkl}^{ABCD} = \sum_{t=1}^{T} \pi_t^X \pi_{it}^{A|X} \pi_{jt}^{B|X} \pi_{kt}^{C|X} \pi_{lt}^{D|X}.$$

The equivalences between these alternative notations are displayed next.

$$P(A = i \mid X = t) \equiv \pi_{it}^{\bar{A}X} \equiv \pi_{it}^{A|X},$$
$$P(B = j \mid X = t) \equiv \pi_{jt}^{\bar{B}X} \equiv \pi_{jt}^{B|X},$$
$$P(C = k \mid X = t) \equiv \pi_{kt}^{\bar{C}X} \equiv \pi_{kt}^{C|X},$$
$$P(D = l \mid X = t) \equiv \pi_{lt}^{\bar{D}X} \equiv \pi_{lt}^{D|X}.$$

In his chapter on choice data, Böckenholt (Chapter 6) uses yet another notational convention to represent the basic LCM as

$$\pi_Y(y) = \sum_{t=i}^{T} \pi_X(t) \pi_{Y|X(t)}(y).$$

In this notational form, the expected joint cooccurrence of the indicator variables on the left of the equation is represented as

$$\pi_Y(y) \equiv \pi_{ijkl}^{ABCD}.$$

As in the previous models, there are $t = 1, \ldots, T$ levels of the latent class variable X_t; thus,

$$\pi_X(t) \equiv \pi_t^X.$$

Also, there are $j = 1, \ldots, J$ indicator variables Y_j. The lower-case y represents the specific joint occurrence, or response pattern (e.g., 0110, 0011), so

$$\pi_{Y_1|X(t)} \equiv \pi_{it}^{\bar{A}X}, \quad \pi_{Y_2|X(t)} \equiv \pi_{jt}^{\bar{B}X},$$

and so forth for all four of the indicator variables, which we have been referring to as A, B, C, and D.

In Chapter 7, Formann and Kohlmann also adopt an alternative set of notations for their presentation of the latent class model. As they note, their notation aids them in highlighting their extension of the latent class model. Their expression of the latent class model can be directly translated into Goodman's notation as

$$\pi(a_s) \equiv \sum_{j=1}^{m} \pi_j \prod_{i=1}^{k} \pi_{i|j}^{a_{si}} (1 - \pi_{i|j})^{1-a_{si}}$$

$$\pi(a_s) \equiv \pi_{ijkl}^{ABCD},$$

$$\pi_j \equiv \pi_t^X,$$

and

$$\pi_{1|j} \equiv \pi_{it}^{\bar{A}X}, \quad \pi_{2|j} \equiv \pi_{jt}^{\bar{B}X}, \quad \pi_{3|j} \equiv \pi_{kt}^{\bar{C}X}, \quad \pi_{4|j} \equiv \pi_{lt}^{\bar{D}X}.$$

In the two chapters that discuss latent class models for longitudinal data (Chapters 10 and 11), alternative notational conventions are also useful in helping focus attention on fitting the basic LCM to model temporal (e.g., change, transition) parameters. In Collins and Flaherty's chapter (Chapter 10), a latent class model is applied to longitudinal data on self-reported drug use collected at two time points in a sample of adolescents. It is easiest to relate Collin and Flaherty's notation to an extended version of the earlier model in which there are six indicator variables:

$$\pi_{ijklmn}^{ABCDEF} = \sum_{t=1}^{T} \pi_t^X \pi_{it}^{\bar{A}X} \pi_{jt}^{\bar{B}X} \pi_{kt}^{\bar{C}X} \pi_{lt}^{\bar{D}X} \pi_{mt}^{\bar{E}X} \pi_{nt}^{\bar{F}X},$$

where the *A*, *B*, and *C* indicator variables are measures of three variables taken at time 1, and the *D*, *E*, and *F* indicator variables are the same three measures, respectively, taken at time 2; Collins and Flaherty refer to these three indicator variables as *I*, *J*, and *K* at time 1 and I', J', and K' at time 2. A series of equality restrictions on the latent class and conditional probabilities (see, e.g., Hagenaars, 1990, especially Chapter 4; McCutcheon, 1987) enables them to define a latent status *A* at time 1 and a latent status *B* at time 2. (Note: their use of latent statuses *A* and *B* are unrelated to our earlier use of *A* and *B* as indicator variables.) Thus, their basic LCM model is

$$\pi_{ijki'j'k'} = \sum_{x=1}^{X} \sum_{a=1}^{A} \sum_{b=1}^{B} \pi_{t}^{X} \pi_{ax}^{\bar{A}X} \pi_{bax}^{\bar{B}AX} \pi_{iax}^{\bar{I}AX} \pi_{jax}^{\bar{J}AX} \pi_{kax}^{\bar{K}AX} \pi_{i'bx}^{\bar{I}BX} \pi_{j'bx}^{\bar{J}BX} \pi_{k'bx}^{\bar{K}BX},$$

where π_x^X is the probability of being in latent class x throughout the entire period of study, $\pi_{ax}^{\bar{A}X}$ is the probability that an observation in latent class x of the latent variable X will be in latent status a of the latent status variable A at time 1, and $\pi_{bax}^{\bar{B}AX}$ is the probability that an observation in latent class x, who was in latent status a at time 1, will be in latent status b at time 2. Thus, with appropriate restrictions, Collins and Flaherty's notation can be equated with the basic "rosetta stone" model as follows:

$$\pi_{ijki'j'k'} \equiv \pi_{ijklmn}^{ABCDEF},$$

$$\pi_{x}^{X} \pi_{ax}^{\bar{A}X} \pi_{bax}^{\bar{B}AX} \equiv \pi_{t}^{X},$$

$$\pi_{iax}^{\bar{I}AX} \equiv \pi_{it}^{\bar{A}X}, \quad \pi_{jax}^{\bar{J}AX} \equiv \pi_{jt}^{\bar{B}X},$$

$$\pi_{kax}^{\bar{K}AX} \equiv \pi_{kt}^{\bar{C}X}, \quad \pi_{i'bx}^{\bar{I}BX} \equiv \pi_{lt}^{\bar{D}X},$$

$$\pi_{j'bx}^{\bar{J}BX} \equiv \pi_{mt}^{\bar{E}X}, \quad \pi_{k'bx}^{\bar{K}BX} \equiv \pi_{nt}^{\bar{F}X}.$$

In their chapter on Markov latent class models (Chapter 11), Langeheine and van de Pol impose a similar set of restrictions for panel data collect at three or more points in time. Instead of using the Greek symbol π to represent all of the probabilities of their probabilistic parameterization, these authors opt for a set of Greek characters, each representing a different kind of probability. The character π is used to indicate the proportion of the sample estimated to be in each of the S "chains" of the Markov model; for example, the Mover–Stayer model ($S = 2$). The character δ is used to indicate the proportion of each Markov chain that is initially found in each of the A latent classes; the character ρ is the proportion used to indicate the measurement properties relating the A

latent classes to the I levels of an indicator variable at sequential time points. Finally, the character τ indicates the proportions used in the transition matrix relating the A latent classes at time 1 to the B latent classes at time 2, the B latent classes at time 2 to the C latent classes at time 3, and so forth. Thus, their basic latent class model for three indicator variables (i.e., time points) is

$$P_{ijk} = \sum_{s=1}^{S}\sum_{a=1}^{A}\sum_{b=1}^{B}\sum_{c=1}^{C} \pi_s \delta^{1}_{a|s} \rho^{1}_{i|as} \tau^{2\,1}_{b|as} \rho^{2}_{j|bs} \tau^{3\,2}_{c|bs} \rho^{3}_{k|cs},$$

where

$$P_{ijk} = \pi^{ABC}_{ijk},$$

and, with a set of equality restrictions similar to those in the Collins and Flaherty Chapter 10,

$$\sum_{a=1}^{A}\sum_{b=1}^{B}\sum_{c=1}^{C} \pi_s \delta^{1}_{a|s} \tau^{2\,1}_{b|as} \tau^{3\,2}_{c|bs} \equiv \pi^{X}_{t}$$

if the S Markov chains are regarded as the T latent classes of interest. Alternatively, we might wish to regard the A latent classes at time 1 as the T latent classes of interest, yielding

$$\sum_{s=1}^{S}\sum_{b=1}^{B}\sum_{c=1}^{C} \pi_s \delta^{1}_{a|s} \tau^{2\,1}_{b|as} \tau^{3\,2}_{c|bs} \equiv \pi^{X}_{t},$$

and so forth. Finally, the conditional probabilities in this model are

$$\rho^{1}_{i|as} \equiv \pi^{\bar{A}X}_{it}, \quad \rho^{2}_{j|bs} \equiv \pi^{\bar{B}X}_{jt}, \quad \rho^{3}_{k|cs} \equiv \pi^{\bar{C}X}_{kt}.$$

The final two chapters that use probabilistic parameterizations – the chapters by Rudas (Chapter 12) and Wedel and DeSarbo (Chapter 13) – use only the latent class, or mixture, probability π^{X}_{t} in a manner that is consistent with the notation of basic latent class model introduced by Goodman. In each instance, the authors of these chapters intend for this proportion to designate the probability that a given observation will be located in one of the T classes, although Rudas' approach clearly indicates that $T = 2$; the proposed model either describes the portion of the population in which an observations is located (i.e., $t = 1$) or it does not (i.e., $t = 2$). In Wedel and DeSarbo's notation, the latent class probability permits T to assume a value greater than 2; the remainder of their basic model, however, permits a wide range of link functions, as they note.

The chapter by Winship, Mare, and Warren (Chapter 15) fully adopts the loglinear parameterization of the basic latent class model. As they note, their usage of the loglinear notation is somewhat unconventional inasmuch as they focus on latent cells rather than latent dimensions. Consequently, there is a somewhat complicated relationship between their parameterization and the usual loglinear parameterization of the latent class model.

Appendix B: Further Readings

This appendix lists a number of books, chapters, and articles that may prove useful to those who wish to learn more about latent class analysis. Although the listing is by no means exhaustive, an effort has been made to include many of the most widely known sources. We have selected a set of general headings with which to separate the various works. Because several of the works can be classified under multiple topics, however, the reader is advised to consider works listed under headings other than that of his or her immediate interest. The bibliographies of the individual contributions are also an excellent source of additional readings on specific topics related to the latent class model.

A. INTRODUCTORY AND OVERVIEW WORKS

Clogg, C. C. (1981). "New developments in latent structure analysis." In D. M. Jackson & E. F. Borgatta (eds.), *Factor Analysis and Measurement in Sociological Research.* Beverley Hills, CA: Sage, pp. 214–80.

Clogg, C. C. (1995). "Latent class models." In G. Arminger, C. C. Clogg, & M. E. Sobel (Eds.), *Handbook of Statistical Modeling for the Social and Behavioral Sciences*. New York: Plenum, pp. 311–59.

McCutcheon, A. L. (1987). *Latent Class Analysis*. Beverly Hills: Sage.

B. HISTORICAL DEVELOPMENTS (CHRONOLOGICAL ORDER)

Lazarsfeld, P. F. (1950). "The logical and mathematical foundation of latent structure analysis." In S. A. Stouffer, L. Guttman, E. A. Suchman, P. F. Lazarsfeld, S. A. Star, & J. A. Clausen (Eds.), *Measurement and Prediction*. Princeton, NJ: Princeton University Press, pp. 362–412.

Lazarsfeld, P. F. (1950). "The interpretation and computations of some latent structures." In S. A. Stouffer, L. Guttman, E.A. Suchman, P. F. Lazarsfeld,

S. A. Star, & J. A. Clausen (eds.), *Measurement and Prediction*. Princeton, NJ: Princeton University Press, pp. 413–72.

Lazarsfeld, P. F., & Henry, N. W. (1968). *Latent Structure Analysis*. Boston: Houghton Mifflin.

Goodman, L. A. (1974a). "Exploratory latent structure analysis using both identifiable and unidentifiable models," *Biometrika*, **61**, 215–31.

Goodman, L. A. (1974b). "The analysis of systems of qualitative variables when some of the variables are unobservable. Part I: modified latent structure approach," *American Journal of Sociology*, **79**, 1197–1259.

Goodman, L. A. (1978). *Analyzing Qualitative/Categorical Data: Log-Linear Models and Latent-Structure Analysis*. J. Magidson (ed). Cambridge, MA: Abt Books.

Haberman, S. J. (1979). *Qualitative Data Analysis*, Vols. 1 & 2. New York: Academic.

Langeheine, R., & Rost, J., Eds. (1988). *Latent Trait and Latent Class Models*. New York: Plenum.

Rost, J., & Langeheine, R., Eds. (1997). *Applications of Latent Trait and Latent Class Models in the Social Sciences*. New York: Waxmann.

C. MEASUREMENT

Clogg, C. C. (1988). "Latent class models for measuring." In R. Langeheine & J. Rost (eds.), *Latent Trait and Latent Class Models*. New York: Plenum. pp. 173–205.

Dayton C. M. (1999). *Latent Class Scaling Analysis*. Thousand Oaks, CA: Sage.

Dayton, C. M., & Macready, G. B. (1980). "A scaling model with response errors and intrinsically unscalable respondents," *Psychometrika*, **45**, 343–56.

Goodman, L. A. (1975). "A new model for scaling response patterns: an application of the quasi-independence concept," *Journal of the American Statistical Association*, **70**, 755–68.

Heinen, T. (1996). *Latent Class and Discrete Latent Trait Models: Similarities and Differences. Thousand Oaks*, CA: Sage.

Lindsay, B., Clogg, C. C., & Grego, J. (1991). "Semiparametric estimation in the Rasch model and related exponential response models, including a simple latent class model for item analysis," *Journal of the American Statistical Association*, **86**, 96–107.

McCutcheon, A. L. (1996). "Multiple group association models with latent variables: an analysis of secular trends in abortion attitudes, 1972–1988." In A. Raftery (ed.), *Sociological Methodology*, 1996. Cambridge, MA: Blackwell.

Uebersax, J. S. (1993). "Statistical modeling of expert ratings on medical treatment appropriateness," *Journal of the American Statistical Association*, **88**, 421–7.

D. CAUSAL MODELS

Hagenaars, J. A. (1990). *Categorical Longitudinal Data: Log-Linear Panel, Trend and Cohort Analysis*. Newbury Park, CA: Sage.

Hagenaars, J. A. (1993). *Loglinear Models with Latent Variables*. Newbury Park, CA: Sage.
Vermunt, Jeroen K. (1997). *Log-Linear Models for Event Histories*. Thousand Oaks, CA: Sage.

E. UNOBSERVED HETEROGENEITY

Everitt, B. S., & Hand, D. J. (1981). *Finite Mixture Models*. New York: Chapman & Hall.
Lindsay, B. G. (1995). *Mixture Models: Theory, Geometry and Applications*. Hayward, CA: Institute of Mathematical Science.
McLachlan, G. J., & Basford, K. E. (1988). *Mixture Models: Inference and Applications to Clustering*. New York: Marcel Dekker.
McLachlan, G., & Peel, D. (2000). *Finite Mixture Models*. New York: Wiley.
Titterington, D. M., Smith, A. F. M., & Makov, U. E. (1985). *Statistical Analysis of Finite Mixture Distributions*. New York: Wiley.
Wedel, M., & Kamakura, W. A. (1998). *Market Segmentation: Conceptual and Methodological Foundations*. Boston: Kluwer Academic.

Appendix C: Selected Software; Webpage

In the past several years, a number of programs that estimate latent class models have appeared. These programs, based on maximum-likelihood estimation, have clearly superceded the earliest programs for latent class analysis: MLLSA (maximum likelihood latent structure analysis) by Clifford Clogg, LCAG (latent classes according to Goodman) by Jacques Hagenaars and Ruud Luijkx, and LAT and Newton by Shelby Haberman. The programs listed herein are a selection of some of the better-known programs that are currently available for solving a variety of latent class models.

WINMIRA

WINMIRA is a Windows-based program that estimates and tests a wide variety of discrete mixture models for categorical variables, including models with both nominal and continuous latent variables. In addition to latent class models, this program can be used to estimate Rasch models and mixed Rasch models with dichotomous and polytomous data.

PANMARK

PANMARK estimates and tests a range of latent class models, although as its name – a combination of panel Markov models – implies, this program was developed to estimate longitudinal latent class models. PANMARK also has a number of features such as its ability to generate multiple sets of starting values, which may be useful in avoiding local maxima, its reporting of asymptotic standard errors for model parameter estimates, and its use of bootstrapping methods for comparing models with different numbers of latent classes.

LATENTGOLD

LatentGOLD is a Windows-based latent class analysis program that is compatible with SPSS. The program estimates a series of LCMs referred to as latent class cluster, latent class factor, and latent class regression models. The basic latent class model is included within the program's LC cluster model; the latent class scaling models are included within the LC factor model; and the LC regression model includes a range of causal models within the latent class context. Latent GOLD also includes a number of useful features such as Bayesian constants that help avoid boundary value parameter estimates, bivariate residual details, the ability to include covariates in the models, and the ability to permit indicator variables to violate conditional independence.

LEM

LEM estimates a wide range of both unconstrained and constrained latent class, loglinear, and association models, as well as several other categorical data models. The program includes a number of desirable features, such as checking model identifiability, providing asymptotic standard errors for parameter estimates, permitting local dependence among indicator variables, advanced estimation algorithms, and others. LEM is accompanied by several example programs to facilitate program setup and analysis.

MPLUS

MPlus estimates a variety of latent class and mixture models, and it combines these models with several other familiar analytic models such as factor analysis, regression analysis (linear and logistic), SEMs and others, such as the usual latent class model, latent profile analysis, mixtures of continuous variables, and growth curve mixtures. Mplus also includes a number of attractive features such as estimation from complex sampling designs, Monte Carlo estimation, and others.

WINLTA

WinLTA is a Windows-based latent class analysis program for estimating both the latent class and the latent transition analysis (LTA) model. The WinLTA program estimates longitudinal, latent class Markov models with a number of useful features, such as permitting multiple (identical) indicators at each wave of data collection, examining multiple groups, placing constraints on the transition matrices, and others.

GLIMMIX

GLIMMIX is a Windows-based latent class analysis program aimed at market researchers for segmentation analysis based on latent class models. In addition to the usual latent class model, GLIMMIX estimates a range of latent class (mixture) regression models and accomodates a wide range of data types, including frequency, pick any/*n*, paired comparisons, conjoint data, and others. A wide range of distributions (e.g., normal, Poisson, gamma, binomial, and multinomial) are also allowed.

DNEWTON

DNewton is a Windows-based latent class analysis program for the estimation of LCMs using the loglinear parameterization. The program implements a stabilized Newton–Raphson algorithm and uses design matrices, enabling the estimation of a wide range of loglinear latent class models.

WEBPAGE

More information about these (and other) programs and where and how to obtain them may be found on the books' webpage at us.cambridge.org/titles/. Moreover, the webpage contains or refers to the main data sets that have been used in this book, along with many program setups for the main models that have been applied to these data sets.

Author Index

Subject Index